W9-AUI-753

Nuclear Structure and Gene Expression

CELL BIOLOGY: A Series of Monographs

EDITORS

Volumes published since 1985

B. G. Atkinson and D. B. Walden (editors). CHANGES IN EUKARYOTIC GENE EXPRESSION IN RESPONSE TO ENVIRONMENTAL STRESS, 1985

Reginald M. Gorczynski (editor). RECEPTORS IN CELLULAR RECOGNITION AND DEVELOPMENTAL PROCESSES, 1986

Govindjee, Jan Amesz, and David Charles Fork (editors). LIGHT EMISSION BY PLANTS AND BACTERIA, 1986

Peter B. Moens (editor). MEIOSIS, 1986

Robert A. Schlegel, Margaret S. Halleck, and Potu N. Rao (editors). MOLECULAR REGULATION OF NUCLEAR EVENTS IN MITOSIS AND MEIOSIS, 1987

Monique C. Braude and Arthur M. Zimmerman (editors). GENETIC AND PERINATAL EFFECTS OF ABUSED SUBSTANCES, 1987

E. J. Rauckman and George M. Padilla (editors). THE ISOLATED HEPATOCYTE: USE IN TOXICOLOGY AND XENOBIOTIC BIOTRANSFORMATIONS, 1987

Heide Schatten and Gerald Schatten (editors). THE MOLECULAR BIOLOGY OF FERTILIZATION, 1989

Heide Schatten and Gerald Schatten (editors). THE CELL BIOLOGY OF FERTILIZATION, 1989

Anwar Nasim, Paul Young, and Byron F. Johnson (editors). MOLECULAR BIOLOGY OF THE FISSION YEAST, 1989

Mary P. Moyer and George Poste (editors). COLON CANCER CELLS, 1990

Gary S. Stein and Jane B. Lian (editors). MOLECULAR AND CELLULAR APPROACHES TO THE CONTROL OF PROLIFERATION AND DIFFERENTIATION, 1991

Victauts I. Kalnins (editor). THE CENTROSOME, 1992

Carl M. Feldherr (editor). NUCLEAR TRAFFICKING, 1992

Christer Sundqvist and Margareta Ryberg (editors). PIGMENT-PROTEIN COMPLEXES IN PLASTIDS: SYNTHESIS AND ASSEMBLY, 1993

David H. Rohrbach and Rupert Timpl (editors). MOLECULAR AND CELLULAR ASPECTS OF BASEMENT MEMBRANES, 1993

Danton H. O'Day (editor). SIGNAL TRANSDUCTION DURING BIOMEMBRANE FUSION, 1993

John K. Stevens, Linda R. Mills, and Judy E. Trogadis (editors). THREE-DIMENSIONAL CONFOCAL MICROSCOPY: VOLUME INVESTIGATION OF BIOLOGICAL SPECIMENS, 1994

Lawrence I. Gilbert, Jamshed R. Tata, and Burr G. Atkinson (editors). METAMORPHOSIS: POSTEMBRYONIC REPROGRAMMING OF GENE EXPRESSION IN AMPHIBIAN AND INSECT CELLS, 1996

R. John Ellis (editor). THE CHAPERONINS, 1996

R. Curtis Bird, Gary S. Stein, Jane B. Lian, and Janet L. Stein (editors). NUCLEAR STRUCTURE AND GENE EXPRESSION, 1997

Nuclear Structure and Gene Expression

Edited by

R. Curtis Bird
Department of Pathobiology
Auburn University
Auburn, Alabama

Gary S. Stein
Department of Cell Biology and Cancer Center
University of Massachusetts Medical Center
Worcester, Massachusetts

Jane B. Lian
Department of Cell Biology and Cancer Center
University of Massachusetts Medical Center
Worcester, Massachusetts

Janet L. Stein
Department of Cell Biology and Cancer Center
University of Massachusetts Medical Center
Worcester, Massachusetts

ACADEMIC PRESS

San Diego London Boston New York Sydney Tokyo Toronto

Front cover photograph: A cluster of serum-starved HeLa S3 cells treated with okadaic acid (19 n*M*) and reacted with a murine anti-cdk1 (p34*cdc*[2]) monoclonal antibody (Oncogene Science, Inc., Uniondale, New York). Strong granular perinuclear activity (red) indicative of increased cdk1 expression, due to okadaic acid induction, is observed in treated cells. Courtesy of S. D. Lenz and R. C. Bird, Department of Pathobiology, Auburn University, Auburn, Alabama.

This book is printed on acid-free paper. ♾

Academic Press, Inc.
15 East 26th Street, 15th Floor, New York, New York 10010, USA
http://www.apnet.com

Academic Press Limited
24-28 Oval Road, London NW1 7DX, UK
http://www.hbuk.co.uk/ap/

Library of Congress Cataloging-in-Publication Data

Nuclear structure and gene expression / edited by R. Curtis Bird.
p. cm. -- (Cell biology series)
Includes bibliographical references and index.
ISBN 0-12-100160-1 (alk. paper)
1. Nuclear matrix. 2. Gene expression. I. Bird, R. Curtis,
II. Series: Cell biology.
QH603.N83N83 1996
574.87'32--dc20 96-35306
CIP

PRINTED IN THE UNITED STATES OF AMERICA
96 97 98 99 00 01 BB 9 8 7 6 5 4 3 2 1

Contents

III Chromatin Structure and Transcription Complexes

6 Nuclear Architecture in Developmental Transcriptional Control of Cell Growth and Tissue-Specific Genes

Gary S. Stein, André J. van Wijnen, Janet L. Stein, Jane B. Lian, and Martin Montecino

7 Transcription by RNA Polymerase II and Nuclear Architecture

Derick G. Wansink, Luitzen de Jong, and Roel van Driel

IV Nuclear mRNP Complexes

8 Protein Deposition on Nascent Pre-mRNA Transcripts

Sally A. Amero and Kenneth C. Sorensen

Contributors

Numbers in parentheses indicate the pages on which the authors' contributions begin.

Sally A. Amero (243), Department of Molecular and Cellular Biochemistry, Stritch School of Medicine, Loyola University Chicago, Maywood, Illinois 60153

Ronald Berezney (35), Department of Biological Sciences, State University of New York at Buffalo, Buffalo, New York 14260

R. Curtis Bird (145), Department of Pathobiology, Auburn University, Auburn, Alabama 36849

Benjamin J. Blencowe (3), Department of Biology and Center for Cancer Research, Massachusetts Institute of Technology, Cambridge, Massachusetts 02139

Luitzen de Jong (87, 215), E. C. Slater Institute, University of Amsterdam, 1018 TV Amsterdam, The Netherlands

Yoshinori Kohwi* (111), The Burnham Institute, La Jolla Cancer Research Center, La Jolla, California 92037

Terumi Kohwi-Shigematsu* (111), The Burnham Institute, La Jolla Cancer Research Center, La Jolla, California 92037

Jane B. Lian (177), Department of Cell Biology and Cancer Center, University of Massachusetts Medical Center, Worcester, Massachusetts 01655

Karin A. Mattern (87), E. C. Slater Institute, University of Amsterdam, 1018 TV Amsterdam, The Netherlands

Martin Montecino† (177), Department of Cell Biology and Cancer Center, University of Massachusetts Medical Center, Worcester, Massachusetts 01655

* Present Address: Department of Cancer Biology, Lawrence Berkeley National Laboratory, University of California, Berkeley, CA 94720.

† Present Address: Department of Molecular Biology, University of Concepción, Casilla 2407, Concepción, Chile.

Jeffrey A. Nickerson (3), Department of Biology, Massachusetts Institute of Technology, Cambridge, Massachusetts 02139

Sheldon Penman (3), Department of Biology, Massachusetts Institute of Technology, Cambridge, Massachusetts 02139

Kenneth C. Sorensen (243), Department of Molecular and Cellular Biochemistry, Stritch School of Medicine, Loyola University Chicago, Maywood, Illinois 60153

Gary S. Stein (177), Department of Cell Biology and Cancer Center, University of Massachusetts Medical Center, Worcester, Massachusetts 01655

Janet L. Stein (177), Department of Cell Biology and Cancer Center, University of Massachusetts Medical Center, Worcester, Massachusetts 01655

Roel van Driel (87, 215), E. C. Slater Institute, University of Amsterdam, 1018 TV Amsterdam, The Netherlands

André J. van Wijnen (177), Department of Cell Biology and Cancer Center, University of Massachusetts Medical Center, Worcester, Massachusetts 01655

Derick G. Wansink (215), E. C. Slater Institute, University of Amsterdam, 1018 TV Amsterdam, The Netherlands

Preface

There is emerging recognition that structural and functional properties of the cell nucleus are interrelated. During the past several years, extensive insight into the complexities of linkage between the components of nuclear architecture and the parameters of transcription, processing of gene transcripts, and DNA replication has accrued. There has been an evolution in our understanding of gene expression from the cataloging of gene regulatory sequences and cognate transcription factors to the pursuit of mechanisms mediating cross talk between promoter domains and the convergence of multiple signaling pathways at multiple promoter elements. These mechanistic questions are being pursued within the context of responsiveness to physiological regulatory signals controlling proliferation, differentiation, and commitment to cell and tissue phenotypic requirements. The involvement of nuclear structure in facilitating regulatory mechanisms is becoming increasingly evident. This has resulted in a model of cell regulation in which subtle changes are evolved through integrated spatial, temporal, structural, and informational response mechanisms.

This volume assimilates the contributions of genome organization as well as of the components of the nuclear matrix to the control of DNA and RNA synthesis. Nuclear domains which accommodate DNA replication and gene expression are considered in relation to short-term developmental and homeostatic requirements as well as to long-term commitments to phenotypic gene expression in differentiated cells. Consideration is given to the involvement of nuclear structure in gene localization as well as to the targeting and concentration of transcription factors. Aberrations in nuclear architecture associated with and potentially functionally related to pathologies are evaluated. Tumor cells are described from the perspective of the striking modifications in both the composition and organization of nuclear components. Concepts as well as experimental approaches which are defining functionality of nuclear morphology are presented.

The challenges we now face are to further define components of nuclear architecture that contribute to the regulation of gene expression and serve as rate-limiting steps in DNA replication and transcriptional control. There is a necessity to resolve the involvement of multiple levels of nuclear structure in establishing the fidelity of physiological responsiveness.

Our current knowledge of promoter organization and the repertoire of transcription factors that mediate activities provide a single dimensional map of options for biological control. We are beginning to appreciate the additional structural and functional dimensions provided by chromatin structure and nucleosome organization, as well as by subnuclear localization and targeting of both genes and transcription factors by specific components of the nuclear matrix. Particularly exciting is the growing body of evidence for dynamic modifications in nuclear structure which parallel modifications in the expression of genes. However, the extent to which nuclear structure regulates and/or is regulated by modifications in gene expression remains to be experimentally established. Interrelationships between the control of chromatin organization and higher order nuclear architecture are not well understood.

It is imperative to determine whether nuclear matrix-mediated subnuclear distribution of actively transcribed genes is responsive to nuclear matrix association of transcriptional and posttranscriptional regulatory factors. And we cannot dismiss the possibility that the association of regulatory factors with the nuclear matrix is consequential to sequence-specific interactions of transcriptionally active genes with the nuclear matrix. As these issues are resolved, we will gain additional insight into the determinants of cause and/or effect relationships which interrelate specific components of nuclear architecture with gene expression at the transcriptional and posttranscriptional levels. However, it is justifiable to anticipate that while nuclear structure–gene expression interrelationships are operative under all biological conditions, situation-specific variations are the rule rather than the exception. As subtleties in the functional components of nuclear architecture are further defined, the significance of nuclear domains to DNA replication, transcription, processing of RNA transcripts, and nuclear–cytoplasmic exchange of regulatory molecules will be additionally understood.

R. Curtis Bird
Gary S. Stein
Jane B. Lian
Janet L. Stein

I

Introduction

1

The Nuclear Matrix: Past and Present

SHELDON PENMAN,[1] BENJAMIN J. BLENCOWE,[1,2] AND JEFFREY A. NICKERSON[1]

Department of Biology[1] and Center for Cancer Research[2]
Massachusetts Institute of Technology
Cambridge, Massachusetts

I. INTRODUCTION

Science that considers the very large or the very small must deal with matters beyond the mind's ability to envision accurately. In physics, at both cosmological and subatomic scales, we simply abandon attempts to visualize and substitute highly abstruse mathematical formulations. Cell biology also must often deal with the very minute but, as a nonmathematical science, it has no recourse to mathematical abstraction. Instead, cell biologists construct the best mental pictures possible from the often limited information available. Unfortunately, it is easy to forget how much our ideas owe to imperfect observing instruments and techniques and to the faulty perceptions to which they may lead. Once incorporated into conventional wisdom, such beliefs, even when quite vague, have remarkable tenacity and acquire vociferous champions.

The heretofore often halting progress in understanding nuclear structure offers a case in point. Our brains cannot really picture the dense complexity contained within the 5-μm spheroid of the eukaryotic nucleus. Indeed, even if a nucleus were expanded 100,000 times to about the size of a basketball, it would still appear so densely crammed as to boggle the imagination. Chromatin would correspond to 60 miles of fine filament, meticulously packed and ordered in the sphere so that specific bits, about one foot long, are accessible for transcription into molecules of RNA. These transcripts are then processed in a series of complex, highly ordered steps by machinery tucked among the dense coils of chromatin. Once every cell cycle, the entire length of chromatin is duplicated and the 120 miles of filament are then folded into 46 bundles or chromosomes. Astonishingly, the scaffolding that established this order is able to disassemble and reassemble into two smaller versions every time the cell divides.

Even at the enormously magnified scale of a basketball, the nucleus presents an unimaginable array of complex structures and machines. Shrunk back to the 5-μm size of the nucleus, its structure would seem miraculous. Oddly, until recently, the common perception of the nuclear interior was of a formless "plasm" with chromatin randomly floating about. Although indirect evidence of an organizing nuclear structure was abundant, opposition to its acceptance was remarkably tenacious. The reluctance to accept the existence of nonchromatin nuclear scaffolding resulted, at least partly, from the utter invisibility of the nuclear matrix in conventional electron microscopy (EM) micrographs.

Two technical advances have made possible microscopy of the nuclear matrix. First, new extraction protocols are able to remove the electron-opaque chromatin without totally devastating the remaining nuclear structure. These procedures afford a biochemically purified nuclear matrix. However, to be effective for microscopy, the procedures must be coupled with embedment-free EM which can image, often with striking clarity, the resulting three-dimensional fiber network of the nuclear matrix.

Individual catalytic processes and their enabling machinery are not uniformly distributed in the nucleus but rather are constrained to highly localized spatial domains. Many of the reactions, functional components, and intermediates in nuclear metabolism have been identified and characterized by the combined application of biochemical and molecular genetic approaches. In particular, the development of *in vitro* systems that faithfully perform DNA replication transcription, and the processing of pre-messenger RNA (pre-mRNA) has allowed the detailed dissection of these processes. Although powerful for the characterization of essential components, a major limitation of *in vitro* systems is their apparent inability to reproduce accurately many important *in vivo* kinetic and regulatory properties. The limitations of *in vitro* systems suggest that important features have been lost. Perhaps the major missing element is the organization of the intact nucleus, whose architecture may integrate and coordinate transcription and processing.

A wide variety of nuclear metabolic domains is emerging. The largest of these are nucleoli, sites of ribosomal RNA synthesis and ribosomal subunit assembly. There are various smaller domains identified by the processes they contain, such as DNA replication or transcription, as well as domains identified by their composition, such as those highly enriched in RNA splicing factors. The structural basis for this segregation of metabolic components has received relatively little attention, perhaps, because it is not easily analyzed in a well-defined *in vitro* system. It is striking that gentle removal of chromatin leaves the domain organization of the nucleus almost unperturbed. This was a compelling reason for postulating a nonchromatin nuclear scaffolding that organizes the domains and, consequently, nuclear metabolism.

The link between nuclear architecture and gene expression may also be an important clue for understanding phenotype-specific gene expression and the alteration of phenotype in malignancy. While regulation of the immediate response of genes to environmental signals is becoming increasingly well understood, far less is known about the mechanisms that set the stable patterns of gene expression in differentiated cells. Structural features of the nucleus, including size, shape, and internal organization, vary with cell type and are radically altered by malignancy. Coincident with these changes are large changes in gene expression. It seems likely that phenotype- and malignancy-specific alterations in nuclear matrix structure and composition are the link between these long-term, tissue-specific states of gene expression, states whose alteration characterizes the malignant cell.

II. THE NUCLEAR MATRIX: HISTORY AND DEFINITION

Early light microscopy showed the cell nucleus as containing only chromatin and nucleoi. The remainder of the nuclear space seemed to be occupied only by a translucent gel which, in the absence of any defining properties, was termed the nuclear sap, nucleoplasm, or karyolymph, terms that, despite their obvious inappropriateness, persist until today. Unfortunately, early electron micrographs did little to contradict this simplistic view of nuclear architecture. The conventional resin-embedded, ultrathin section shows only the nuclear contents that happen to be exactly at the surface of the section. The images show masses of chromatin, dense nucleoli, and some vague, initially uninterpretable structures. However, with more sophisticated selective staining techniques, even the thin section reveals a more highly structured nucleus than had heretofore been imagined.

A. Fibrogranular Ribonucleoprotein Network of the Intact Nucleus

Careful study of electron micrographs showed that, in addition to the conspicuous dense chromatin, the nucleus contained a fibrogranular network in the interchromatin space between patches of condensed heterochromatin. The ethylenediaminetetraacetic acid (EDTA)-regressive stain selectively accentuated RNA over DNA and showed that this material was distinct from chromatin and instead contained RNA (Bernhard, 1969; Monneron and Bernhard, 1969; Petrov and Bernhard, 1971). This RNA- or, more precisely, ribonucleoprotein (RNP)-containing network occurs throughout the nuclear interior except in regions of dense heterochromatin.

The RNP network revealed by the EDTA-regressive technique is made up of both granules and irregular fibers. These were named according to their spatial relationship to the chromatin; thus they are called peri- and interchromatin granules and fibers. As early as 1978, it was clear that removing chromatin left the RNP fibrils of this network in place, although they could be destroyed by subsequent ribonuclease (RNase) A digestion (Herman *et al.,* 1978). Experiments combining the EDTA-regressive stain with autoradiography to detect nascent (nonnucleolar) RNA indicated that the interchromatin granules and perichromatin fibrils are sites of active heterogeneous nuclear RNA (hnRNA) synthesis (Fakan *et al.,* 1976; Bachellerie *et al.,* 1975; Fakan and Hughes, 1989). Both structures contain polyadenylated RNA, hnRNP proteins, small nuclear RNPs (snRNPs), and non-snRNP splicing factors such as Ser-Arg (SR) proteins (Visa *et al.,* 1993; Spector *et al.,* 1991; Fakan *et al.,* 1984; Puvion *et al.,* 1984). Despite the deficiencies of early electron micrographs, the images led Fawcett to recognize that terms such as nuclear sap were no longer appropriate. He suggested nuclear matrix as a preferable substitute (Fawcett, 1966). Although many names have subsequently been used to identify this structure, we prefer the designation nuclear matrix as referring to the actual nuclear structure and not simply to the product of a particular isolation technique.

B. Development of Matrix Isolation Protocols

Salt extraction of nuclei leaves insoluble nuclear components, and this process formed the basis for the first matrix isolation procedures. Studies performed as early as 1942 (Mayer and Gulick, 1942) identified a subfraction of nuclear proteins that resisted extraction with high ionic strength solutions. Although these proteins were clearly of interest, their relative insolubility made them difficult to analyze. Similar high ionic strength extractions are now used in conjunction with nuclease digestion to isolate the nuclear matrix, which is largely free of chromatin. The biochemical analysis of the matrix, however, is still plagued by the insolubility of its structural components.

In their pioneering studies, Berezney and Coffey (1974, 1975, 1977) developed a protocol employing nuclease digestion and elevated ionic strength for the isolation of the nuclear matrix. Their procedure removed chromatin from isolated rat liver nuclei by deoxyribonuclease (DNase) I digestion, followed by removal of the cleaved chromatin by 2*M* NaCl extraction. The aim was to remove the last vestige of DNA and the procedure did provide a remnant structure free of chromatin, but the preservation of nuclear matrix ultrastructure was another matter.

Since the time of the early work of Berezney and Coffey, many nuclear matrix isolation protocols have been employed. Some are variants on the Berezney–Coffey method using different salts, ionic strengths, nuclear isolation procedures, or enzymes (e.g., Kaufman *et al.,* 1981; Capco *et al.,* 1982; Hodge *et al.,* 1977; Adolph, 1980; Stuurman *et al.,* 1990). Others remove chromatin from the nucleus by more novel means using detergents such as lithium 3,5-diiodosalicylate (LIS) (Mirkovitch *et al.,* 1984) or using electrophoretic fields with agarose-embedded nuclei (Jackson and Cook, 1985, 1986, 1988).

C. Comparison of Nuclear Matrix Protocols

While the existence of the nuclear matrix is no longer a contentious matter, there is less than unanimity as to its exact nature. The basic conundrum is to determine the prior nature of the matrix, which can be neither seen nor measured until after some rather drastic purification procedures. The problem is not new in some scientific disciplines; it is relatively rare in biology.

One path through a thicket of alternative preparation methods and conclusions about the nuclear matrix is to marry resinless section EM with biochemical determinations and apply reasonable criteria as to what a suitable outcome should be (Capco *et al.,* 1982; Fey *et al.,* 1986). A procedure that produced a highly disrupted matrix would be judged inferior to one that yielded an organized structure. Such judgments are, admittedly, initially somewhat subjective but, in the end, the approach must justify itself by yielding consistent, predictive results. Central to this strategy is the resinless section microscopy technique, which, in contrast to conventional thin-section microscopy, allows the whole structure to be imaged with high contrast and in three dimensions (Capco *et al.,* 1984; Penman, 1995). Procedures are judged according as to how organized a structure they provide.

Using resinless section microscopy to evaluate fractionation procedures, Penman and co-workers developed relatively gentle methods which effectively removed chromatin and soluble material from nuclei while leaving an apparently well-preserved nuclear structure (Capco *et al.,* 1982; Fey *et al.,* 1986; He *et al.,* 1990). In this nuclear matrix preparation procedure, whole cells were extracted first with Triton X-100 in a buffer of physiologic pH and ionic strength. This effectively removed membranes and soluble proteins. A stronger double-detergent combination of Tween 40 and deoxycholate in a low ionic strength buffer then removed the cytoskeleton except for the intermediate filaments. The remaining membrane-free nuclei, still connected at the surface lamina to the intermediate filaments, were then digested with DNase I and extracted with 0.25 *M* ammonium sulfate. These steps removed more than 97% of nuclear DNA and essentially all the histones. The resulting structure was termed an RNA-containing nuclear matrix because it retained about 75% of nuclear RNA. Ultrastructural preservation was judged by criteria such as these: Did the structure fill the nuclear space uniformly? Was the nuclear lamina visible and intact? Was the internal matrix attached to the lamina? Was fine structure still visible? According to these criteria, the nuclear matrix structure yielded by the DNase I 0.25 *M* ammonium sulfate procedure was better preserved than that provided by previous protocols. Most important, the composition of the protein showed highly reproducible features never seen before, such as a strong dependence on cell type and state of differentiation. Variants of the Penman protocol have been developed that are equivalent, as judged by ultrastructural preservation and protein composition (Fey *et al.,* 1986; Fey and Penman, 1988).

Much deeper insight into matrix structure is afforded by further extracting with 2 *M* NaCl the previous 0.25 *M* $(NH_4)_2SO_4$ nuclear matrix preparation. This is not the same as extracting the nucleus directly with 2 *M* salt, as in previous procedures; the stepwise extraction is critical.

D. Operational Definitions of the Nuclear Matrix

A bugaboo long plaguing nuclear matrix studies has been the operational definition. It has been suggested that the nuclear matrix is defined by a particular isolation protocol. The term was sometimes meant pejoratively to suggest that the matrix was essentially the product of the preparation method—or the investigator's imagination.

Today, the existence of a nonchromatin nuclear structure can be inferred simply from the spatial organization of nuclear metabolism and from the survival of this spatial organization after chromatin removal by many different protocols. Additionally, as we have discussed above (Section II,A), the RNA-containing structure that is at least part of the nuclear matrix can be imaged without any fractionation protocol by using

EDTA-regressive staining techniques. Although protocols in different laboratories isolate nuclear matrix preparations with differing degrees of preservation, our goal remains to understand the structure that organizes nuclear metabolism in living cells. It is that living structure that we define as the nuclear matrix and that we strive to understand.

III. ULTRASTRUCTURE OF THE NUCLEAR MATRIX

A. Principal Regions of the Nuclear Matrix

The nuclear matrix appears to be composed of at least two structurally distinct regions: the nuclear lamina, a protein shell constructed primarily of the lamin proteins A, B, and C (Gerace *et al.,* 1984; Krohne *et al.,* 1987; Franke, 1987; Fisher *et al.,* 1986; Georgatos *et al.,* 1994), and the internal nuclear matrix. However, truncating the matrix at the lamina is somewhat arbitrary. The nuclear matrix is firmly connected to the extensive network of intermediate filaments occupying the cytoplasmic space and anchoring on the outer surface of the nuclear lamina. This can be seen most clearly in a whole-mount electron micrograph (Figs. 1 and 2). The nuclear matrix and intermediate filaments are integrated into a single cell-wide structure that retains the overall geometry and appearance of the intact cell. We have referred to this architectural skeleton as the nuclear matrix–intermediate filament complex, implying that there are two separate constituents. It is well to remember that assigning a separate existence to the intermediate filaments is, in part, a historical accident; the filaments were observed in the cytoplasm long before the nuclear matrix was identified. It is not unlikely that the matrix and intermediate filaments will eventually be considered a single, integrated entity—the nuclear matrix–intermediate filament scaffold (NM–IF).

B. Nuclear Matrix–Intermediate Filament Scaffold

The intermediate filaments of adjacent cells are coupled to adjacent cells through the desmosomes (Fig. 2) and form, in epithelia at least, a

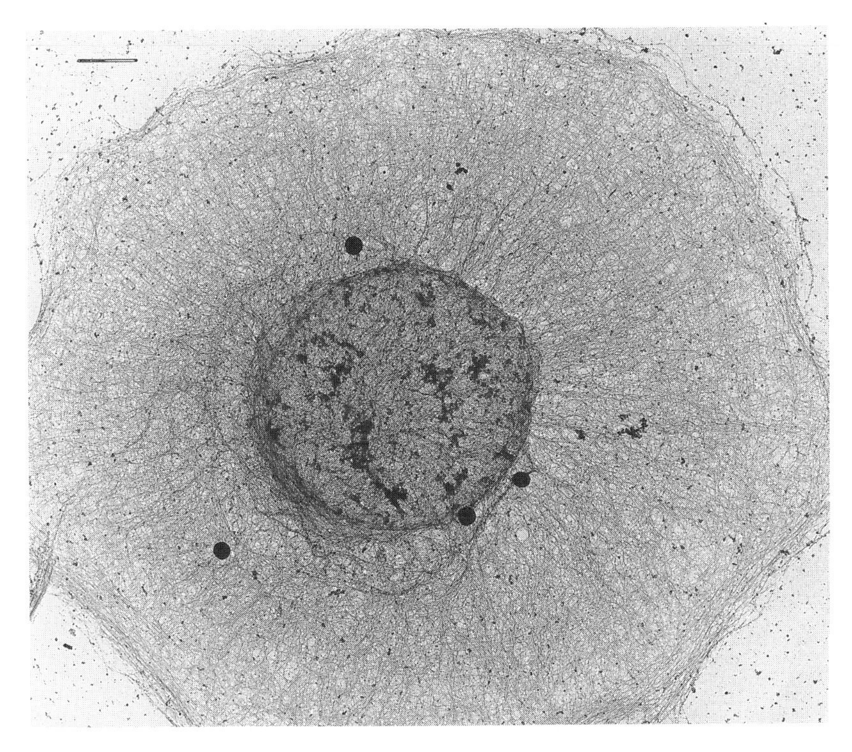

Fig. 1. The NM–IF scaffold. Colon cells grown on nickel grids were extracted with 0.5% Triton X-100 to remove soluble proteins. Chromatin was removed by DNase I digestion and 0.25 *M* ammonium sulfate extraction. The resulting nuclear matrix–intermediate filament scaffold was fixed, dehydrated, and critical point dried. In this whole-mount micrograph the cell has not been sectioned, and we are viewing the entire structure. The fine filamentous web formed by the intermediate filaments is connected to the internal nuclear matrix through connections to the nuclear lamina. The internal nuclear matrix is not visible but is masked by the dense lamina. Bar: 2 μm.

more or less continuous tissue-wide armature or scaffold. This armature, essentially a multicellular NM–IF, underlies tissue architecture and serves as a link between cell organization and issue morphology. At the level of the individual cell, the intermediate filaments mechanically couple the nucleus to the plasma membrane and probably contribute to the form and position of the nucleus. This may explain how nuclear shape is coupled to cell shape (Sims *et al.,* 1992; Hansen and Ingber, 1992; Ingber, 1993). Donald Ingber has proposed cellular tensegrity models for

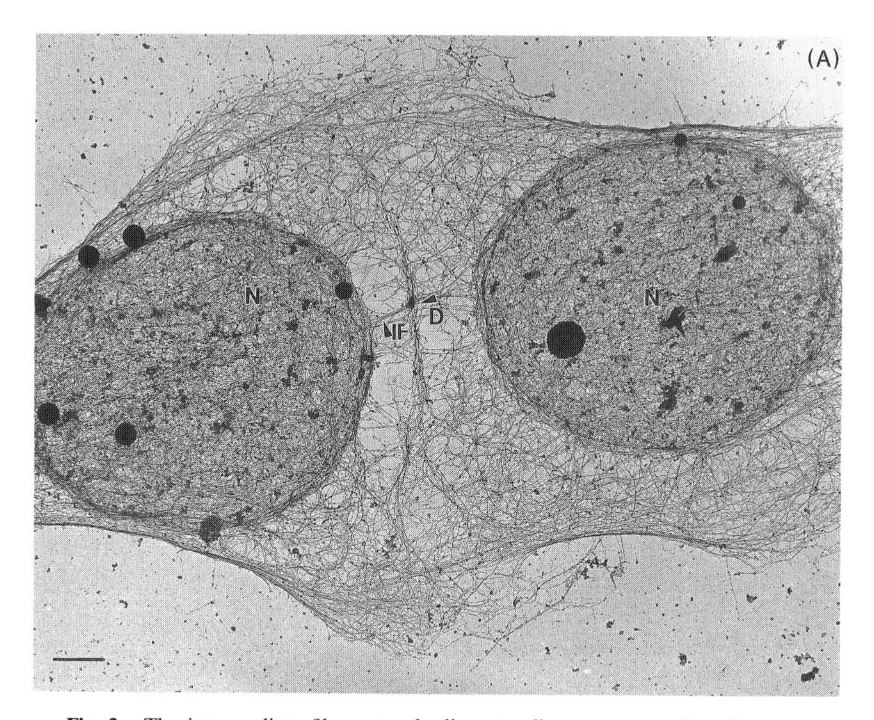

Fig. 2. The intermediate filaments of adjacent cells are connected at desmosomes. Colon cells were grown for whole-mount EM on nickel grids. Cells were detergent-extracted before chromatin was removed with DNase I and 0.25 *M* ammonium sulfate to uncover the nuclear matrix–intermediate filament scaffold. The structure was then fixed, dehydrated, and critical point dried. (A) Whole-mount electron micrograph of two adjacent colon cells. The intermediate filaments (IF) are seen connected to the nucleus (N) and the desmosomes (D) at the cell periphery. Bar: 1 μm. (B) A higher-magnification view of the same colon cells showing that the intermediate filaments (IF) of adjacent cells are interconnected at desmosomes (D). Bar: 0.5 μm.

mechanical signal transduction (Ingber and Folkman, 1989; Ingber *et al.*, 1994) that can explain the observed effects of cell shape on growth and gene expression (Folkman and Moscona, 1978; Ben-Ze'ev *et al.*, 1980; Ingber *et al.*, 1994). This model requires nuclear structure to be mechanically linked through the cytoskeleton to the exterior of the cell. The NM–IF is an excellent candidate for this function. In tissue, the multicellular NM–IF presumably couples internal nuclear structures and

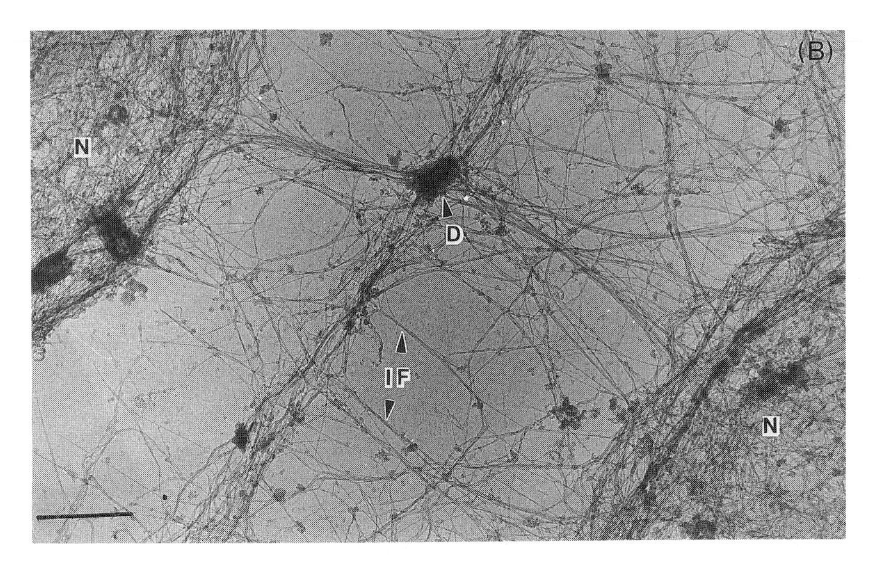

Fig. 2. (*Continued*)

nuclear metabolism to overall tissue form, affording a means of spatially regulating gene expression that does not require invoking intangible morphogen gradients.

The tissue-wide coupling of nuclei through the NM–IF may also explain a profound but often overlooked puzzle in cell differentiation: what signals modulate gene expression over the extent of a tissue (e.g., altering skin from the palm to the back of the hand)? Such modulation is often governed by another tissue (e.g., the endoderm in the case of skin), but the question then is how a tissue pattern is established and maintained in the signaling cells. The tissue-wide NM–IF scaffolding, evident in nuclear matrix preparations made from model tissues, may have evolved precisely to meet these needs of tissue pattern as well as the requirements of cell structure. Consequently, we might expect such cell organization to be a property of metazoans (and presumably metaphyta) and absent from or greatly altered in simple unicellular eukaryotes. The most primitive eukaryotes we have examined so far are dinoflagellates which apparently lack an intermediate filament scaffolding attached to the nucleus (unpublished observations, 1993).

C. Nuclear Matrix Surface or Nuclear Lamina

Conventional electron micrographs of the cell nucleus in intact cells show a double membrane forming the outer nuclear boundary (reviewed in Nigg, 1989). These membranes disappear as such in the initial Triton X-100 detergent extraction and a single proteinaceous layer, the nuclear lamina, forms the outer boundary of the nuclear matrix. It appears that the outer nuclear membrane is completely dissolved. In contrast, the inner nuclear membrane apparently consists of lipids intercalated into the lamin network of the nuclear lamina. The lamina remains after the lipids are extracted. The nuclear pores that normally traverse the envelope remain embedded in the nuclear lamina (Fig. 3). The attachment of pores to a nuclear lamina was first proposed by Aaronson and Blobel (1974), and the structure is often referred to as the pore–lamina complex (Fisher *et al.*, 1982; Gerace *et al.*, 1978, 1984). The intermediate filaments appear to be joined to the lamina at pores by filamentous crossbridges (Carmo-Fonseca *et al.*, 1988) whose protein composition has not yet been identified.

D. Nuclear Matrix RNA

In addition to its structural proteins, the nuclear matrix has a large amount of hnRNA that is closely associated. When prepared using RNase inhibitors, the matrix retains more than 70% of total nuclear RNA (Fey *et al.*, 1986; Fey and Penman, 1988; He *et al.*, 1990). It is not completely clear whether this RNA, or some subset of it, should be considered a component of the nuclear matrix structure. Certainly it has a role in chromatin organization because digesting nucleus with RNase A or pretreating cells with transcription inhibitors causes a massive collapse of the chromatin structure (Nickerson *et al.*, 1989).

IV. CORE FILAMENTS AND METABOLIC CENTERS OF THE NUCLEAR MATRIX

Analysis of nuclear matrix structure has lagged behind awareness of its many metabolic functions. Serious technical as well as intellectual

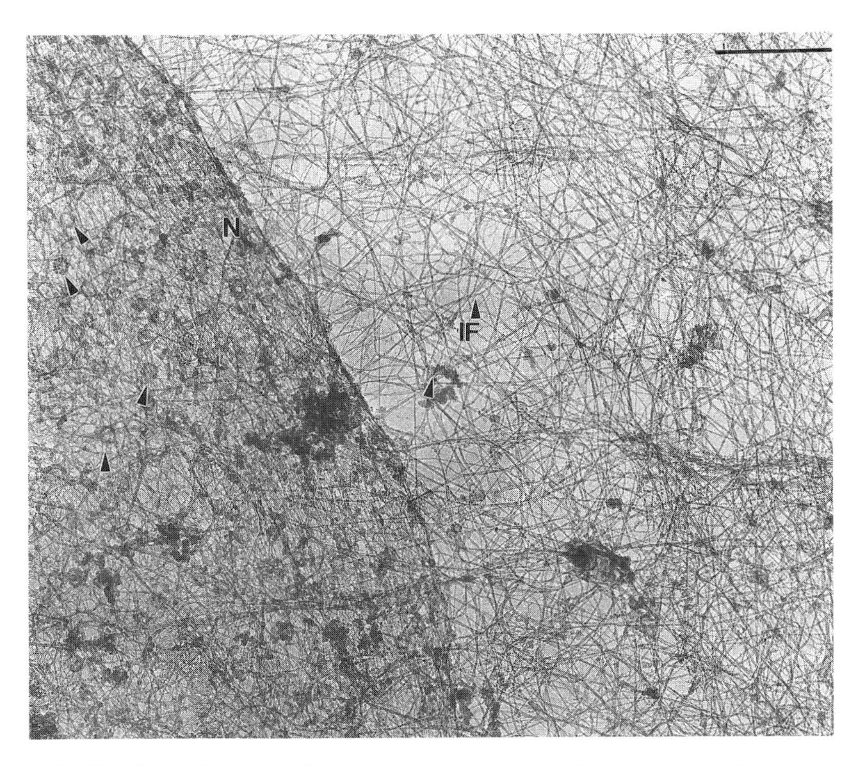

Fig. 3. The nuclear lamina has embedded nuclear pores. Colon cells grown for whole-mount EM on nickel grids were first extracted with 0.5% Triton X-100. Chromatin was removed by DNase I digestion and 0.25 *M* ammonium sulfate extraction. The resulting nuclear matrix–intermediate filament scaffold was fixed, dehydrated, and critical point dried. The intermediate filaments (IF) are connected to the nuclear lamina (N) through connections to the nuclear lamina, which displays numerous pores (arrowheads). In the unextracted nucleus (not shown) the nuclear envelope covers the lamina, leaving only the pores exposed. Bar: 0.5 μm.

obstacles had to be overcome, and only now are we making some progress in structure studies. So far, the evidence suggests that the nuclear matrix consists of various metabolically active sites enmeshed in and positioned by an armature of anastomosing 10-nm "core" filaments (He *et al.*, 1990). A standard 0.25 *M* $(NH_4)_2SO_4$ nuclear matrix preparation (Capco *et al.*, 1982; Fey *et al.*, 1986) is further extracted with 2 *M* NaCl to reveal the underlying network of core filaments. Enmeshed in the filament network are numerous dense, granular, and morphologically varied bodies that are the sites of much of nuclear metabolism (Fig. 4). At higher magnification, the nuclear core filaments of human cells are heterogeneous in size,

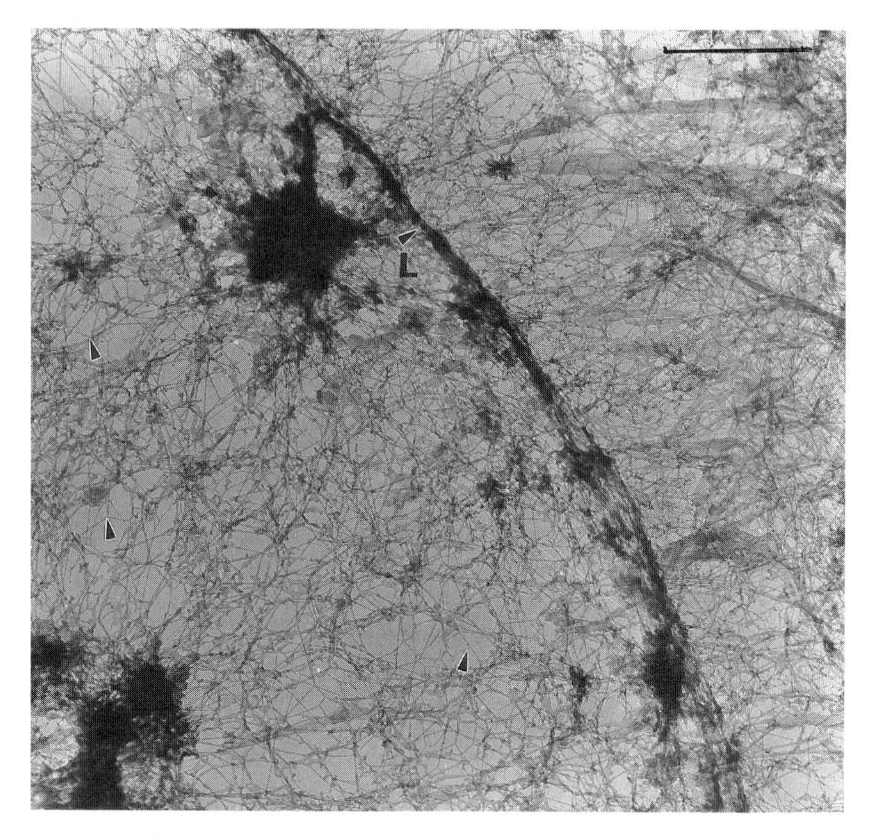

Fig. 4. The core filaments of the HeLa nuclear matrix. HeLa cells grown in suspension were detergent-extracted to remove soluble proteins and then digested with pure DNase I and extracted with 0.25 *M* ammonium sulfate to remove chromatin and uncover the nuclear matrix. The matrix was then further extracted with 2 *M* NaCl to reveal the core filaments of the nuclear matrix. The structures were fixed, sectioned, and prepared for EM by the resinless technique (Capco *et al.*, 1984). The core filament network, shown at high magnification, is connected to the nuclear lamina (L). The web of core filaments is intimately connected to and suspends a dense mass. Some similar masses stain with antibodies against RNA splicing factors, and these may correspond to speckled domains. The arrowheads point to the smooth junctions between core filaments. Bar: 0.2 μm.

with two prevalent classes whose mean diameters are 9 nm and 13 nm. The filaments are organized in a network that is highly branched and anastomose smoothly, with no obvious junction structures (Fig. 4). This arrangement of core filaments and enmeshed dense bodies has been found in human HeLa, MCF-7, SiHa, and CaSki cells, as well as in rat hepatocytes, and is probably universal in metazoan cells.

Nuclear filaments have also been reported by Jackson and Cook, (1988), who use a very different procedure to remove part of the chromatin under isotonic conditions. The unusual nature of their procedure makes it difficult to compare their results directly with those obtained with the 0.25 *M* $(NH_4)_2SO_4$ nuclear matrix preparation (He *et al.,* 1990). They observed 10-nm filaments which showed regular striations not seen in the salt-eluted matrix preparation. The demonstration of 10-nm core filaments by two radically different matrix preparation procedures strongly supports the hypothesis that these filaments are a basic component of nuclear matrix structure.

V. DENSE BODIES, FLUORESCENCE SPECKLES, AND FUNCTIONAL DOMAINS OF THE NUCLEUS

A remarkable conclusion from recent studies is that the great majority of nuclear antigens stain with punctate fluorescence patterns. Parallel EM studies have shown that the speckles correspond to the dense bodies seen in Fig. 4. The core filaments support and position the enmeshed, granular dense bodies, which vary widely in their overall morphology and functions. The largest of these appear to be nucleolar remnants surviving the 2 *M* extraction. Nucleoli are the most prominent and best-studied nuclear domains. They are known to be the sites of ribosomal RNA synthesis, processing, and assembly into ribosomal subunits, although strong hints of other functions are emerging. The known structural and functional properties of the nucleolus have been extensively reviewed elsewhere (Fischer *et al.,* 1991; Scheer *et al.,* 1993).

Several types of nonnucleolar matrix domains have been defined based on criteria such as size, number per nucleus, and presence of specific nuclear constituents. These include structures involved in DNA replication, as well as distinct matrix-associated structures involved in the metabolism of pre-mRNA such as transcript foci, speckled domains, coiled bodies, and tracks. These nuclear domains, labeled for fluorescence optical microscopy or by gold bead conjugated probes for resinless EM, clearly correspond to the dense bodies seen in resinless section electron micrographs of nuclear matrix structure.

A. DNA Replication Foci

Among the earliest nonnucleolar domains to be studied are the replication foci. Eukaryote DNA replication takes place in a few hundred discrete replication foci, each of which is thought to contain some of the estimated 50,000 replication origins activated during S phase (Nakamura *et al.,* 1986; Nakayasu and Berezney, 1989). The size and distribution of these foci change in an ordered fashion during S phase of the cell cycle (D'Andrea *et al.,* 1983; Goldman *et al.,* 1984; Hatton *et al.,* 1988; O'Keefe *et al.,* 1992). These sites also contain factors involved in DNA modification and cell cycle control. To date, these factors include DNA polymerase α (Bensch *et al.,* 1982), replication protein A (a 70-kDa subunit) Cardoso *et al.,* 1993), DNA ligase (Lasko *et al.,* 1990), proliferating cell nuclear antigen (PCNA) (Bravo and MacDonald-Bravo, 1987), DNA methyltransferase (Leonhardt *et al.,* 1992), cyclin A, and cyclin-dependent kinase 2 (cdk2) (Cardoso *et al.,* 1993).

The site of DNA synthesis is marked specifically by incorporating 5′-bromodeoxyuridine triphosphate (BrdUTP) into nascent DNA which is then detected with anti-BrdU antibodies. The foci are labeled exclusively (Nakamura *et al.,* 1986; Nakayasu and Berezney, 1989; O'Keefe *et al.,* 1992). The extraction of chromatin leaves remnant DNA and some replication factors unchanged in spatial arrangement, indicating coupling to the nuclear matrix (Pardoll *et al.,* 1980; Berezney *et al.,* 1982; Dijkwel *et al.,* 1986; Jackson and Cook, 1986; Razin, 1987; Foster and Collins, 1985; Tubo and Berezney, 1987a,b; Vaughn *et al.,* 1990; Nakayasu and Berezney, 1989). Resinless section microscopy has been used to visualize these "replication factories" within the matrix–filament network (Hózak *et al.,* 1993).

B. Transcription Foci

Pulse labeling of cells with 5′-bromouridine 5′-triphosphate reveals that transcription by RNA polymerase II occurs in over 100 well-defined areas distributed throughout the nucleoplasmic space except for the nucleoli (Jackson *et al.,* 1993; Wansink *et al.,* 1993). Consistent with previous observations that the earliest pulse-labeled transcripts associate with the nuclear matrix, both the number and the nuclear distribution of the transcript domains are unchanged after removal of chromatin and extraction of the nuclei in 0.25 *M* salt. These sites are not generally coincident with the larger and less abundant speckled domains, where splicing factors are concentrated (Wansink *et al.,* 1993).

C. Coiled Bodies

There are approximately one to five coiled bodies per cell nucleus. These structures have a highly conserved morphology and have been detected in organisms as diverse as mammals and plants. Coiled bodies were initially identified by the Spanish cytologist Ramón y Cajál at the turn of the century (Ramón y Cajál, 1903, 1909) and rediscovered several times during the past three decades (Monneron and Bernhard, 1969; Hardin *et al.,* 1969; Seite *et al.,* 1982; Lafarga *et al.,* 1983; Carmo-Fonseca *et al.,* 1992). Autoimmune patient sera were discovered which selectively label these structures. These autoimmune sera recognize a protein of 80 kDa, p80 coilin, highly concentrated in coiled bodies but also present in lower concentration throughout the nucleoplasm, excluding the nucleoli (Raska *et al.,* 1990, 1991; Andrade *et al.,* 1991). The p80 coilin sequence (Andrade *et al.,* 1991) bears some similarity to a protein concentrated in snRNP-rich structures, snurposomes, in the amphibian oocyte (Wu *et al.,* 1993). Coiled bodies are sites of concentration of each of the major snRNAs involved in pre-mRNA splicing and also of several snRNP-associated proteins (Fakan *et al.,* 1984; Eliceiri and Ryerse, 1984; Raska *et al.,* 1991; Carmo-Fonseca *et al.,* 1991a,b; Huang and Spector, 1992; Blencowe *et al.,* 1993). In addition to snRNP components, coiled bodies contain elevated concentrations of both the 65-kDa and 35-kDa subunits of the non-snRNP splicing factor U2AF (Carmo-Fonseca *et al.,* 1991a; Zamore and Green, 1991; Zhang *et al.,* 1992).

Immunofluorescence staining of p80 coilin is unaltered in matrix preparations compared to unextracted cells, indicating that the coiled body structure and p80 coilin are associated with the nuclear matrix (our unpublished findings). Also, U2snRNA remains associated with coiled bodies after chromatin extraction in a low-salt buffer but, unlike p80 coilin, the majority of U2 is lost on extraction at higher salt concentrations (Carmo-Fonseca *et al.*, 1991a). Although these structures contain spliceosomal snRNP components and a subset of non-snRNP splicing factors, they do not appear to be sites of pre-mRNA splicing, and their function and significance remain unknown.

D. Speckled Domains of RNA Splicing Factors

Mammalian nuclei typically contain 20–50 domains which stain with probes specific for pre-mRNA splicing factors in a characteristic speckled staining pattern. These speckled domains tend to be concentrated in a plane just below the midline and parallel to the growth surface of cultured cells (Carter *et al.*, 1993). Conventional electron micrographs show the domains corresponding to a network of interconnecting structures, often referred to as interchromatin granules and perichromatin fibrils (Perraud *et al.*, 1979; Spector *et al.*, 1993, 1991; Fakan *et al.*, 1984; see the following discussion). These structures are preferentially stained by the EDTA-regressive method (Bernhard, 1969). Resinless section electron micrographs show interchromatin granules and perichromatin fibrils that are, in reality, dense bodies enmeshed in the extensive network of nuclear matrix core filaments (reviewed in Nickerson and Penman, 1992).

Nuclear splicing factor speckles were first detected by the staining patterns of autoimmune patient sera which recognize protein or RNA components of snRNPs (Perraud *et al.*, 1979; Lerner *et al.*, 1981; Spector *et al.*, 1983). More recently, speckled domains (and coiled bodies) were simultaneously detected with nuclease-stable antisense probes specific for each snRNA involved in splicing, in combination with monoclonal antibodies (MAbs) specific for snRNP proteins and the snRNA trimethylguanosine (m3G) cap structure (Carmo-Fonseca *et al.*, 1992; Huang and Spector, 1992). The speckled domains are also highly enriched in non-snRNP splicing factors belonging to the SR family (Roth *et al.*, 1990; Fu and Maniatis, 1990; Spector *et al.*, 1991). A group of SR proteins share a common phospho epitope structure detected by a MAb104 (Zahler *et al.*, 1992). These proteins have one or two RNA recognition motifs (RRMs) at the N terminus and a domain rich in serine–arginine diamino acid repeats at the C terminus. Antibodies which recognize one or more proteins belonging to the MAb104-reactive family show pronounced speckled staining patterns that precisely coincide with the speckled staining patterns obtained with antibodies to snRNPs. Other proteins containing an SR domain, which are not of the MAb104 class, are also concentrated in speckles. These include the U1 snRNP-associated 70-kDa protein (Verheijen *et al.,* 1986), a 64-kDa autoantigen with sequence similarities to the 65-kDa subunit of U2AF (Imai *et al.,* 1993), and two different *Drosophila* alternative splicing factors: *suppressor of white apricot* $su(w^a)$ and *transformer (tra)* (Li and Bingham, 1991). In addition, we have recently identified three high molecular weight proteins that are related to the SR family of proteins and which are present in speckled domains (Blencowe *et al.,* 1995).

E. RNA Transcript "Tracks"

An important function of the nuclear matrix is the transport of RNA molecules within the nucleus. Using probes that distinguish the localization of a particular gene from its corresponding transcript, and that distinguish intron from exon sequences within the transcript, it was shown that certain viral and cellular pre-mRNAs are localized within curvilinear tracks (Lawrence *et al.,* 1989; Huang and Spector, 1991; Xing *et al.,* 1993). Both the quantitative and qualitative appearance of individual RNA tracks was preserved in a matrix preparation that removed 95% of chromatin as well as bulk protein and phospholipid material (Xing and Lawrence, 1991). These tracks emerge from a single focus of transcription, coincident with the gene, and then usually extend out toward the nuclear periphery. Consistent with the splicing taking place within the track, intron sequences are detected in only a relatively small portion of the track, whereas exon sequences are detected throughout the track (Xing *et al.,* 1993). Significantly, tracks defined by some transcripts intersect a speckled domain, whereas other tracks do not.

VI. NUCLEAR MATRIX AND CELL PHENOTYPE

Perhaps the most intriguing aspect of nuclear matrix function is its apparent role in regulating gene expression. In higher eukaryotes, every cell contains the complete genome of the organism but expresses only a precisely defined subset of genes. It is the expression of this set of genes that determines and defines the stable cell phenotype. Phenotype-specific genes are expressed at precisely the right place and moment in development and can remain stably turned on for the life of the organism. The mechanism responsible for this process remains one of the great puzzles of biology.

The cell structure regulating phenotype-specific gene expression probably has a cell type-specific protein composition. Such cell type specificity eluded researchers until an effective matrix isolation procedure was developed. Studies on the nuclear matrix showed that although chromatin proteins are largely invariant from cell to cell, the protein composition of the nuclear matrix depends on the differentiation state and varies markedly with cell type. This was first observed by Fey and Penman (1988) in a study of the nuclear matrix protein composition of human cell lines. Examining nuclear matrix proteins by two-dimensional gel electrophoresis, they identified nuclear matrix proteins that were present in every cell type and others that were present only in cells derived from one tissue type. These observations were extended by Stuurman (Stuurman *et al.*, 1989, 1990), who found a similar dependence of nuclear matrix composition cell type in murine embryonal cells and tissues. Additional studies by Getzenberg and Coffey (1990; Partin *et al.*, 1993) showed that different cell types within the rat prostate express distinct, although overlapping, sets of nuclear matrix proteins.

Changes in nuclear matrix protein composition have also been observed during development. The fetal rat calvarial osteoblast has become an important *in vitro* developmental system for studying phenotype-specific gene expression (Stein *et al.*, 1994). In culture, fetal rat osteoblasts differentiate through three well-defined developmental stages, each with a unique and characteristic pattern of gene activity. Each developmental stage has a characteristic pattern of nuclear matrix protein composition (Dworetzky *et al.*, 1990). At least two of these osteoblast nuclear matrix proteins bind regulatory elements in the developmentally regulated osteocalcin gene (Bidwell *et al.*, 1993; Dworetzky *et al.*, 1992). Because the nuclear matrix is the only nuclear component with a markedly phenotype-specific composition and because it has contacts with gene sequences, it is reasonable to think that the nuclear matrix participates in the regulation and coordination of gene expression.

VII. THE NUCLEAR MATRIX IN MALIGNANCY

Pathological diagnosis is based on the inevitable conformation changes of the malignant cell. The most marked of these cellular alterations of nuclear size, shape, and organization accompany malignancy (Kamel *et al.*, 1990; Underwood, 1990). Although these nuclear alterations have been an intrinsic part of pathodiagnosis for perhaps a century, their cause and their biochemical significance are not understood. However, nuclear matrix studies may offer some insight. Because the matrix determines nuclear form, it is not surprising that malignancy-related alterations in nuclear structure should reflect a change in nuclear matrix composition and structure (Partin *et al.*, 1993; Pienta and Coffey, 1992).

Because nuclear matrix protein composition reflects cell type and state of differentiation (Fey and Penman, 1988; Stuurman *et al.*, 1989; Dworetzky *et al.*, 1990), the changes in these properties by malignant transformation should induce changes in matrix proteins. Several laboratories have reported that certain tumors have nuclear matrix proteins which are not present in the corresponding normal tissue. In the first such study, it was reported that human prostate tumors have a nuclear matrix protein not present in normal prostate tissue or in benign prostatic hyperplasia (Partin *et al.*, 1993). Several malignancy-specific nuclear matrix proteins in human infiltrating ductal carcinoma have been identified which are not present in normal breast tissue (Khanuja *et al.*, 1993). Six nuclear matrix proteins have been reported in human colon adenocarcinoma tumor samples which were absent from normal colon tissue (Keesee *et al.*, 1994). Such malignancy-specific nuclear matrix proteins, if their existence can be confirmed, may have considerable diagnostic value and may help to explain the changes in nuclear structure that accompany malignancy.

The finding of malignancy-specific nuclear matrix proteins may at first seem puzzling. As we have seen, nuclear matrix protein composition changes with phenotype. This has been most clearly established in the developing rat calvarial osteoblast. Studies on rat osteosarcoma cells suggest that these tumor cells are expressing matrix proteins normally expressed only at earlier developmental stages in the osteoblast lineage (Bidwell *et al.*, 1994). Thus, the changes in matrix protein composition which are reported in preliminary studies of human tumor tissue may eventually be explained as a regression to earlier stages in the development of the normal tissue.

VIII. CONCLUSION

Biochemical and molecular genetic analysis has identified and characterized many of the molecules and reactions important in nucleic acid metabolism. These approaches, however, provide little information about the architectural organization of metabolism in the cell. Indeed, the cell is far more than a vessel holding a solution of interacting molecules. The substructure within the cell imposes an essential spatial order on metabolism and provides an important level of regulation. Most important, a critical but currently much ignored role of cells in the organism is the establishment of tissue and organ architecture. Patterns of cell change over the topography of an organ cannot be understood in terms of current biochemistry alone. "Morphogen gradients" would convey insufficient spatial information and so are inadequate for specifying the complex three-dimensional pattern of tissue architecture. When the nucleus was thought to be a gel, it was impossible to envision how cell and tissue architecture might affect gene function. The coupling of gene expression to nuclear structure and thereby through the NM–IF to tissue-wide architecture provides a first indication of gene control systems that could respond to spatial signals.

A more complete understanding of nucleic acid metabolism in the living cell will require the combined application of biochemistry and structural cell biology. Studies of cell metabolism and cell architecture have not often been conducted together, partly because very different technical skills are required. In addition, different philosophies and temperaments may be required. Still, neither approach alone can answer all the important questions we can now pose about nuclear metabolism. It is clear that continuing cooperation, collaboration, and communication among disciplines will be essential if we are to characterize and understand the architecture of nuclear metabolism.

REFERENCES

Aaronson, P. P., and Blobel, G. (1974). On the attachment of the nuclear pore complex. *J. Cell Biol.* **62,** 746–754.

Adolph, K. (1980). Organization of chromosomes in HeLa cells: Isolation of histone-depleted nuclei and nuclear scaffolds. *J. Cell Sci.* **42,** 291–304.

Andrade, L. E., Chan, E. K., Raska, I., Peebles, C. L., Roos, G., and Tan, E. M. (1991). Human autoantibody to a novel protein of the nuclear coiled body: Immunological characterization and cDNA cloning of p80-coilin. *J. Exp. Med.* **173,** 1407–1419.

Bachellerie, J. P., Puvion, E., and Zalta, J. P. (1975). Ultrastructural organization and biochemical characterization of chromatin–RNA–protein complexes isolated from mammalian cell nuclei. *Eur. J. Biochem.* **58,** 327–337.

Bensch, K., Tanaka, S., Hu, S.-Z., Wang, T., and Korn, D. (1982). Intracellular localization of human DNA polymerase α with monoclonal antibodies. *J. Biol. Chem.* **257,** 8391–8396.

Ben-Ze'ev, A., Farmer, S., and Penman, S. (1980). Protein synthesis requires cell-surface contact while nuclear events respond to cell shape in anchorage-dependent fibroblasts. *Cell (Cambridge, Mass.)* **21,** 365–372.

Berezney, R., and Coffey, D. S. (1974). Identification of a nuclear protein matrix. *Biochem. Biophys. Res. Commun.* **60,** 1410–1417.

Berezney, R., and Coffey, D. S. (1975). Nuclear protein matrix: Association with newly synthesized DNA. *Science* **189,** 291–292.

Berezney, R., and Coffey, D. S. (1977). Nuclear matrix: Isolation and characterization of a framework structure from rat liver nuclei. *J. Cell Biol.* **73,** 616–637.

Berezney, R., Basler, J., Buchholtz, L. A., Smith, H. C., and Siegel, A. J. (1982). Nuclear matrix organization and DNA replication. *In* "The Nuclear Envelope and Nuclear Matrix" (G. G. Maul, ed.), pp. 183–197. Alan R. Liss, New York.

Bernhard, W. (1969). A new procedure for electron microscopical cytology. *J. Ultrastruct. Res.* **27,** 250–265.

Bidwell, J. P., van Wijnen, A. J., Fey, E. G., Dworetzky, S., Penman, S., Stein, J. L., Lian, J. B., and Stein, G. S. (1993). Osteocalcin gene promoter-binding factors are tissue-specific nuclear matrix components. *Proc. Natl. Acad. Sci. U.S.A.* **90,** 3162–3166.

Bidwell, J. P., Fey, E. G., van Wijnen, A. J., Penman, S., Stein, J. L., Lian, J. B., and Stein, G. S. (1994). Nuclear matrix proteins distinguish normal diploid osteoblasts from osteosarcoma cells. *Cancer Res.* **54,** 28–32.

Blencowe, B. J., Carmo-Fonseca, M., Behrens, S.-E., Lührmann, R., and Lamond, A. I. (1993). Interaction of the human autoantigen p150 with splicing snRNPs. *J. Cell Sci.* **105,** 685–697.

Blencowe, B. J., Nickerson, J. A., Issner, R., Penman, S., and Sharp, P. A. (1994). Association of nuclear matrix antigens with exon-containing splicing complexes. *J. Cell Biol.* **127,** 593–607.

Bravo, R., and MacDonald-Bravo, H. (1987). Existence of two populations of cyclin/proliferating cell nuclear antigen during the cell cycle: Association with DNA replication sites. *J. Cell Biol.* **105,** 1549–1554.

Capco, D. G., Wan, K. M., and Penman, S. (1982). The nuclear matrix: Three-dimensional architecture and protein composition. *Cell (Cambridge, Mass.)* **29,** 847–858.

Capco, D. G., Krocmalnic, G., and Penman, S. (1984). A new method for preparing embedment-free sections for transmission electron microscopy: Applications to the cytoskeletal framework and other three-dimensional networks. *J. Cell Biol.* **98,** 1878–1885.

Cardoso, M. C., Leonhardt, H., and Nadal-Ginard, B. (1993). Reversal of terminal differentiation and control of DNA replication: Cyclin A and Cdk2 specifically localize at subnuclear sites of DNA replication. *Cell (Cambridge, Mass.)* **74,** 979–992.

Carmo-Fonseca, M., Cidadao, A. J., and David-Ferreira, J. F. (1988). Filamentous cross-bridges link intermediate filaments to the nuclear pore complexes. *Eur. J. Cell. Biol.* **45,** 282–290.

Carmo-Fonesca, M., Tollervey, D., Pepperkok, R., Barabino, S. M. L., Merdes, A., Brunner, C., Zamore, P. D., Green, M. R., Hurt, E., and Lamond, A. I. (1991a). Mammalian nuclei contain foci which are highly enriched in components of the pre-mRNA splicing machinery. *EMBO J.* **10,** 195–206.

Carmo-Fonseca, M., Pepperkok, R., Sproat, B. S., Ansorge, W., Swanson, M. S., and Lamond, A. I. (1991b). In vivo detection of snRNP-rich organelles in the nuclei of mammalian cells. *EMBO J.* **10,** 1863–1873.

Carmo-Fonseca, M., Pepperkok, R., Carvalho, M. T., and Lamond, A. I. (1992). Transcription-dependent colocalization of the U1, U2, U4/U6, and U5 snRNPs in coiled bodies. *J. Cell Biol.* **117,** 1–14.

Carter, K. C., Bowman, D., Carrington, W., Fogarty, K., McNeil, A., Fay, F. S., and Lawrence, J. B. (1993). A three-dimensional view of precursor messenger RNA metabolism within the mammalian nucleus. *Science* **259,** 1330–1335.

D'Andrea, A. D., Tantravahi, U., LaLande, M., Perle, M A., and Latt, S. A. (1983). High resolution analysis of the timing of replication of specific DNA sequences during S phase of mammalian cells. *Nucleic Acids Res.* **11,** 4753–4774.

Dijwel, P. A., Wenink, P. W., and Poddighe, J. (1986). Permanent attachment of replication origins to the nuclear matrix in BHK-cells. *Nucleic Acids Res.* **14,** 3241–3249.

Dworetzky, S., Wright, K. L., Fey, E. G., Penman, S., Lian, J. B., Stein, J. L., and Stein, G. S. (1992). Sequence-specific DNA binding proteins are components of a nuclear matrix attachment site. *Proc. Natl. Acad. Sci. U.S.A.* **89,** 4178–4182.

Dworetzky, S. I., Fey, E. G., Penman, S., Lian, J. B., Stein, J. L., and Stein, G. S. (1990). Progressive changes in the protein composition of the nuclear matrix during osteoblast differentiation. *Proc. Natl. Acad. Sci. U.S.A.* **87,** 4605–4609.

Eliceiri, G. L., and Ryerse, J. S. (1984). Detection of intranuclear clusters of Sm antigens with monoclonal anti-Sm antibodies by immunoelectron microscopy. *J. Cell. Physiol.* **121,** 449–451.

Fakan, S., and Hughes, M. E. (1989). Fine structural ribonucleoprotein components of the cell nucleus visualized after spreading and high resolution autoradiography. *Chromosoma* **98,** 242–249.

Fakan, S., Puvion, E., and Spohr, G. (1976). Localization and characterization of newly synthesized nuclear RNA in isolated rat hepatocytes. *Exp. Cell. Res.* **99,** 155–164.

Fakan, S., Lesser, G., and Martin, T. E. (1984). Ultrastructural distribution of nuclear ribonucleoproteins as visualized by immunocytochemistry on thin sections. *J. Cell Biol.* **98,** 358–363.

Fawcett, D. W. (1966). "An Atlas of Fine Structure: The Cell, Its Organelles and Inclusions," pp. 2–3. Saunders, Philadelphia.

Fey, E. G., and Penman, S. (1988). Nuclear matrix proteins reflect cell type of origin in cultured human cells. *Proc. Natl. Acad. Sci. U.S.A.* **85,** 121–125.

Fey, E. G., Krochmalnic, G., and Penman, S. (1986). The non-chromatin substructures of the nucleus: The ribonucleoprotein (RNP)-containing and RNP-depleted matrices analyzed by sequential fractionation and resinless section electron microscopy. *J. Cell Biol.* **102,** 1654–1665.

Fischer, D., Weisenberger, D., and Scheer, U. (1991). Assigning functions to nucleolar structures. *Chromosoma* **101,** 133–140.

Fisher, D. Z., Chaudhary, N., and Blobel, G. (1986). cDNA sequencing of nuclear lamins A and C reveals primary and secondary structural homology to intermediate filament proteins. *Proc. Natl. Acad. Sci. U.S.A.* **83,** 6450–6454.

Fisher, P. A., Berrios, M., and Blobel, G. (1982). Isolation and characterization of a proteinaceous subnuclear fraction composed of nuclear matrix, peripheral lamina and nuclear pore complexes from embryos of *Drosophila melanogaster. J. Cell Biol.* **92,** 674–686.

Folkman, J., and Moscona, A. (1978). Role of cell shape in growth control. *Nature* (*London*) **273,** 345–349.

Foster, K. A., and Collins, J. M. (1985). The interrelation between DNA synthesis rates and DNA polymerases bound to the nuclear matrix in synchronized HeLa cells. *J. Biol. Chem.* **260,** 4229–4235.

Franke, W. W. (1987). Nuclear lamins and cytoplasmic intermediate filament proteins: A growing multigene family. *Cell* (*Cambridge, Mass.*) **48,** 3–4.

Fu, X. D., and Maniatis, T. (1990). Factor required for mammalian spliceosome assembly is localized to discrete regions in the nucleus. *Nature* (*London*) **343,** 437–441.

Georgatos, S. D., Meier, J., and Simos, G. (1994). Lamins and lamin-associated proteins. *Curr. Opin. Cell Biol.* **6,** 347–353.

Gerace, L., Blum, A., and Blobel, G. (1978). Immunocytochemical localization of the major polypeptides of the nuclear pore complex-lamina fraction-interphase and mitotic distribution. *J. Cell Biol.* **79,** 546–566.

Gerace, L., Comeau, C., and Benson, M. (1984). Organization and modulation of the nuclear lamina structure. *J. Cell Sci., Suppl.* **1,** 137.

Getzenberg, R. H., and Coffey, D. S. (1990). Tissue specificity of the hormonal response in sex accessory tissues is associated with nuclear matrix protein patterns. *Mol. Endocrinol.* **4,** 1336–1342.

Goldman, M. A., Holmquist, G. P., Gray, M. C., Caston, L. A., and Nag, A. (1984). Replication timing of genes and middle repetitive sequences. *Science* **224,** 686–692.

Hansen, L. R., and Ingber, D. E. (1992). Regulation of nucleocytoplasmic transport by mechanical forces transmitted through the cytoskeleton. *In* "Nuclear Trafficing" (C. Feldherr, ed.), pp. 71–78. Academic Press, San Diego, CA.

Hardin, J. H., Spicer, S. S., and Greene, W. B. (1969). The paranucleolar structure, accessory body of Cajál, sex chromatin, and related structures in nuclei of rat trigeminal neurons: A cytochemical and ultrastructural study. *Anat. Rec.* **164,** 403–432.

Hatton, K. S., Dhar, V., Gahn, T. A., Brown, E. H., Mager, D., and Schildkraut, C. L. (1988). Temporal order of replication of multigene families reflects chromosomal location and transcriptional activity. *Cancer Cells* **6,** 335–340.

He, D., Nickerson, J. A., and Penman, S. (1990). The core filaments of the nuclear matrix. *J. Cell Biol.* **110,** 569–580.

Herman, R., Weymouth, L., and Penman, S. (1978). Heterogeneous nuclear RNA-protein fibers in chromatin-depleted nuclei. *J. Cell Biol.* **78,** 663–674.

Hodge, L. D., Mancini, P., Davis, F. M., and Heywood, P. (1977). Nuclear matrix of HeLa S3 cells. Polypeptide composition during adenovirus infection and in phases of the cell cycle. *J. Cell Biol.* **72,** 194–208.

Hozák, P., Hassan, A. B., Jackson, D. A., and Cook, P. R. (1993). Visualization of replication factories attached to a nucleoskeleton. *Cell* **73,** 361–373.

Huang, S., and Spector, D. L. (1991). Nascent pre-mRNA transcripts are associated with nuclear regions enriched in splicing factors. *Genes Dev.* **5,** 2288–2302.

Huang, S., and Spector, D. L. (1992). U1 and U2 small nuclear RNAs are present in nuclear speckles. *Proc. Natl. Acad. Sci. U.S.A.* **89,** 305–308.

Imai, H., Chan, E. K. L., Kiyosawa, K., Fu, X.-D., and Tan, E. M. (1993). Novel nuclear autoantigen with splicing factor motifs identified with antibody from hepatocellular carcinoma. *J. Clin. Invest.* **92,** 2419–2462.

Ingber, D. E. (1993). Cellular tensegrity: Defining new rules of biological design that govern the cytoskeleton. *J. Cell Sci.* **104,** 467–475.

Ingber, D. E., and Folkman, J. (1989). Tension and compression as basic determinants of cell form and function: Utilization of a cellular tensegrity mechanism. *In* "Cell Shape Determinants: Regulation and Regulatory Role" (W. Stein and F. Bronner, eds.), pp. 1–32. Academic Press, Orlando, FL.

Ingber, D. E., Dike, L., Hansen, L., Karp, S., Liley, H., Maniotis, A., McNamee, H., Mooney, D., Plopper, G., Sims, J., and Wang, N. (1994). Cellular tensegrity: Exploring how mechanical changes in the cytoskeleton regulate cell growth, migration, and tissue pattern during morphogenesis. *Int. Rev. Cytol.* **150,** 173–224.

Jackson, D. A., and Cook, P. R. (1985). Transcription occurs at nucleoskeleton. *EMBO J.* **4,** 919–925.

Jackson, D. A., and Cook, P. R. (1986). Replication occurs at a nucleoskeleton. *EMBO J.* **5,** 1403–1410.

Jackson, D. A., and Cook, P. R. (1988). Visualization of a filamentous nucleoskeleton with a 23nm axial repeat. *EMBO J.* **7,** 3667–3678.

Jackson, D. A., Hassan, A. B., Errington, R. J., and Cook, P. R. (1993). Visualization of focal sites of transcription within human nuclei. *EMBO J.* **12,** 1059–1065.

Kamel, H. M., Kirk, J., and Toner, P. G. (1990). Ultrastructural pathology of the nucleus. *Curr. Top. Pathol.* **82,** 17–89.

Kaufman, S. H., Coffey, D. S., and Shaper, J. H. (1981). Considerations in the isolation of rat liver nuclear matrix, nuclear envelope, and pore complex lamina. *Exp. Cell Res.* **132,** 105–123.

Keesee, S. K., Meneghini, M., Szaro, R. P., and Wu, Y.-J. (1994). Nuclear matrix proteins in colon cancer. *Proc. Natl. Acad. Sci. U.S.A.* **91,** 1913–1916.

Khanuja, P. S., Lehr, J. E., Soule, H. D., Gehani, S. K., Noto, A. C., Choudhury, S., Chen, R., and Pienta, K. J. (1993). Nuclear matrix proteins in normal and breast cancer cells. *Cancer Res.* **53,** 3394–3398.

Krohne, G., Wolin, S. L., McKeon, F. D., Franke, W. W., and Kirschner, M. W. (1987). Nuclear lamin LI of *Xenopus laevis:* cDNA cloning, amino acid sequence and binding specificity of a member of the lamin B subfamily. *EMBO J.* **6,** 3801–3808.

Lafarga, M., Hervas, J. P., Santa-Cruz, M. C., Villegas, J., and Crespo, D. (1983). The "accessory body" of Cajál in the neuronal nucleus. A light and electron microscopic approach. *Anat. Embryol.* **166,** 19–30.

Lasko, D. D., Tomkinson, A. E., and Lindahl, T. (1990). Biosynthesis and intracellular localization of DNA ligase I. *J. Biol. Chem.* **265,** 12618–12622.

Lawrence, J. B., Singer, R. H., and Marselle, L. M. (1989). Highly localized tracks of specific transcripts within interphase nuclei visualized by in situ hybridization. *Cell* (*Cambridge, Mass.*) pp. 493–502.

Leonhardt, H., Page, A. W., Weier, H. U., and Bestor, T. H. (1992). A targeting sequence directs DNA methyltransferase to sites of DNA replication in mammalian nuclei. *Cell* (*Cambridge, Mass.*) **71,** 865–873.

Lerner, E. A., Lerner, M. R., Janeway, L. A., and Steitz, J. A. (1981). Monoclonal antibodies to nucleic acid containing cellular constituents: Probes for molecular biology and autoimmune disease. *Proc. Natl. Acad. Sci. U.S.A.* **78,** 2737–2741.

Li, H., and Bingham, P. M. (1991). Arginine/serine-rich domains of the su(wa) and tra RNA processing regulators target proteins to a subnuclear compartment implicated in splicing. *Cell (Cambridge, Mass.)* **67,** 335–342.

Mayer, D. T., and Gulick, A. (1942). The nature of the proteins of cellular nuclei. *J. Biol. Chem.* **46,** 433–440.

Mirkovitch, J., Mirault, M., and Laemmli, U. K. (1984). Organization of the higher order chromatin loop: Specific DNA attachment sites on nuclear scaffold. *Cell (Cambridge, Mass.)* **39,** 223–232.

Monneron, A., and Bernhard, W. (1969). Fine structural organization of the interphase nucleus in some mammalian cells. *J. Ultrastruct. Res.* **27,** 266–288.

Nakamura, H., Morita, T., and Sato, C. (1986). Structural organization of replicon domains during DNA synthetic phase in the mammalian nucleus. *Exp. Cell Res.* **165,** 291–297.

Nakayasu, H., and Berezney, R. (1989). Mapping replication sites in the eukaryotic cell nucleus. *J. Cell Biol.* **108,** 1–11.

Nickerson, J. A., and Penman, S. (1992). The nuclear matrix: Structure and involvement in gene expression. *In* "Molecular and Cellular Approaches to the Control of Proliferation and Differentiation" (G. Stein and J. Lian, eds.), pp. 343–380. Academic Press, San Diego, CA.

Nickerson, J. A., Krochmalnic, G., Wan, K. M., and Penman, S. (1989). Chromatin architecture and nuclear RNA. *Proc. Natl. Acad. Sci. U.S.A.* **86,** 177–181.

Nigg, E. A. (1989). The nuclear envelope. *Curr. Opin. Cell Biol.* **1,** 435–440.

O'Keefe, R. T., Henderson, S. C., and Spector, D. L. (1992). Dynamic organization of DNA replication in mammalian cell nuclei: Spatially and temporally defined replication of chromosome specific a-satellite DNA sequences. *J. Cell Biol.* **116,** 1095–1100.

Pardoll, D. M., Vogelstein, B., and Coffey, D. S. (1980). A fixed site of DNA replication in eukaryotic cells. *Cell* **19,** 527–536.

Partin, A. W., Getzenberg, R. H., CarMichael, M. J., Vindivich, D., Yoo, J., Epstein, J. I., and Coffey, D. S. (1993). Nuclear matrix protein patterns in human benign prostatic hyperplasia and prostate cancer. *Cancer Res.* **53,** 744–746.

Penman, S. (1995). Rethinking cell structure. *Proc. Natl. Acad. Sci. U.S.A.* **92,** 5251–5257.

Perraud, M., Gioud, M., and Monier, J. C. (1979). Intranuclear structures of monkey kidney cells recognised by immunofluorescence and immuno-electron microscopy using anti-ribonucleoprotein antibodies. *Ann. Immunol. (Paris)* **130C,** 635–647.

Petrov, P., and Bernhard, W. (1971). Experimentally induced changes of extranucleolar ribonucleoprotein components of the interphase nucleus. *J. Ultrastruct. Res.* **35,** 386–402.

Pienta, K. J., and Coffey, D. S. (1992). Nuclear-cytoskeletal interactions: Evidence for physical connections between the nucleus and cell periphery and their alteration by transformation. *J. Cell. Biochem.* **49,** 357–365.

Puvion, E., Virion, A., Assens, C., Leduc, E. H., and Jeanteur, P. (1984). Immunocytochemical identification of nuclear structures containing snRNPs in isolated rat liver cells. *J. Ultrastruct. Res.* **87,** 180–189.

Ramón y Cajál, S. (1903). Un sencillo metodo de coloracion selectiva del reticulo protoplasmico y sus efectos en los diversos organos nerviosos. *Trab. Lab. Invest. Biol.* **2,** 129–221.

Ramón y Cajál, S. (1909). "Histologie du Système Nerveux de l'Homme et des Vertébrés," Vols. I and II. Reprinted by Consejo Superior de Investigaciones Cientificas, Madrid, 1952.

Raska, I., Ochs, R. L., Andrade, L. E., Chan, E. K., Burlingame, R., Peebles, C., Gruol, D., and Tan, E. M. (1990). Association between the nucleolus and the coiled body. *J. Struct. Biol.* **104,** 120–127.

Raska, I., Andrade, L. E., Ochs, R. L., Chan, E. K., Chang, C. M., Roos, G., and Tan, E. M. (1991). Immunological and ultrastructural studies of the nuclear coiled body with autoimmune antibodies. *Exp. Cell Res.* **195,** 27–37.

Razin, S. V. (1987). DNA interactions with the nuclear matrix and spatial organization of replication and transcription. *Bioessays* **6,** 19–23.

Roth, M. B., Murphy, C., and Gall, J. G. (1990). A monoclonal antibody that recognizes a phosphorylated epitope stains lampbrush chromosome loops and small granules in the amphibian germinal vesicle. *J. Cell Biol.* **111,** 2217–2223.

Scheer, U., Thiry, M., and Goessens, G. (1993). Structure, function and assembly of the nucleolus. *Trends Cell Biol.* **3,** 236–241.

Seite, R., Pebusque, M. J., and Vio-Cigna, M. (1982). Argyrophilic proteins on coiled bodies in sympathetic neurons identified by Ag-NOR procedure. *Biol. Cell* **46,** 97–100.

Sims, J. R., Karp, S., and Ingber, D. E. (1992). Altering the cellular mechanical force balance results in integrated changes in cell, cytoskeletal and nuclear shape. *J. Cell Sci.* **103,** 1215–1222.

Spector, D. L., Schrier, W. H., and Busch, H. (1983). Immunoelectron microscopic localization of snRNPs. *Biol. Cell* **49,** 1–10.

Spector, D. L., Fu, X.-D., and Maniatis, T. (1991). Associations between distinct pre-mRNA splicing components and the cell nucleus. *EMBO J.* **10,** 3467–3481.

Spector, D. L., Lark, G., and Huang, S. (1992). Differences in snRNP localization between transformed and non-transformed cells. *Mol. Biol. Cell* **3,** 555–569.

Stein, G. S., van Wijnen, A., Stein, J. L., Lian, J. B., Bidwell, J. P., and Montecino, M. (1994). Nuclear architecture supports integration of physiologically regulatory signals for transcription of cell growth and tissue-specific genes during osteoblast differentiation. *J. Cell. Biochem.* **55,** 4–15.

Stuurman, N., van Driel, R., de Jong, L., Meijne, A. M., and van Renswoude, J. (1989). The protein composition of the nuclear matrix of murine P19 embryonal carcinoma cells is differentiation-stage dependent. *Exp. Cell Res.* **180,** 460–466.

Stuurman, N., Meijne, A. M., van der Pol, A. J., de Jong, L., van Driel, R., and van Renswoude, J. (1990). The nuclear matrix from cells of different origin. Evidence for a common set of matrix proteins. *J. Biol. Chem.* **265,** 5460–5465.

Tubo, R. A., and Berezney, R. (1987a). Identification of 100 and 150S DNA polymerase-primase megacomplexes solubilized from the nuclear matrix of regenerating rat liver. *J. Biol. Chem.* **262,** 5857–5865.

Tubo, R. A., and Berezney, R. (1987b). Nuclear matrix-bound DNA primase. Elucidation of an RNA priming system in nuclear matrix isolated from regenerating rat liver. *J. Biol. Chem.* **262,** 6637–6642.

Underwood, J. C. (1990). Nuclear morphology and grading of tumours. *Curr. Top. Pathol.* **82,** 1–15.

Vaughn, J. P., Dijkwel, P. A., Mullenders, L. H. F., and Hamlin, J. L. (1990). Replication forks are associated with the nuclear matrix. *Nucleic Acids Res.* **18,** 1965–1969.

Verheijen, R., Kuijpers, H., Vooijs, P., van Venrooij, W., and Ramaekers, F. (1986). Distribution of the 70k U1 RNA-associated protein during interphase and mitosis: Correlation with other U RNP particles and proteins of the nuclear matrix. *J. Cell Sci.* **86,** 173–190.

Visa, N., Puvion-Dutilleul, F., Harper, F., Bachellerie, J. P., and Puvion, E. (1993). Intranuclear distribution of poly(A) RNA determined by electron microscope in situ hybridization. *Exp. Cell Res.* **208,** 19–34.

Wansink, D. G., Schul, W., van der Kraan, I., Steensel, B., van Driel, R., and de Jong, L. (1993). Fluorescent labeling of nascent RNA reveals transcription by RNA polymerase II in domains scattered throughout the nucleus. *J. Cell Biol.* **122,** 283–293.

Wu, Z., Murphy, C., Wu, C. H., Tsvetkov, A., and Gall, J. G. (1993). Snurposomes and coiled bodies. *Cold Spring Harbor Symp. Quant. Biol.* **58,** 747–754.

Xing, Y., and Lawrence, J. B. (1991). Preservation of specific RNA distribution within the chromatin-depleted nuclear substructure demonstrated by *in situ* hybridization coupled with biochemical fractionation. *J. Cell Biol.* **112,** 1055–1063.

Xing, Y., Johnson, C. V., Dobner, P. R., and Lawrence, J. B. (1993). Higher level organization of individual gene transcription and splicing. *Science* **259,** 1326–1330.

Zahler, A. M., Lane, W. S., Stolk, J. A., and Roth, M. B. (1992). SR proteins: A conserved family of pre-mRNA splicing factors. *Genes Dev.* **6,** 837–847.

Zamore, P. D., and Green, M. R. (1991). Biochemical characterization of U2 snRNP auxiliary factor: An essential pre-mRNA splicing factor with a novel intranuclear distribution. *EMBO J.* **10,** 207–214.

Zhang, M., Zamore, P. D., Carmo-Fonseca, M., Lamond, A. I., and Green, M. R. (1992). Cloning and intracellular localization of the U2 small nuclear ribonucleoprotein auxiliary factor small subunit. *Proc. Natl. Acad. Sci. U.S.A.* **89,** 8769–8773.

II

Nuclear Matrix Structure and Function

2

The Nuclear Matrix: Mapping Genomic Organization and Function in the Cell Nucleus

RONALD BEREZNEY

Department of Biological Sciences
State University of New York at Buffalo
Buffalo, New York

I. INTRODUCTION

In the past, understanding of the relationships between nuclear architecture and function was compromised by both the structural complexities of the cell nucleus and the limitations of the methods of analysis. Recent findings, however, have greatly stimulated interest in this area and offer the promise of major breakthroughs in the near future. In one series of investigations, discrete domains of genomic organization and function were detected *in situ* using high-resolution and computer-aided microscopic approaches. These studies indicate that the genome and its associated processes of DNA replication, transcription, RNA splicing, and transport events are compartmentalized within discrete functional regions inside the mammalian cell nucleus.

Another important step in elucidating the relationships between nuclear architecture and function occurred with the isolation and characterization of the nuclear matrix (Berezney and Coffey, 1974). Besides its relevance for understanding details of nuclear structure in correlation with function, the isolated nuclear matrix offers a powerful *in vitro* approach for dissecting the molecular components involved in genomic organization and function. For example, appropriate nuclear matrix *in vitro* systems are available for studying DNA replication, transcription, and RNA splicing. The isolated nuclear matrix also represents a valuable starting point for cloning and sequencing novel proteins potentially involved in genomic organizational and/or functional processes. This chapter will focus on the basic properties of the nuclear matrix in relation to nuclear architecture, function, and regulation. Nuclear matrix isolation, nuclear matrix proteins, and DNA replication sites associated with the nuclear matrix will also be briefly summarized.

II. *IN SITU* NUCLEAR MATRIX

Until recently, our view of the cell nucleus was highly influenced by classical light microscopic observations of the 19th and early 20th centuries. These studies revealed three main structural components in the cell nucleus: (1) the chromatin, (2) the nucleolus, and (3) the nucleoplasm, nuclear sap, or karyolymph, whose translucent appearance led to the belief that this was a clear fluid or gel.

The application of electron microscopy to the cell nucleus from the late 1950s to the mid-1970s led to a new view of the cell nucleus. These studies revealed for the first time heterogeneous structures in the interchromatinic regions between condensed chromatin and led Fawcett (1966) to state: "The chromatin and nucleolus are dispersed in a matrix traditionally called the nuclear sap or karyolymph. . . . Although these terms are still in use, they seem inappropriate designations for the congeries of submicroscopic granules of varying size and density that comprise the interchromosomal material of the interphase nucleus. . . . The terms nuclear matrix or nuclear ground substance are preferable."

Unfortunately, this new view of the cell nucleus was not widely accepted, particularly among molecular biologists. To them this new model was too complicated and did not seem to add anything significant to our understanding of genetic processes—merely speculations. What was clearly lacking in this new electron microscopic view was a global perspective of the overall nuclear architecture and how this might relate to genomic organization and function. Despite the significant progress in defining the details of individual granules and fibrils in the nucleus, the relationship of these structures to genomic functions was still unclear; most important, there was no concept of how this might all fit together into an organized framework.

A. The Nuclear Matrix in Whole Cells

Figure 1a shows the basic morphology of the *in situ* rat liver nucleus using thin-sectioned electron microscopy. Dense chromatin (heterochromatin) is found along the nuclear periphery, along the nucleolar periphery, and at other regions in the nuclear interior. Between the dense chromatin areas are the interchromatinic regions, which contain at least two main structural components: the diffuse chromatin, or euchromatin, and the nonchromatin nuclear matrix composed of nonchromatin granules (i.e., the interchromatin and perichromatin granules) and proteinaceous fibrils of 30–50 Å (Bernhard and Granboulan, 1963; Smetana *et al.*, 1971). It was generally difficult, however, to distinguish between diffuse chromatin fibers sectioned in various directions and the nonchromatin granules and fibers of the nuclear matrix.

An important breakthrough occurred with the development of the ethylenediaminetetraecetic acid (EDTA)-regressive staining procedure (Bernhard, 1969; Monneron and Bernhard, 1969). By preferentially bleaching the DNA-containing chromatin structures while maintaining significant contrast of the nonchromatin (especially RNA-containing) structures, a new view of the cell nucleus was revealed (Fig. 1b). An extensive nonchromatin fibrogranular nuclear matrix is observed within

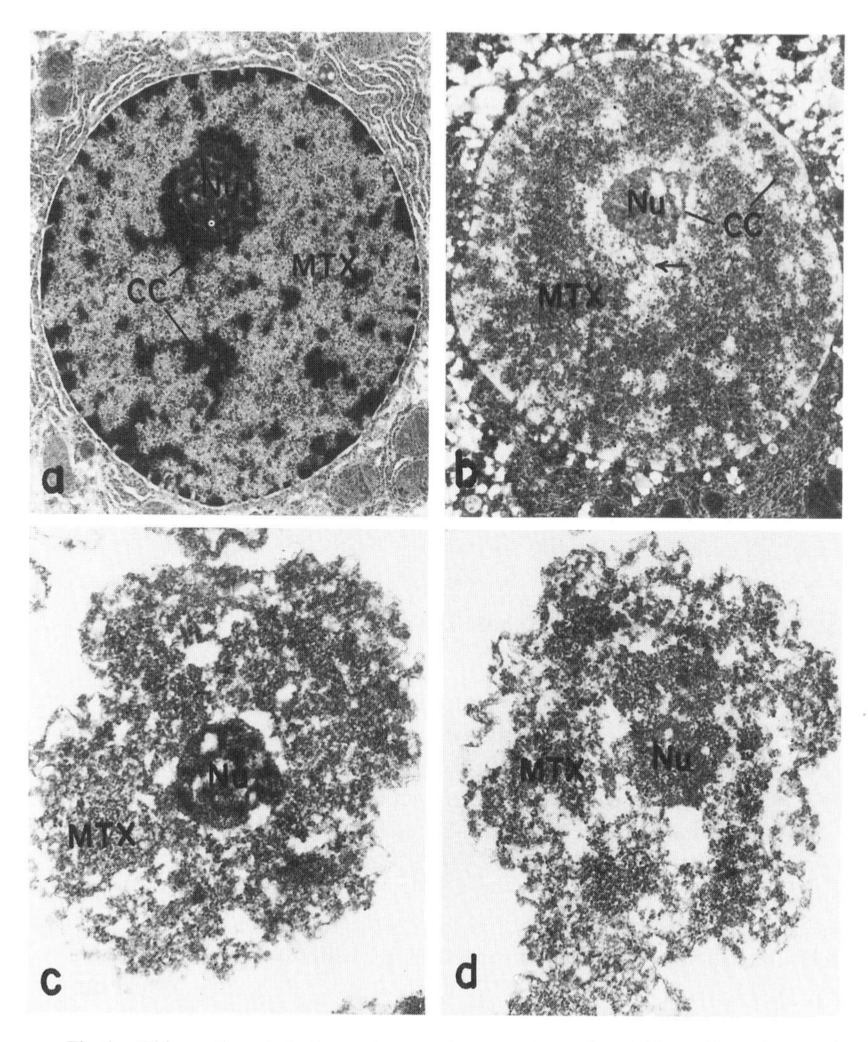

Fig. 1. Thin-sectioned electron micrograph comparison of nuclei in rat liver tissue and isolated nuclear matrix by standard staining with uranyl acetate and lead citrate and EDTA-regressive staining. CC, condensed chromatin (heterochromatin); MTX, matrix region; NU, nucleolus. Note the similarity in structure of the isolated matrix and the *in situ* matrix. (a) Rat liver, standard staining; (b) rat liver, EDTA-regressive staining; (c) isolated rat liver nuclear matrix, standard staining; (d) isolated rat liver nuclear matrix, EDTA-regressive staining.

the interchromatinic regions, while the dense chromatin regions show little or no staining. The nuclear pore complexes are especially densely stained (see arrows in Fig. 2), and channels of nuclear matrix material appear to radiate from the nuclear interior to the nuclear pore complexes. This was also observed in earlier studies (Monneron and Bernhard, 1969; Franke and Falk, 1970) and may correspond to passageways for regulated transport of RNA transcripts to the nuclear pores (Blobel, 1985).

B. Chromatin Domains

Besides the nuclear matrix, the interchromatinic regions contain the diffuse chromatin, or euchromatin. Specific staining of DNA in these regions by Derenzini *et al.* (1977, 1978) showed that the euchromatin occupies discrete regions between the nonchromatin nuclear matrix structures. The studies of Derenzini *et al.* (1977, 1978) further revealed that the diffuse chromatin is only a minor component of the visible structure in the interchromatinic region. The major components are the ribonucleoprotein (RNP)-rich nonchromatin fibrils and granules that compose the nuclear matrix.

Taken together, these ultrastructural findings suggest a potentially close *in situ* association between actively replicating and transcribing chromatin (euchromatin) and nuclear matrix components inside the functioning cell nucleus. One mode of interaction between the nuclear matrix and chromatin occurs at the DNA loop attachment sites (Gasser and Laemmli, 1987). Other levels of association between chromatin and nuclear matrix components are feasible and will require further investigations of specific nuclear matrix proteins and their possible interactions with chromatin.

C. Ribonucleoprotein Domains

Electron microscopic analyses of nuclear matrix structures observed within the interchromatinic regions of the nucleus such as interchromatin granules, perichromatin granules, and fibrils stress the RNP nature of these components (Bernhard and Granboulan, 1963; Monneron and

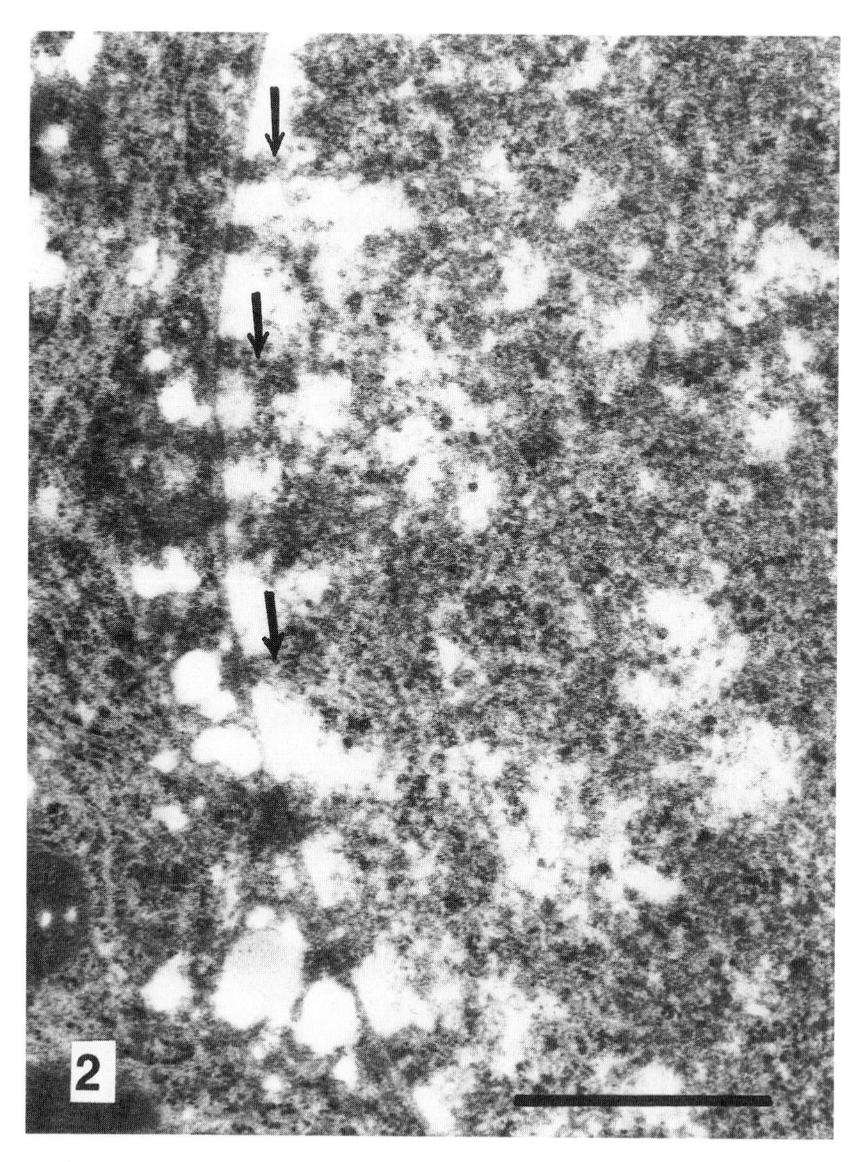

Fig. 2. Thin-sectioned electron micrograph of an *in situ* rat liver nucleus in the area of the nuclear pore complex by the EDTA-regressive staining procedure of Bernhard. In this micrograph the interior *in situ* matrix appears to be contiguous with the nuclear pore complexes of the surrounding nuclear envelope. Arrows indicate the "channels" of matrix materials radiating toward the nuclear pore complexes. Bar: 1 μm.

Bernhard, 1969), and the EDTA-regressive staining pattern of the nucleus is generally interpreted as an enhancement of these presumptive RNP domains (Bernhard, 1969; Monneron and Bernhard, 1969). This view is further supported by high-resolution electron microscopic autoradiography performed on EDTA-regressively stained sections (Puvion and Moyne, 1981; Fakan and Bernhard, 1971), where newly transcribed extranucleolar RNA is preferentially localized along the borders (where the perichromatin fibrils are concentrated) and within the interior of the interchromatinic regions (where the interchromatin granules are concentrated).

Pulse-chase experiments demonstrated that the newly transcribed RNA migrates through the interchromatinic region as RNP particles or fibrils toward the nuclear periphery (Fakan and Bernhard, 1973; Nash *et al.*, 1975; Fakan *et al.*, 1976). This suggests that a network of RNP particles and/or fibrils (nuclear matrix structures) extends from the sites of transcription to final release through the nuclear pore complexes. These strands or channels of nuclear matrix structures may correspond to dynamic assembly lines for the coordinate transcription, splicing, processing, and transport of RNA (Berezney, 1984).

D. Functional Topography and Dynamics of the Nuclear Matrix

Our current knowledge of *in situ* nuclear structure leads us to view the mammalian cell nucleus as a three-dimensional collage of condensed chromatin (heterochromatin), interchromatinic regions (euchromatin and nonchromatin nuclear matrix structures), nucleolar compartments, and a surrounding double-membraned nuclear envelope containing nuclear pore complexes. At least two main structural domains are postulated to exist within the *in situ* matrix: chromatin domains and RNP domains. This dual aspect of nuclear matrix organization inside the nucleus has important implications. It immediately suggests heterogeneity of the matrix at both the organizational and functional levels. The chromatin domains of the matrix (e.g., DNA loop attachment sites) are presumably quite different from RNP domains with respect to individual components and their organization.

Functionally, the nuclear matrix can be divided into chromatin domains of replication and/or transcription versus domains of posttranscriptional RNA processing. At the level of replication and transcription, the matrix may provide a structural milieu for the expression of functional loops of chromatin and the organization of active replicational and/or transcriptional complexes. The potentially continuous nature of the matrix system also suggests that these two distinct structural (chromatin and RNP structures) and functional (DNA replication, transcription, and RNA processing and transport) domains may be linked. In this manner, the matrix may play an important integrative and regulatory role for cascade and assembly-line events.

The dynamic properties of nuclear architecture have been reviewed (Berezney, 1979, 1984). Briefly, the cell nucleus responds to physiological changes by a variety of morphological alterations. One of the most common alterations involves a change in the size of the nucleus. This has been strikingly observed in nuclear transplantation and cell fusion experiments in which cell nuclei activated to synthesize DNA show enormous enlargement (Graham *et al.,* 1966; Harris, 1968, 1970). Moreover, the size of the cell nucleus is constantly changing during the cell cycle. The HeLa cell nucleus, for example, doubles in volume between G_1 and the end of S phase (Maul *et al.,* 1972).

In addition to overall size changes, treatment of cells with drugs that inhibit various biosynthetic processes often results in striking perturbations in nuclear shape and internal morphology (Berezney, 1979). Many of these drugs are carcinogens, and it has been known since the time of the pioneering studies by 19th-century cytopathologists that cancer cells are often characterized by alterations in the shape and internal morphology of their cell nuclei.

By applying appropriate directional forces to cells, Ingber and colleagues (1994) observed coordinate changes in overall cell and nuclear shape in relation to mechanical signal transduction. Altering cell shape affected growth, indicating a close relationship between genomic function and the three-dimensional organization of the cell nucleus and the dynamic nature of nuclear as well as cytoplasmic architecture.

The cell nucleus is a dynamic structure capable of rapid and striking changes in organization. We propose that the basis for these changes is a nuclear matrix structure that is in a state of constant flux and is highly dependent on nuclear function. In this view, overall nuclear organization is maintained despite the constant alterations in interactions at the local level. It is these changes at the local level (e.g., replication, transcription, or RNA-splicing sites), however, which accumulate and lead to changes in overall nuclear architecture.

III. ISOLATED NUCLEAR MATRIX

In 1974, Berezney and Coffey, following previous studies of residual nuclear structures (for references, see Berezney and Coffey, 1976; Berezney, 1979), reported the first characterization of nuclear matrix isolated from rat liver tissue. Subsequent studies by this group (Berezney and Coffey, 1977) and many others (for references, see Berezney, 1979, 1984; Shaper *et al.*, 1978) led to the characterization of nuclear matrices from a wide variety of eukaryotic cells from unicellular organisms to man.

A. Isolation Procedures

Most procedures for nuclear matrix isolation are based on the original protocols reported by Berezney and Coffey (1974, 1977). Briefly, morphologically intact nuclei are isolated and subjected to a series of treatments involving nuclease digestion, salt extractions, and detergent (Triton X-100) extraction. Conditions are chosen so that nuclear structures are maintained throughout the extraction protocol, including the final nuclear matrix fraction, despite the removal of most chromatin and protein and disruption of the nuclear membranes with detergent.

A major modification for nuclear matrix isolation was introduced by Laemmli's group (Mirkovitch *et al.*, 1984), who used the chaotropic agent and detergent lithium 3,5-diiodosalicylate (LIS) instead of salt solutions for extraction. This preparation has been termed the nuclear scaffold to distinguish it from its salt-extracted counterpart and has been widely used for the study of specific DNA sequences associated with the nuclear matrix (Gasser and Laemmli, 1987). In a comprehensive study of the HeLa nuclear matrix, it was found that nuclear matrices prepared by the LIS method with heat stabilization were strikingly similar to their salt-extracted counterparts in polypeptide profiles and in both ultrastructure and biochemical composition (Belgrader *et al.*, 1991a). In contrast, nuclear matrices prepared by salt extraction in the absence of heat or chemical stabilization showed typical nuclear matrix morphology and chemical composition, while the corresponding LIS-prepared nuclear matrices were highly disrupted and virtually devoid of internal nuclear matrix structure (Belgrader *et al.*, 1991a).

Other preparations of these types of structures have been termed nucleoskeletons, nuclear ghosts, nuclear cages, and residual nuclear structures, to name just a few. The term nuclear matrix, however, is used by the vast majority of the more than 1000 papers published in this field. Cook and associates use the term nucleoid for nuclear matrices that have intact, supercoiled DNA associated with them (McCready *et al.*, 1980). The term DNA-rich nuclear matrices has also been used to describe these preparations (Berezney and Buchholtz, 1981a). Cook's group has also developed a novel approach to studying nuclear matrix organization. They embed cells in agarose beads, nuclease digest the chromatin with restriction enzymes, and electroelute the chromatin at isotonic salt concentrations (Jackson and Cook, 1985). The remaining nuclear structures are called nucleoskeletons.

Nuclear matrices isolated from a wide range of organisms and cell types have a number of characteristic features that are summarized in Table I. Many of the major architectural features of the intact nucleus are maintained in isolated nuclear matrices despite the removal of 75–90% of the

TABLE I

Properties of Isolated Nuclear Matrix

Property	Comments
Isolation	Wide range from lower eukaryotes to humans
Structure	Tripartite structure consisting of pore complex lamina, residual nucleoli, and fibrogranular internal matrix
Composition	Contains 10–25% of total nuclear protein and tightly bound DNA and RNA
Polypeptides	Heterogeneous profile of nonhistone proteins including cell type and differentiation state-specific proteins, as well as common proteins termed nuclear matrins. Nuclear lamina has simpler profile, with lamins A, B, and C (60–70 kDa) predominating
Cell cycle	Cell cycle dynamics of nuclear matrix in interphase cells remains to be elucidated. It is likely that components of interphase nuclear matrix (e.g., DNA attachment sites) are preserved in mitotic cells as chromosome scaffold
Functions	While precise role(s) of nuclear matrix in nuclear function remain to be elucidated, its proposed role as site for organization and regulation of replication, transcription, and RNA splicing and processing is supported by vast array of functional properties associated with isolated nuclear matrix

total nuclear protein and virtually all of the chromatin. The isolated matrices also contain large amounts of tightly bound RNA, lesser amounts of DNA (depending on the degree of nuclease digestion), and only trace amounts of lipids if nonionic detergent extraction (e.g., Triton X-100) is performed. Protein is the predominant macromolecular component. A multitude of different proteins are found, with enrichment of the higher molecular weight nonhistone proteins in the nucleus and depletion of many of the lower molecular weight ones, especially the histones (Berezney, 1979, 1984). Three of the major proteins are lamins A, B, and C, which migrate between 60 and 70 kDa on sodium dodecyl sulfate–polyacrylamide gel electrophoresis (SDS–PAGE) and are the most abundant components of the surrounding residual nuclear envelope or nuclear lamina.

Laemmli and co-workers (Aldolph *et al.*, 1977) first demonstrated that isolated chromosomes extracted with nuclease and high salt maintain a residual protein chromosomal structure termed the chromosome scaffold. Further studies showed that the chromosomal DNA loops are attached to the scaffold structure (Paulson and Laemmli, 1977). Because the DNA loops are attached to the nuclear matrix in interphase cells, it has been widely suggested that at least certain components of the interphase nuclear matrix (e.g., the DNA attachment sites) are maintained in mitotic cells as components of the chromosome scaffold. Despite this widely held belief, our knowledge of the precise relationships between the proteins composing the interphase matrix and the chromosome scaffold is very limited.

B. Morphology of Isolated Nuclear Matrix

Three main structural regions typically compose the isolated nuclear matrix: a surrounding residual nuclear envelope or nuclear lamina containing morphologically recognizable nuclear pore complexes, residual components of nucleoli, and an extensive fibrogranular internal matrix. The last structure is believed to represent residual components of the *in situ* nuclear matrix structures observed in whole cells (Berezney and Coffey, 1977). When EDTA-regressive staining is used as a criterion, a similarity is seen between the fibrogranular internal matrix of the isolated rat liver nuclear matrix and the *in situ* nuclear matrix visualized in intact tissue (Fig. 1b,d). The virtual identity of the electron microscopic images of the isolated nuclear matrices following EDTA-regressive staining and the images obtained by conventional fixation and staining (Fig. 1c,d) further supports the derivation of the internal matrix from the *in situ* nuclear matrix structure in intact liver tissue.

A major consideration in preparing nuclear matrices that optimally resemble the *in situ* nuclear matrix is the procedure used for isolation of nuclei. Procedures involving exposure of nuclei to hypotonic solutions often lead to major perturbations in internal nuclear structure. These

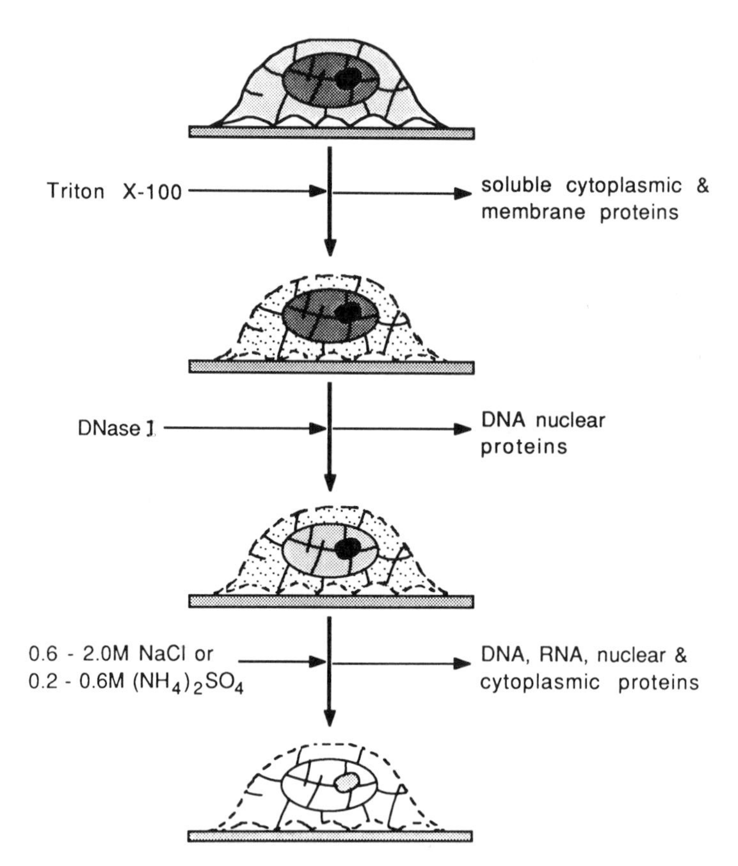

Fig. 3. Nuclear matrix prepared by *in situ* extraction of cells. Whole cells grown on coverslips can be directly extracted to isolate the nuclear matrix (*bottom*). Remnants of the cell surface and cytoskeleton help to anchor this *in situ* nuclear matrix to the coverslips. These preparations are especially valuable for cytochemical and immunolocalization studies but are also useful for biochemical studies.

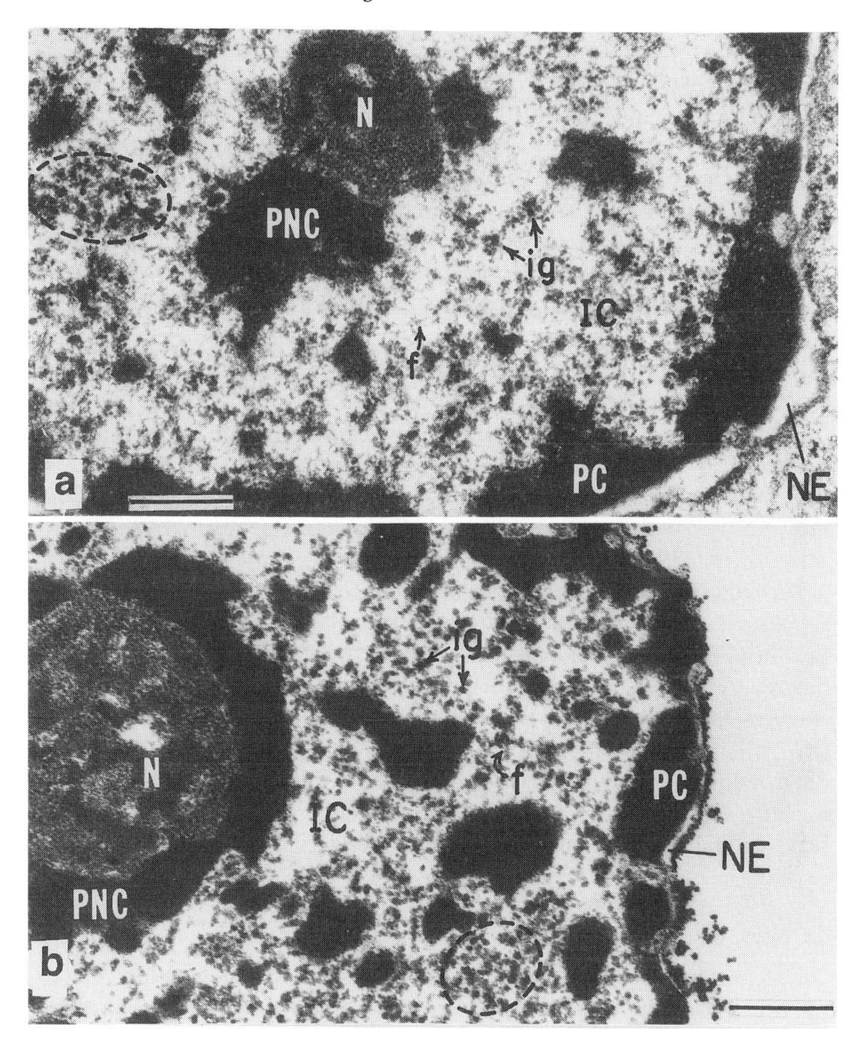

Fig. 4. Electron micrograph of sections of rat liver nuclei *in situ* in liver tissue (a) and in the isolated nuclear fraction (b). The isolated rat liver nuclei maintain many of the structural features characteristic of nuclei *in situ.* These include the NE, nuclear envelope; N, nucleolus; PNC, perinucleolar condensed chromatin; PC, peripheral condensed chromatin; and IC, interchromatinic areas. The IC contain electron-dense particles 150–250 Å in diameter termed interchromatin granules (ig), as well as less electron-dense fibrous material (f). Clusters of interchromatin granules are enclosed by a broken line. Magnification: ×44,000. From Berezney and Coffey (1977). Reproduced from *The Journal of Cell Biology,* 1977, **73,** pp. 616–637, by copyright permission of The Rockefeller University Press.

structural alterations are then maintained in the isolated nuclear matrix. Unfortunately, hypotonic swelling of cells is an important step in isolating pure, intact nuclei from most cells grown in culture. Some conditions of nuclei isolated by hypotonic swelling, however, appear to work better than others, as illustrated in the study of Belgrader *et al.* (1991a) for isolation of the HeLa nucleus and nuclear matrix. In general, however, the nuclear matrix morphology of cells grown in culture is best studied by direct extraction of cells grown on coverslips. The laboratories of Deppert (Staufenbiel and Deppert, 1984) and Penman (Fey *et al.*, 1986; Nickerson *et al.*, 1990) have pioneered the development of these so-called *in situ* nuclear matrix preparations, which are particularly valuable for electron microscopic studies as well as for immunolocalization studies. A diagrammatic illustration of a typical *in situ* matrix preparation is shown in Fig. 3. Numerous studies have documented that such *in situ* preparations offer the advantage of better maintenance of nuclear morphology, with biochemical and functional properties very similar to those of nuclear matrices prepared from isolated nuclei.

Perhaps the most widely characterized nucleus at the electron microscopic level, both in intact tissue and as isolated nuclei, is the rat liver nucleus. With the use of isotonic and hypertonic sucrose solutions, procedures have been developed which maintain the ultrastructure of the isolated rat liver nucleus while allowing a high degree of purification (Blobel and Potter, 1966; Berezney *et al.*, 1972; Berezney, 1974). This is illustrated in Fig. 4, where the ultrastructure of a typical isolated rat liver nucleus is strikingly similar to the *in situ* nuclear morphology observed in rat liver tissue. Aside from the maintenance of the general disposition of

---

Fig. 5. (a) Electron micrograph of a section through the nuclear matrix revealing the internal structural components of the matrix. RN, residual nucleolus; IM, internal matrix framework; RE, residual nuclear envelope layer. Note the empty spaces surrounding the residual nucleoli and along the periphery (arrows). These may correspond to regions previously occupied by the perinucleolar and peripheral condensed chromatin in untreated nuclei (compare with Fig. 7a and 7b). Magnification: ×21,000. (b) Higher magnification electron micrograph of a section of the nuclear matrix in the region of the residual nuclear envelope. A close association of the IM with the RE is evident (white arrows). A residual nuclear pore complex structure is projecting through the RE layer (black arrows). Regions of the IM (enclosed in the broken line) resemble clusters of ICG seen in both isolated and *in situ* nuclei (compare with the regions enclosed by broken lines in Fig. 4a and 4b). Magnification: ×20,000. From Berezney and Coffey (1977). Reproduced from *The Journal of Cell Biology*, 1977, **73**, pp. 616–637, by copyright permission of The Rockefeller University Press.

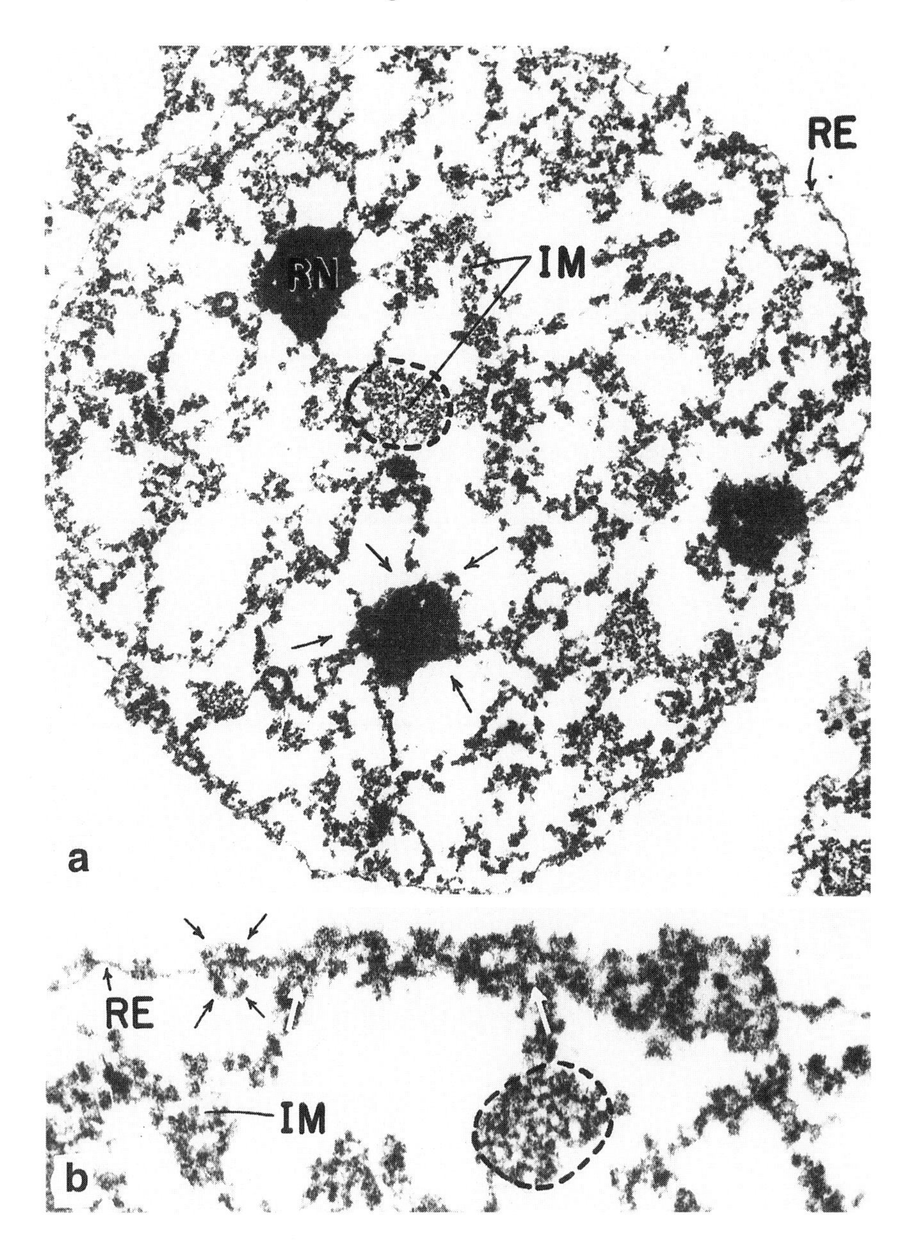
RE
RN
IM
a
RE
IM
b

condensed chromatin and interchromatinic regions, there is a remarkable maintenance of interchromatin granules (ICG) (including the characteristic clusters of ICG) and the less discrete-appearing fibrous material, which is nonetheless characteristic of the *in situ* matrix.

Preparation of nuclear matrices from such morphologically well-maintained rat liver nuclei enabled us to compare directly structural details at the electron microscopic level of the isolated nuclear matrix in relation to the nucleus *in situ.* Figures 5 and 6 show the tripartite structure (nuclear lamina, residual nucleus, and internal matrix) at lower and higher magnifications. Large empty regions, especially surrounding the nucleoli, suggest that these regions were formerly occupied by condensed chromatin domains. The fibrogranular structure of the isolated nuclear matrix is similar to the fibrogranular structures of the *in situ* matrix. This includes the appearance of discrete ICG (often appearing as clusters in the isolated nuclear matrix), as well as the less discrete and less densely staining matrix protein filaments (MPF).

The fibrous components of *in situ* and isolated nuclear matrices are difficult to visualize by standard thin-sectioning electron microscopy, but the poor quality of imaging does not mean that they do not exist or that they are not important. Indeed, these wispy, poorly staining MPF structures (30–50 Å in diameter) have been consistently identified in

---

Fig. 6. (a) High magnification of the nuclear matrix in the area of the residual nucleolus. The residual nucleolus (RN) appears to be continuous with the internal matrix framework (IM). Empty spaces (arrows) surrounding the RN may correspond to regions previously occupied by the perinucleolar condensed chromatin (compare with Figure 4a and 4b). Note that these empty areas are bordered by the nucleolus and the IM. Magnification: ×102,000. (b) High-magnification section of a region in the interior of an isolated nucleus. Distinct areas of condensed chromatin (C) are seen in the upper-center and lower-left regions of the micrograph. The regions between these C areas are the interchromatinic regions, which contain electron-dense interchromatin granules (ig) and less electron-dense fibrous structures (f). Magnification: ×116,000. (c) High magnification of a section through the IM of the nuclear matrix. This residual framework structure consists of electron-dense matrix particles (mp) and matrix fibers (f) which are very similar to the interchromatinic structures of isolated as well as *in situ* liver nuclei (compare with Figs. 9b and 7a and 7b). Magnification: ×116,000. (d,e) High-magnification sections through the residual nuclear envelope layer of the nuclear matrix. Distinct residual nuclear pore complex structures are observed which still retain their characteristic annular structure (arrows). Central granules are often visible in tangenital sections through the residual pore complex structures (white arrow in e). Magnification: ×158,000. From Berezney and Coffey (1977). Reproduced from *The Journal of Cell Biology,* 1977, **73,** pp. 616–637, by copyright permission of The Rockefeller University Press.

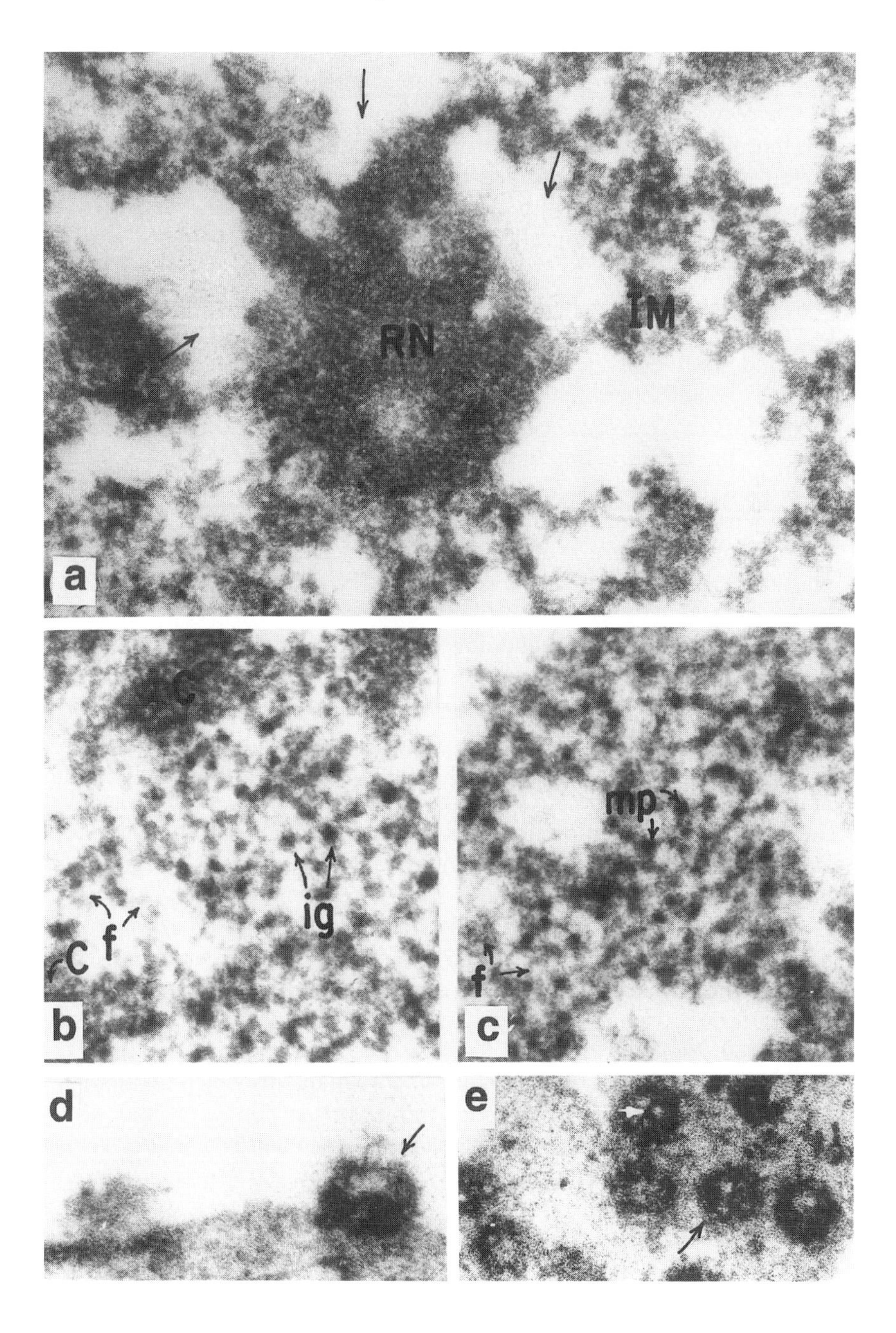
a
RN
IM
b
C
ig
f
C
c
mp
f
d
e

this region of the nucleus (Bernhard and Granboulan, 1963) but have been virtually ignored in most studies of nuclear organization. In one of the very few studies, however, Smetana *et al.* (1971) suggested that the fine filaments (MPF) are protein structures because they were digested with pepsin. In contrast, the ICG were not affected by pepsin. However, predigestion with pepsin made these granules sensitive to ribonuclease (RNase) digestion, indicating the RNP nature of the ICG.

The classical studies of Monneron and Bernhard (1969) using EDTA-regressive staining also identified 30- to 50-Å fibers as a basic component of various extrachromatinic structures. These include perichromatin fibers and granules, coiled bodies, fibers associated with the nuclear pore complexes, and fibers that interconnect ICG. Monneron and Bernhard (1969) further stressed the similar staining properties of the fibers and suggested a structural link between perichromatin fibers and ICG. These ill-defined, hard-to-visualize filaments that are maintained in an appropriately prepared nuclear matrix preparation may therefore represent a fundamental structural component of the nuclear matrix. Indeed, Comings and Okada (1976) first termed the MPF matrixin; they were later called matricin (Berezney, 1980).

While standard thin-sectioning electron microscopy and EDTA-regressive staining enable the visualization of a fibrogranular structure in whole cells and isolated nuclear structures, the structural information obtained is limited. In particular, each section is only a very thin slice through the entire nucleus. Thus the potentially three-dimensional organization formed by the ICG and the MPF is not addressed. Moreover, the structural information obtained about the MPF is extremely limited, as the tracing of individual filaments would be severely compromised by the thinness of the sections.

As a step toward improving this situation, Penman and co-workers have used whole-mount and resinless thick-section electron microscopy to study the nuclear matrix (Capco *et al.,* 1982; Fey *et al.,* 1986; Nickerson *et al.,* 1990). A complex three-dimensional network of filaments with associated granular structures was observed. Further fractionation revealed a three-dimensional network of core filaments which the Penman group has interpreted as the core framework structure around which other components are assembled in the complete nuclear matrix structure (He *et al.,* 1990). The core filaments averaged 80–100 Å in diameter and were structurally distinct from cytoplasmic filament systems. Jackson and Cook (1988) have also reported a filament system in the nucleoskeleton obtained without high-salt extraction.

We have observed similar extended filaments of approximately 100 Å in more spread-out regions of whole-mount preparations of isolated rat liver nuclear matrix (Fig. 7). As summarized previously, conventional thin-sectioning electron microscopy has demonstrated that the internal matrix is composed predominantly of ICG that are enmeshed in less electron-dense, fibrous-like material. Filaments with a minimum diameter of 30–50 Å, the MPF, can be observed extending from this material. The possible identity of the MPF with the core filaments identified by Penman and co-workers (He *et al.,* 1990) is an important area of future investigation.

In earlier studies of nuclear subfractionation, some investigators found that procedures related to those used for nuclear matrix isolation (nuclease, salt, and detergent) could also lead to so-called empty nuclear matrices which contained the surrounding nuclear lamina with nuclear pore complexes but were devoid of internal matrix structure. After some initial confusion, it became apparent that the internal nuclear matrix is much more sensitive to extraction than the surrounding nuclear lamina (Kaufmann *et al.,* 1981). This has led to more optimized preparations for both nuclear matrix with well-preserved internal matrix structure and nuclear lamina free of internal matrix components (Smith *et al.,* 1984; Belgrader *et al.,* 1991a). If nuclei are digested with RNase A and extracted for nuclear matrix in the presence of sulfhydryl reducing agents such as dithiothreitol, the internal matrix is destabilized and empty matrices consisting exclusively of nuclear lamina are obtained. Preparation of nuclear matrix in the absence of RNase and dithiothreitol leads to typical tripartite matrices with an elaborate internal matrix structure.

IV. NUCLEAR MATRIX PROTEINS

Nuclear matrix proteins are the nonhistone proteins comprising the nuclear matrix following nuclease, salt, and detergent extraction of isolated cell nuclei (see Section III,B on nuclear matrix isolation). While virtually all known nuclear functions are associated with this proteinaceous nucleoskeletal structure (see Section IV,A on the functional properties of nuclear matrices), our knowledge of the proteins which compose the nuclear matrix is very limited. There is no doubt, however, that a detailed molecular analysis of the individual nuclear matrix proteins is

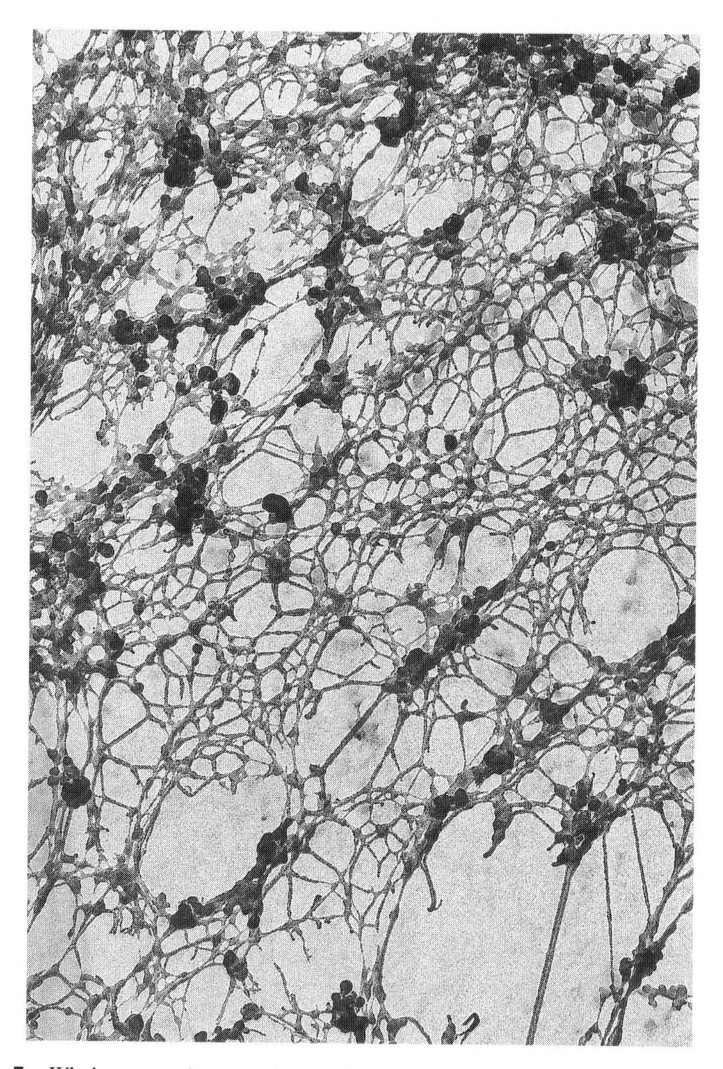

Fig. 7. Whole-mount electron micrograph of an isolated rat liver nuclear matrix spread on an aqueous surface reveals a fibrous network structure. The specimen was critically point dried and rotary shadowed with platinum–palladium. The delicate matrix lacework is considerably disrupted in the absence of critical point drying.

of paramount importance for deciphering the structural organization and molecular properties of the nuclear matrix and its functions.

A. General Properties of Nuclear Matrix Proteins

Initial studies of the nuclear matrix proteins suggested that they represent a major component of the acidic nonhistone proteins of the cell nucleus (Berezney and Coffey, 1977). This conclusion was based on a combination of classical fractionation procedures (the nuclear matrix proteins are predominantly soluble in alkaline) and total amino acid analysis, which revealed a remarkable similarity of the previously characterized residual acidic nuclear protein fraction extracted from mammalian cell nuclei (Steele and Busch, 1963). Two-dimensional PAGE studies, however, clearly indicate that there are many proteins in the nuclear matrix with overall basic charge, as well as a large population of acidic proteins (Peters and Comings, 1980; Nakayasu and Berezney, 1991; Fig. 8).

A major problem in identifying proteins native to the nuclear matrix is that nuclear matrices prepared from tissue culture cells are invariably contaminated with large amounts of cytoskeletal proteins, particularly the intermediate filament proteins (Capco *et al.,* 1982; Staufenbiel and Deppert, 1983; Verheijen *et al.,* 1986; Belgrader *et al.,* 1991a). Naturally, any *in situ* nuclear matrix preparation behaves this way. Fey and Penman (1988) have circumvented this problem by using an extraction procedure which separates intermediate filament proteins from the true nuclear matrix components. In contrast, the polypeptide profiles of nuclei isolated from tissues such as rat liver are likely to reflect largely the true nuclear proteins.

The first reports of nuclear matrix proteins separated on one-dimensional SDS-PAGE stressed the predominance of three major proteins between 60 and 70 kDa (Berezney and Coffey, 1974, 1977; Comings and Okada, 1976). It was apparent from those one-dimensional gels, however, that while the 60- to 70-kDa triplet was a major component, many other protein bands were also present, especially at molecular weights ranging from approximately 50,000 to more than 200,000 (Berezney and Coffey, 1977). At about the same time, Aaronson and Blobel (1975) reported that the nuclear pore complex–lamina fraction was also composed of

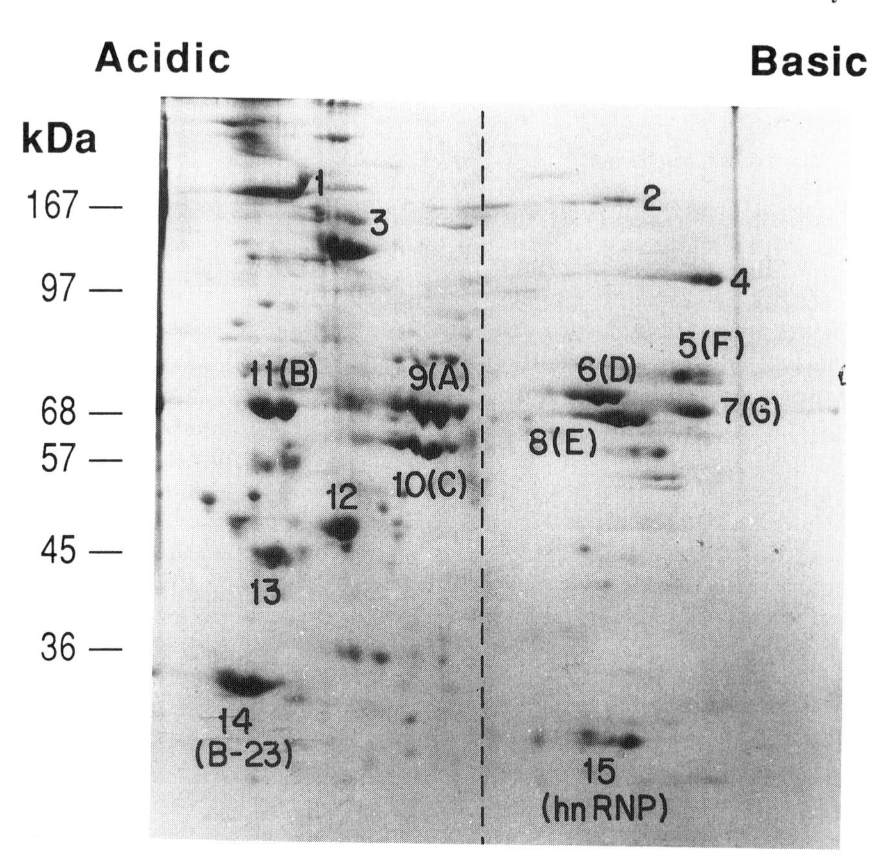

Fig. 8. Two-dimensional PAGE of rat liver nuclear matrix proteins. Total rat liver matrix protein was run on a nonequilibrium pH gradient in the first dimension and on SDS–PAGE in the second dimension. The major proteins detected with Coomassie blue staining were numerically labeled (1–15), including one minor spot (protein 2) and another spot (protein 1) which often stained less intensely. The major proteins included lamins A, B, and C, the nucleolar protein B-23, and residual hnRNP proteins. The remaining eight major components appear to represent hitherto uncharacterized major proteins of the cell nucleus. Because these proteins stained the internal fibrogranular matrix by immunofluorescence, they have been termed nuclear matrins to distinguish them from the nuclear lamins along the nuclear periphery (Nakayasu and Berezney, 1991). A group of nonlamin proteins which migrated as more basic components, but in the same molecular mass range (60–75 kDa) as the lamins, are also labeled matrins D, E, F, and G. Molecular mass markers are indicated in kilodaltons. From Nakayasu and Berezney (1991).

three major proteins between 60 and 70 kDa, which were ultimately termed nuclear lamins A, B, and C by Gerace and Blobel (1980).

A controversy then developed as to whether the 60- to 70-kDa proteins were strictly components of the surrounding nuclear lamina or were also present in the internal nuclear matrix (Berezney, 1980; Gerace and Blobel, 1980). It turned out that both groups were right. Nuclear lamins A, B, and C are major components of the 60- to 70-kDa proteins, but there are also other proteins in this molecular weight range which are present in nuclear matrix preparations but not present in purified nuclear lamina. This was apparent only on two-dimensional SDS–PAGE (Berezney, 1984). Study of these other 60- to 70-kDa proteins has now become a major focus of our research (see Section IV,B on the nuclear matrins). To complete the story, reports indicate that nuclear lamins are also found in the interior of the nuclear matrix (Goldman *et al.,* 1992; Bridger *et al.,* 1993; Moir *et al.,* 1994, 1995).

Two-dimensional analyses of nuclear matrix proteins performed by several different groups all stress the high degree of complexity of these polypeptide profiles (Peters *et al.,* 1982; Fey and Penman, 1988; Stuurman *et al.,* 1990; Nakayasu and Berezney, 1991). Using [^{35}S]methionine labeling for detection, Fey and Penman (1988) have detected over 200 proteins in the nuclear matrix. Stuurman *et al.* (1990) have also found enormous complexity in the two-dimensional profiles with the sensitive silver procedure. Despite this complexity, these studies are already providing valuable information. For example, the total nuclear matrix proteins can be separated into two major classes: those which are found in a variety of cell lines (common matrix proteins) and those which are cell type, hormonal, or differentiation state-dependent (Fey and Penman, 1988; Dworetzky *et al.,* 1990; Getzenberg and Coffey, 1990; Stuurman *et al.,* 1990).

The advantage of using two-dimensional PAGE to identify specific nuclear matrix proteins is further illustrated by the recent use of this approach to define differences in nuclear matrix proteins between normal and cancer cells grown in culture or in tumor tissues (Getzenberg *et al.,* 1991; Partin *et al.,* 1993; Pienta and Lehr, 1993a; Khanuja *et al.,* 1993; Bidwell *et al.,* 1994; Keesee *et al.,* 1994). The potential application of these findings to the diagnosis, prognosis, and possibly treatment of human cancers is worthy of further studies (see, e.g., Pienta and Lehr, 1993b).

Our overall knowledge of the proteins which comprise the nuclear matrix structure is very limited. One general property that needs further investigation is the phosphorylation of nuclear matrix proteins. Previous studies indicate that many nuclear matrix proteins are phosphorylated (Allen *et al.*, 1977; Henry and Hodge, 1983). There are only a few reports, however, dealing with the phosphorylation or biosynthesis and turnover of specific nuclear matrix proteins in relation to the cell cycle and its regulation (Milavetz and Edwards, 1986; Halikowski and Liew, 1987; Zhelev *et al.*, 1990).

The nuclear matrix itself contains a variety of protein kinases (Table III), at least some of which are capable of phosphorylating endogenous nuclear matrix proteins (Sikorska *et al.*, 1988; Sahyoun *et al.*, 1984; Capitani *et al.*, 1987; Tawfic and Ahmed, 1994). Payrastre *et al.* (1992) reported differential localization of phosphoinositide kinases, diacylglycerol kinase, and phospholipase C in the nuclear matrix. Because protein kinase C is also present (Capitani *et al.*, 1987), a complete regulatory signal transduction pathway may exist in the cell nucleus, with the nuclear matrix serving an important role in arranging and sequestering the individual components of the regulatory components. In this regard, Coghlan *et al.* (1994) found a nuclear matrix-associated protein (AKAP 95) that demonstrated high-affinity binding sites for the regulatory subunit of type II cyclic adenosine monophosphate (cAMP)-dependent protein kinase. It is very likely that these findings represent only the tip of the iceberg in terms of regulatory factors associated with the nuclear matrix.

Use of antibodies raised against nuclear matrix proteins offers great promise. In various studies, however, it has been difficult to identify the specific nuclear matrix antigen(s) responsible for the immunodecoration or to extend the work beyond the initial characterization (Chaly *et al.*, 1984; Lehner *et al.*, 1986; Noaillac-Depeyre *et al.*, 1987; Turner and Franchi, 1987). Progress is being made, however. The groups of Smith *et al.* (1985) and Penman (Nickerson *et al.*, 1992; Wan *et al.*, 1994) have gone beyond the initial identification to demonstrate potential roles of their respective nuclear matrix-associated proteins in RNA splicing (Smith *et al.*, 1989; Blencowe *et al.*, 1994).

One very significant development is the demonstration by Brinkley and co-workers (Zeng *et al.*, 1994a,b; He *et al.*, 1995) that the nuclear mitotic apparatus protein (NuMA) is a nuclear matrix protein. The propensity of this large protein (>200 kDa) to form a long coiled-coil structure (Compton *et al.*, 1992; Yang *et al.*, 1992) suggests an important role of NuMA in the organization of the nuclear matrix—possibly as a major component of the proposed nuclear matrix core filaments (He *et al.*, 1990).

B. Nuclear Matrins

Studies in our laboratory are concentrating on the major proteins of the nuclear matrix which are common at least among mammalian cells. Using a two-dimensional PAGE system, we have detected in rat liver nuclear matrix about 12 major Coomassie blue-stained proteins along with over 50 minor spots (Fig. 8). Polyclonal antibodies were then generated to individual matrix proteins excised from the two-dimensional gels. Antibodies to known nuclear proteins revealed that five of the major Coomassie blue-stained proteins correspond to lamins A, B, and C, the nucleolar protein B-23, and core heterogeneous nuclear RNP (hnRNP) proteins. The remaining eight proteins were termed nuclear matrins because they are components of the internal nuclear matrix, as revealed by immunofluorescence, and are distinct from the nuclear lamins, as indicated by one- and two-dimensional peptide maps (Nakayasu and Berezney, 1991). A survey of the literature showed no definite relationship of these proteins (matrins, 3, 4, D, E, F, G, 12, and 13) to any other known nuclear proteins (Nakayasu and Berezney, 1991). A summary of these major nuclear matrix proteins, including their identification as nuclear lamins or matrins, is presented in Table II.

Within the eight matrins examined, six formed three "pairs" of related proteins based on antibody cross-reactivity and peptide mapping (matrins D–E, F–G, and 12–13). This suggests that the nuclear matrins may compose a broad family of hitherto undefined proteins in the nucleus, with potential subfamilies indicated by the various protein-pair homologs. This leads us to propose that the term nuclear matrins be adopted as a general designation of nuclear matrix-associated proteins (determined via immunoblots) which are also components of the internal nuclear matrix (determined via immunofluorescence in whole cells versus cells from which the nuclear matrix had been extracted).

We are currently studying the molecular and functional properties of the individual nuclear matrins. As an initial step, we have screened nuclear matrix proteins for DNA-binding activity on Southwestern blots. Using one-dimensional SDS–PAGE, we have shown that the nuclear matrix is enriched in the higher molecular weight DNA-binding proteins found in total rat liver nuclear matrix proteins (Hakes and Berezney, 1991). Approximately 12 major DNA-binding proteins with apparent molecular weights exceeding 40,000 were detected on the one-dimensional blots. Further studies indicated that these proteins preferentially

TABLE II

Major Coomassie Blue-Stained Nuclear Matrix Proteins Derived from Two-Dimensional PAGE

Nuclear matrix protein designation	Approximate size (kDa)	Acidic or basic[b]	Protein identity[c]
1[a]	190	A	Matrin
3	125	A	Matrin
4	105	B	Matrin
5 (F)	77	B	Matrin
6 (D)	72	B	Matrin
7 (G)	68	B	Matrin
8 (F)	66	B	Matrin
9 (A)	68	A	Lamin
10 (B)	66	A	Lamin
11 (C)	62	A	Larnin
12	48	A	Matrin
13	42	A	Matrin
14	34	A	B-23
15	30	B	hnRNP

[a] Designation of protein 1 as a major matrix protein is tentative because it often stains as a more minor component.

[b] Designation as acidic (A) or basic (B) is based on migration across a nonequilibrium pH gradient gel. The dotted lines across the gels indicate the position of pH 7.0. Additional studies with equilibrium pH gradients have verified this general characterization (data not shown).

[c] The identity of the matrix proteins was evaluated by two-dimensional immunoblots to known nuclear proteins. Proteins in the nuclear matrix fraction which were not identified as previously known proteins are termed the nuclear matrins to distinguish them from the well-known nuclear lamins of the nuclear periphery.

bound DNA when competing with excess RNA (Hakes and Berezney, 1991). Two-dimensional Southwestern blots were then performed to identify the specific DNA-binding proteins (Hakes and Berezney, 1991). Approximately 12 distinct spots were detected, including lamins A and C (but not B); matrins D, E, F, G, and 4 (but not 3); and an unidentified protein of about 48 kDa.

As a step toward the further characterization of the nuclear matrins, we have been screening cDNA expression libraries, and have cloned and sequenced two distinct nuclear matrix proteins termed matrin 3 and matrin cyp. Matrin 3 is an acidic nuclear matrix protein (125 kDa; see Table II) that stained the nuclear interior with a fibrogranular pattern typical of the nuclear matrins (Nakayasu and Berezney, 1991; Belgrader *et al.,* 1991b). Analysis of matrin 3 human and rat cDNA sequences indicated a high degree of conservation. The lack of homologies with other proteins or functional motifs, however, gave few clues to possible function (Belgrader *et al.,* 1991b). The presence of over 40 potential phosphorylation sites in the predicted amino acid sequence of matrin 3, however, suggests the possible role of protein phosphorylation in matrin 3 function.

Studies of the matrin 3 gene(s) are in progress as a prelude to investigating matrin 3 function. Preliminary results suggest the presence of at least two closely related genes and the possibility of alternatively spliced transcripts (Mortillaro *et al.,* 1993; Somanathan *et al.,* 1995).

We have cloned and sequenced another nuclear matrix protein termed matrin cyp. Antibodies to matrin cyp recognized an approximately 100-kDa protein that was located in the nucleus and highly enriched in the nuclear matrix (Mortillaro and Berezney, 1995). Immunofluorescence microscopy confirmed the predominantly nuclear localization of matrin cyp. Analysis of the predicted amino acid coding sequence for matrin cyp revealed several distinct domains: (1) a 170-amino acid, amino-terminal region that showed high identity with cyclophilins, a family of ubiquitous proteins that bind to the immunosuppressant drug cyclosporin A and catalyze the conversion of *cis*-proline to *trans*-proline—an event important for protein folding (Fischer *et al.,* 1989; Gething and Sambrook, 1992); (2) an acidic serine-rich region similar to the nuclear localization signal binding (NLS-binding) domains initially described for Nopp140 (Meier and Blobel, 1992); and (3) a series of serine-arginine (SR) repeats throughout the carboxyl half of the protein which are also present in several splicing factors (Zahler *et al.,* 1993) and have been demonstrated to target proteins to splicing factor-rich nuclear speckles (Li and Bingham, 1991).

The presence of SR repeats in matrin cyp is consistent with the colocalization, at the splicing component-rich nuclear speckles, of anti-matrin cyp antibodies with antibodies to splicing components (Spector, 1984; Mortillaro and Berezney, 1995). Moreover, there was no detectable change in the localization of matrin cyp at these sites following extraction of cells for nuclear matrix (Mortillaro and Berezney, 1995).

A fusion protein containing the cyclophilin domain of matrin cyp expressed peptidylprolyl isomerase (PPIase) activity, which was completely abolished by cyclosporin A (Mortillaro and Berezney, 1995). In preliminary experiments, we have detected cyclosporin-sensitive PPIase activity in soluble nuclear matrix protein extracts and are in the process of determining whether matrin cyp, either by itself or with other proteins, is responsible for this activity in the nuclear matrix. Studies on the possible role of this novel nuclear matrix cyclophilin in RNA splicing and/or in assembly or chaperone events at splicing sites are in progress.

It is likely that the next few years will see the elucidation of many nuclear matrix proteins using this molecular cloning approach. This will provide fundamental information about a family of proteins which are of obvious significance for nuclear organization and likely function but have until recently defied analysis.

V. NUCLEAR MATRIX FUNCTIONS

The nuclear matrix was first identified in whole cells as that region of the nucleus where the actively functioning chromatin (euchromatin) is located, along with the nonchromatin fibrogranular matrix structures (Berezney and Coffey, 1977; Berezney, 1984). It is therefore not surprising that isolated nuclear matrices, which show a structural correspondence to the *in situ* defined structures, have a vast array of functional properties associated with them. Table III summarizes many of these major properties along with sample references.

It is important to stress that while it is not surprising to see this multitude of functional properties—ranging from DNA loop attachment sites, to DNA replication, to transcriptional associations, to RNA transcripts, to RNA splicing, to viral associations and their associated functions, and to a vast number of regulatory proteins involved in the functioning and regulation of these properties (e.g., steroid hormone receptors, oncogene proteins, tumor suppressors, heat shock proteins, calmodulin-binding proteins, protein kinases)—the true significance of these associations remains to be determined. Initial results suggest that the isolated nuclear matrix provides a potentially powerful *in vitro* approach for studying the molecular biology of higher order nuclear structure and function.

Because only a few studies have been performed on many of these properties, more studies are needed to define more clearly the nature of the associations and the actual role(s) of nuclear matrix structure in these processes. What is needed for each property is a detailed description of the associated function, the molecular constituents involved, and

TABLE III

Functional Properties Associated with Isolated Nuclear Matrix

Property	Reference
DNA loop attachment site sequences	Gasser and Laemmli (1987)
DNA-binding proteins	Hakes and Berezney (1991)
DNA topoisomerase II	Fernandes and Catapano (1991)
Replicating DNA	van der Velden and Wanka (1987)
Replication origins	(Dijkwel *et al.* (1986)
DNA polymerase α and primase	Tubo and Berezney (1987b)
Other replicative factors	Tubo and Berezney (1987a)
Active gene sequences	Zehnbauer and Vogelstein (1985)
RNA polymerase II	Lewis *et al.* (1984)
Transcriptional regulatory proteins	Feldman and Nevins (1983)
hnRNA and RNP	Verheijen *et al.* (1988)
Preribosomal RNA and RNP	Ciejek *et al.* (1982)
snRNA and RNP	Harris and Smith (1988)
RNA splicing	Zeitlin *et al.* (1987)
Spliceosome-associated proteins	Blencowe *et al.* (1994)
Steroid hormone receptor binding	Rennie *et al.* (1983)
Viral DNA and replication	Smith *et al.* (1985)
Viral premessenger RNA	Mariman *et al.* (1982)
Viral proteins	Covey *et al.* (1984)
Carcinogen binding	Gupta *et al.* (1985)
Oncogene proteins	Eisenman *et al.* (1985)
Tumor suppressors	Mancini *et al.* (1994)
Heat shock proteins	Reiter and Penman (1983)
Calmodulin-binding proteins	Bachs and Carafoli (1987)
HMG-14 and HMG-17 binding	Reeves and Chang (1983)
ADP-ribosylation	Cardenas-Corona *et al.* (1987)
Protein phosphorylation	Allen *et al.* (1977)
Protein kinase A	Sikorska *et al.* (1988)
Protein kinase B	Sahyoun *et al.* (1984)
Protein kinase C	Capitani *et al.* (1987)
Protein kinase CK2	Tawfic and Ahmed (1994)
Histone acetylation and deacetylation	Hendzel *et al.* (1991, 1994)
Protein disulfide oxidoreductase	Altieri *et al.* (1993)
Reversible size changes	Wunderlich and Herlan (1977)

the relationship of the *in vitro* function to *in situ* associations. This last evaluation is most difficult and may require continued studies of the *in vitro* associations until enough is known to plan appropriate experiments at the level of whole cells.

A major development in the last several years which bodes well for future studies is the multitude of immunofluorescence studies demonstrating a variety of functional domains in the nucleus of intact cells (Spector, 1993), including territories for specific chromosomes (Manuelidis, 1985, 1990; Hadlaczky *et al.*, 1986; Cremer *et al.*, 1993), DNA replication sites (Nakamura *et al.*, 1986; Nakayasu and Berezney, 1989; Fox *et al.*, 1991; Kill *et al.*, 1991; O'Keefe *et al.*, 1992; Neri *et al.*, 1992; Hassan and Cook, 1993; Hozák *et al.*, 1993), and transcriptional and RNA splicing sites (Huang and Spector, 1991; Jackson *et al.*, 1993; Wansink *et al.*, 1993).

In a particularly elegant series of experiments, Lawrence and coworkers (1989) demonstrated that abundantly newly transcribed RNA often forms visible "tracks" in the nucleus that presumably extend from sites of transcription toward sites of further processing and transport. This suggested that the newly transcribed RNA was somehow posttranscriptionally ordered in the nucleus. The maintenance of these gene transcript track patterns following extraction for nuclear matrix (Xing and Lawrence, 1991) implicates the nuclear matrix in this structural organization. The Lawrence group has extended these studies to show specific orientation of the tracks with respect to sites of transcription and gene splicing for the fibronectin gene using intron-containing and spliced transcripts (Xing *et al.*, 1993).

A major conclusion of these studies is that transcription and RNA splicing may be a coordinated process in which the nuclear matrix plays a fundamental role in the organization. Jiménez-García and Spector (1993), using different approaches, also conclude that transcription and splicing may be temporally and spatially linked in the cell nucleus. Similarly, the speckled sites where splicing factors and small nuclear RNPs (snRNPs) are concentrated (Spector *et al.*, 1983; van Eekelen *et al.*, 1982) and the sites of DNA replication (Nakayasu and Berezney, 1989) are strikingly maintained following extraction of cells on coverslips for nuclear matrix.

The host of regulatory factors which are in the nucleus and nuclear matrix-associated (Table III) further supports the model of the nuclear matrix as a dynamic entity involved in structural organization, function, and regulation of the genome. Additional findings include the association of a plethora of transcription factors with nuclear matrix (van Wijnen *et al.*, 1993) and the presumptive role of specific nuclear matrix proteins in RNA splicing *in vitro* (Smith *et al.*, 1989; Blencowe *et al.*, 1994).

Section V presents a more detailed description of one of the best-studied functions of the nuclear matrix: DNA replication. In particular, experiments will be described that are designed to "bridge the gap" between *in vitro* matrix systems and replication *in situ*.

VI. NUCLEAR MATRIX AND DNA REPLICATION

Each enormous molecule of eukaryotic chromosomal DNA is divided into hundreds to thousands of independent subunits of replication termed replicons (Hand, 1978). Replication proceeds bidirectionally within each replicon subunit. Individual replicons are further organized into families or clusters of tandemly repeated subunits which replicate as a unit at particular times in S phase (Hand, 1975, 1978; Lao and Arrighi, 1981; Fig. 9A). Up to 100 or more replicons may be organized into each replicon cluster, with an estimated average size of approximately 25 (Hand, 1978; Painter and Young, 1976). The numerous reports that specific DNA sequences are duplicated at precise times within the S phase of eukaryotic cells (Goldman *et al.*, 1984; Hatton *et al.*, 1988) further supports the conclusion that replicon cluster synthesis is temporally and spatially regulated along the chromosomal DNA molecule.

While the existence of replicon subunits, their bidirectional replication, and their organization into functional replicon clusters are well documented, the mechanistic and molecular basis for these fundamental properties remains a long-standing mystery. Even less well understood is what controls the exquisite spatial and temporal patterns of replicon cluster synthesis during S phase. It is remarkable that the approximately 50,000 to 100,000 individual replicons that make up the typical mammalian genome are programmed to replicate once and only once in a precisely choreographed process. This implies a great deal of structural order underlying DNA replication in the cell nucleus. Somehow the molecular details of replication are integrated within the complex three-dimensional organization of the cell nucleus. As summarized in Section V,A, the key player in this process may be the nuclear matrix.

A. Replicating DNA on the Nuclear Matrix

Numerous studies of *in vivo* replicated DNA associated with isolated nuclear matrix have led to a radically new view of DNA replication

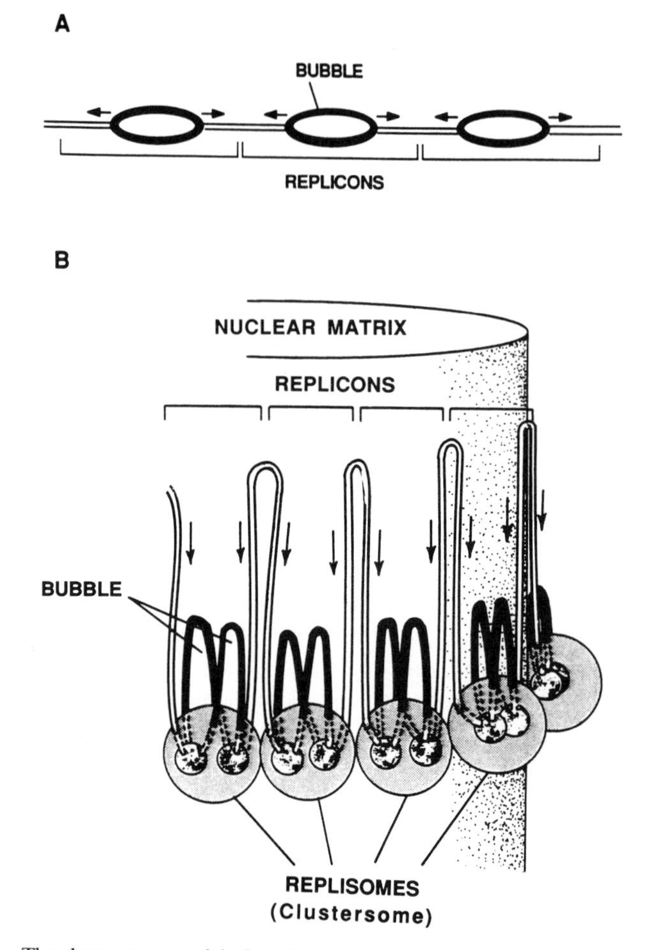

Fig. 9. The clustersome model of nuclear matrix-associated DNA replication. (A) Replicon cluster on linear DNA. Bidirectional replication along three tandemly arranged replicons in a hypothetical replicon cluster is illustrated along a linear DNA molecule. The arrows show the directions of the growing replicational bubbles. Unduplicated DNA is shown in white; duplicated DNA (replicational bubbles) in black. (B) Replicon cluster on nuclear matrix-attached DNA loops. The DNA of the same replicon cluster shown in (A) is now arranged in a series of loops attached to the nuclear matrix at fixed replicational sites. Each of these sites, known as replisomes, also contains the apparatus for copying the DNA. Copying occurs when DNA is reeled across the matrix-bound replisomes, as shown by the arrows. Unduplicated DNA loops are shown in white, duplicated DNA loops (replicational bubbles) in black. Groups of replisomes cluster together to form a higher order assembly for replicon cluster synthesis termed the clustersome. From Tubo and Berezney (1987b).

inside the cell nucleus (Berezney and Coffey, 1975; Dijkwel *et al.*, 1979; McCready *et al.*, 1980; Pardoll *et al.*, 1980; Berezney and Buchholtz, 1981b; Berezney, 1984). It is envisioned that replicating DNA loops corresponding to individual replicon subunits are bound to the nuclear matrix. Bidirectional replication then occurs by the reeling of DNA at the two ends of the loops through matrix-bound replisomes (Fig. 9B). Topographical organization of the replicating DNA loops and the associated replisomes into functional clusters or clustersomes may then provide the basis for replicon clustering.

Consistent with this clustersome model, DNA polymerase, primase, and other replicative components have been found to be associated with isolated nuclear matrix (Table IV; Smith and Berezney, 1980; Jones and Su, 1982; Foster and Collins, 1988; Collins and Chu, 1987; Tubo and Berezney, 1987a,b,c; Paff and Fernandes, 1990) or with the nucleoskeleton obtained via electroelution of chromatin fragments following restriction enzyme digestion under isotonic buffer conditions (Jackson and Cook, 1986a,b,c). The *in vitro* synthesis of Okazaki-sized DNA fragments (Smith and Berezney, 1982), density shift experiments which indicate that the matrix-bound synthesis continues replication along *in vivo*-initiated DNA strands (Tubo *et al.*, 1985), the striking replicative and prereplicative association of DNA polymerase, primase, and other replicative components with the nuclear matrix (Smith and Berezney, 1983; Tubo and Berezney, 1987a,b), and the adenosine triphosphate (ATP)-stimulated processive synthesis by the matrix-bound polymerase (Tubo *et al.*, 1987) all point toward a replicative-related role of these matrix-bound activities.

The clustersome model further predicts that the replicational machinery (replisomes) for many individual replicons is correspondingly clustered at nuclear matrix-bound sites (Fig. 9B). As a step toward testing this aspect of the model, we developed methods to extract the matrix-

TABLE IV

Properties of Nuclear Matrix-Bound DNA Synthesis

Replicative-dependent association of DNA polymerase α and primase
Okazaki fragments continue synthesis at *in vivo* forks
ATP-stimulated processive synthesis requires nuclear matrix attachment
Organization into large megacomplexes (100–150S) is replication dependent

bound replicational complexes (Fig. 10; Tubo and Berezney, 1987b). Most of the matrix-bound DNA polymerase α and primase activities were released in the form of discrete megacomplexes sedimenting on sucrose gradients at approximately 100S and 150S. In contrast, complexes extracted from nuclei during nuclear matrix preparation sedimented at about 8–10S, which is typical of DNA polymerase–primase complexes

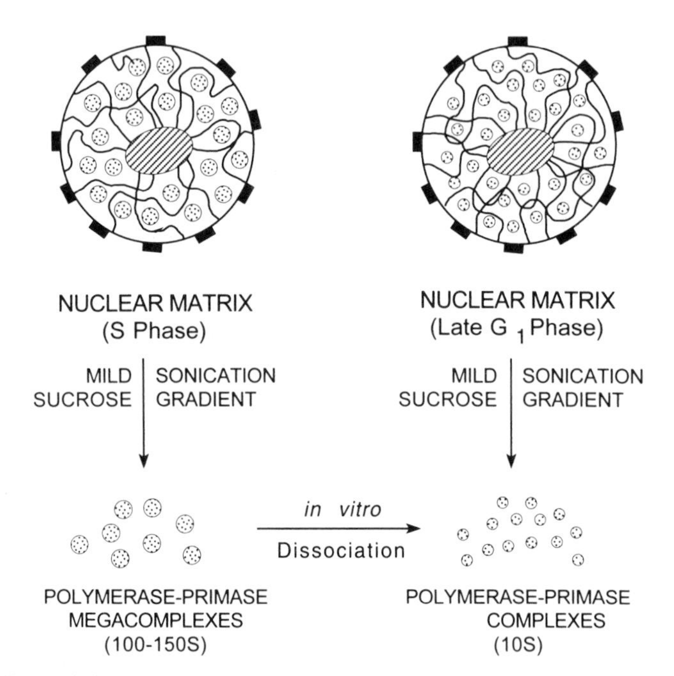

Fig. 10. Isolation of DNA polymerase–primase megacomplexes from rat liver nuclear matrix. Nuclear matrix was isolated from rat liver nuclei prepared at different times following partial hepatectomy (Tubo and Berezney, 1987b). DNA polymerase α and primase activities were effectively released from the isolated matrices by mild sonication and resolved on sucrose gradients. During active replication of the liver cells (22 hr posthepatectomy), most of the enzyme activity sedimented as large 100–150S complexes (megacomplexes). Just before the onset of replication in the regenerating liver (12 hr posthepatectomy), the DNA polymerase α and primase activities were found predominantly in 10S complexes. The corresponding dissociation of the megacomplexes to 8–10S complexes following release from the matrix structure led to the suggestion that the megacomplexes represent clusters of 10S complexes and thus may be components of the hypothetical clustersomes that we proposed are attached to the nuclear matrix in intact cells (see Tubo and Berezney, 1987b).

purified from cells. The rapid conversion of the megacomplexes into the more typically sized 10S complexes following release from the matrix structure suggested that the megacomplexes were composed of clusters of 10S complexes and thus might represent the *in vitro* equivalent of the predicted clustersome (Fig. 10; Tubo and Berezney, 1987b).

B. Visualizing Replication Sites in Mammalian Cells

A model for the arrangement of these putative nuclear matrix-bound clustersomes is shown in Figs. 9B and 10. While our biochemical results supported this model (see Section VI,A), the possibilities of rearrangements or aggregations during nuclear and/or nuclear matrix isolation could not be completely ruled out. What was needed was a method to visualize directly the sites of DNA replication in the nuclei of whole cells.

With this in mind, we developed a permeabilized mammalian cell system to study the incorporation of biotin-11–deoxyuridine 5′-triphosphate (biotin-11–dUTP) into newly replicated DNA. The sites of biotinylated, newly synthesized DNA were then directly visualized by fluorescence microscopy following reaction with Texas Red–streptavidin (Nakayasu and Berezney, 1989). Discrete granular sites of replication were observed. The number of replication granules per nucleus (150 to 300) and their size (0.4–0.8 μm in diameter) are consistent with each replication granule being the site of synthesis of a replicon cluster. At any given time in S phase, one would anticipate that thousands of replicons would be active and arranged in up to several hundred clusters.

The discrete nature of the individual replication granules is more apparent at higher magnification (Fig. 11a). Many of the granules appear to have a somewhat elongated or ellipsoid-like shape. In addition, the characteristic size and shape of the individual replication granules remained the same while the fluorescence intensity progressively increased in pulse periods ranging from 2 to 60 min (Nakayasu and Berezney, 1989). The size of the individual replication granules, therefore, is not determined by the amount of DNA replicated but rather is an inherent organizational property of each replication site. These results strongly support the previously proposed clustersome model. It is proposed, therefore, that each replication granule corresponds to an *in vivo* clustersome (Fig. 9).

To what extent are the *in situ* nuclear patterns of DNA replication maintained following nuclear matrix isolation? To address this question, biotin–dUTP was first incorporated into permeabilized cultured cells (e.g., 3T3 fibroblasts or PtK1 cells), followed by *in situ* extraction for nuclear matrix (Nakayasu and Berezney, 1989). Alternatively, nuclear matrix structures were prepared and followed by *in vitro* incorporation of biotin-dUTP via the nuclear matrix-bound DNA synthesis system. DNA replication granules observed on the nuclear matrix were virtually identical in size and number to those in cells (Fig. 11a,b).

Direct information on the *in vivo* sites of DNA replication was obtained by incorporated 5-bromodeoxyuridine (BrdU) into cultured cells, and immunofluorescence localization was performed following reaction with monoclonal antibodies to BrdU and fluorescein isothiocyanate (FITC) or Texas Red-conjugated secondary antibodies. Despite the lower sensitivity of this *in vivo* approach, we observed significant similarity in the size and number of replication granules compared to those observed in permeabilized cells (Nakayasu and Berezney, 1989). This provides corroborative evidence for the clustersome model and demonstrates that the permeabilized cell approach is useful for studying the organization of replication sites in the cell nucleus. Several other studies of *in vivo* replication sites in mammalian cells have found similar results (Kill *et al.*, 1991; Neri *et al.*, 1992; Fox *et al.*, 1991; O'Keefe *et al.*, 1992; Hassan and Cook, 1993). In addition, these sites have been visualized with electron microscopy, with results consistent with the immunofluorescence patterns (Hozák *et al.*, 1993). Similar sites have been detected following reconstruction of nuclear structures *in vitro* in *Xenopus* egg extracts (Blow and Laskey, 1986).

C. Visualizing Replication Sites in Three Dimensions

A major limitation of these studies is that essentially two-dimensional information is obtained about what is in fact a three-dimensional system. We are using laser scanning confocal fluorescence microscopy following incorporation of BrdU into cultured mammalian cells to obtain real, three-dimensional information about the spatial organization of the replication sites in the cell nucleus. Figure 12a–j shows a series of confocal sections through the nucleus of a 3T3 cell which was replicating its DNA in early to mid-S phase. A typical pattern of replication granules is

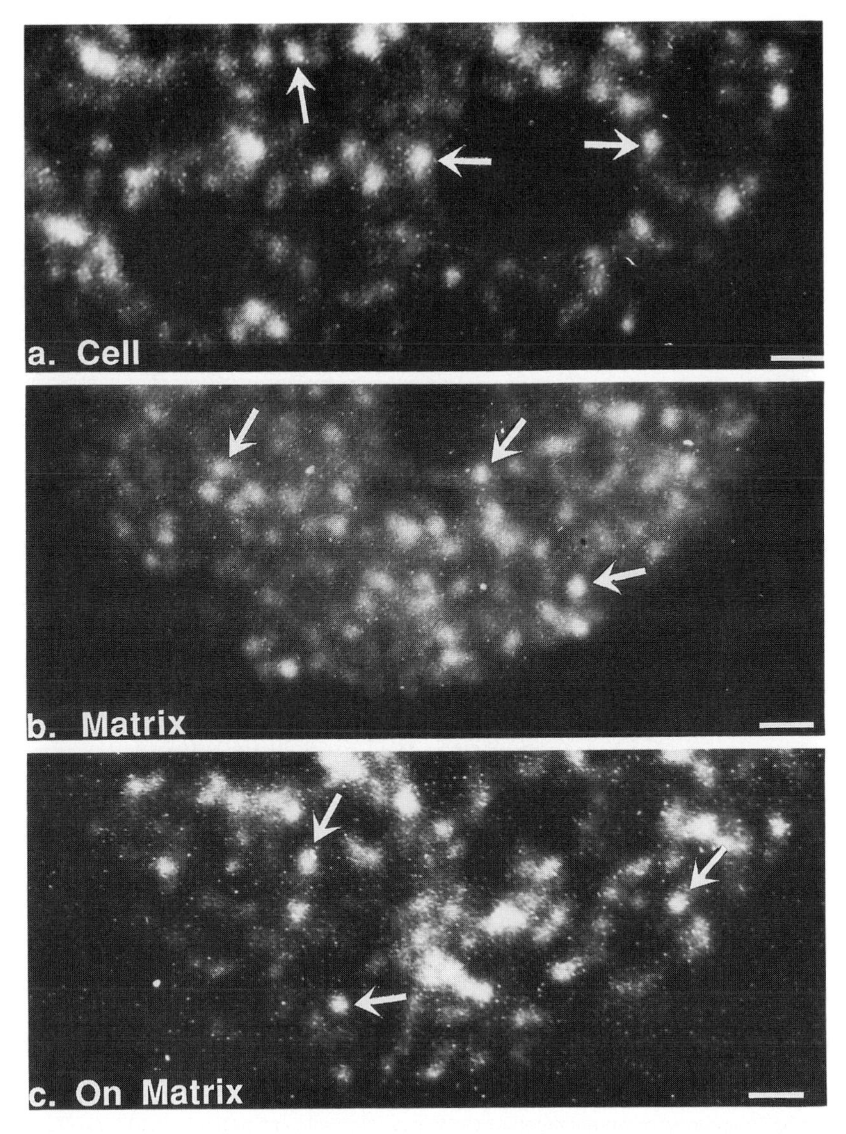

Fig. 11. Replication granules (clustersomes) in permeabilized cells and after extraction of cells for nuclear matrix. (a) PtK1 cells on coverslips were permeabilized with 0.04% Triton X-100 and incubated with a DNA synthesis medium containing biotin-11–dUTP at 37° C for 5 min (Nakayasu and Berezney, 1989). (b) Following incorporating of biotin-11–dUTP, the cells were extracted for nuclear matrix. (c) PtK1 cells were first extracted for nuclear matrix. DNA synthesis was then carried out on nuclear matrix-attached DNA fragments in the presence of biotin-11–dUTP. The sites of DNA synthesis were visualized under the fluorescence microscope following incubation with Texas Red-conjugated streptavidin. Similar granular sites (clustersomes) were detected in all cases (arrows). Bar: 1 μm.

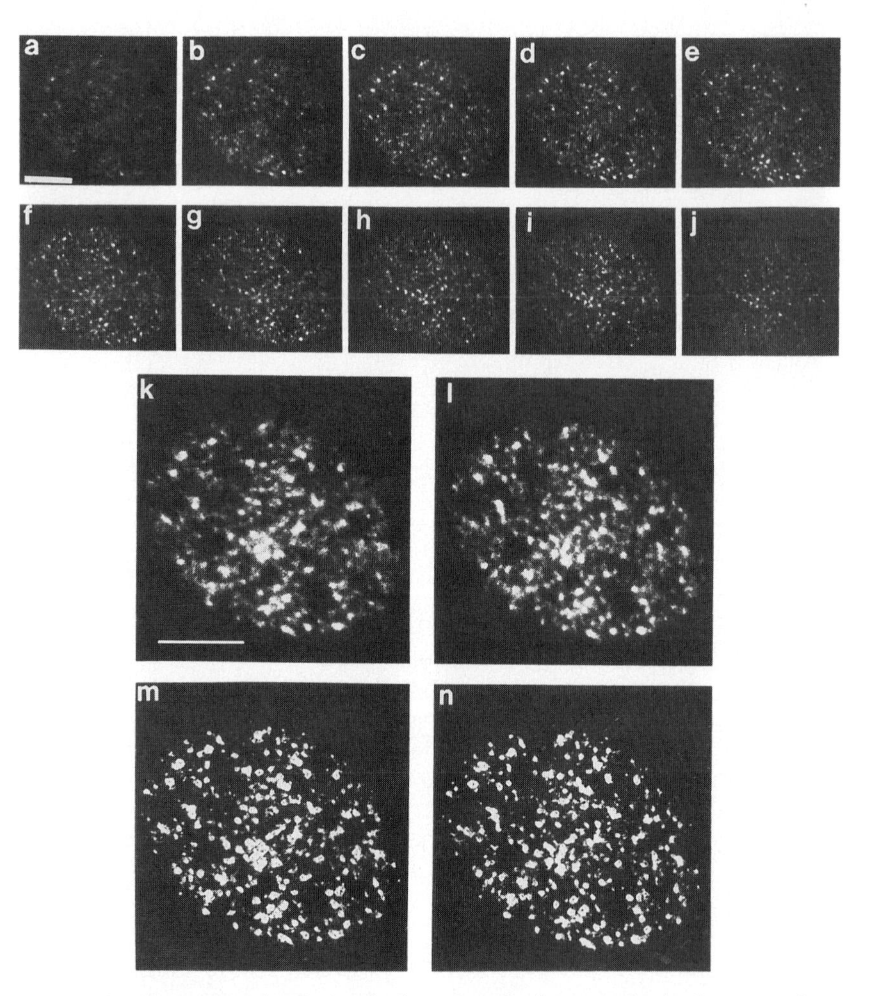

Fig. 12. Three-dimensional visualization of replication granules by laser scanning confocal microscopy. (a–j) Ten representative optical sections (0.3-μm intervals) of a total of 17 sections through the nucleus of 3T3 cells following a 60-min *in vivo* pulse with BrdU and processing for immunofluorescence staining with antibodies to BrdU and FITC-conjugated secondary antibodies. Bar: 10 μm. (k,l) Stereo pair for the three-dimensionally reconstructed image derived from the 17 individual sections. (m,n) The same three-dimensionally reconstructed nucleus showing the contours of the individual replication granules obtained using a multidimensional image analysis system. Bar: 10 μm (k–n).

apparent. Figure 12k–l is a stereo pair of the three-dimensionally reconstructed image derived from the complete series of confocal microscopic sections through this nucleus. Aside from demonstrating the three-dimensional organization, this approach allows the individual sites of replication and their shapes to be seen with enhanced clarity and more details (compare Figs. 13 and 11). Greater heterogeneity in the size and shape of individual clustersomes is observed compared to those visualized with epifluorescence microscopy.

More precise and quantitative information from these three-dimensional images is being obtained by applying multidimensional computer imaging analysis. In our approach, the analysis and processing software uses an enhancement algorithm to suppress the background while intensifying the replication sites. A suitable threshold is then used to obtain the contours of individual sections (Ballard and Brown, 1982; Duda and Hart, 1973). Three-dimensional replication sites are built up from these contours by three-dimensional connected component analysis (Samarabandu *et al.,* 1994). The visualization software renders these volume contours as wire frame diagrams on the screen (Fig. 12m,n). It is important to note that the *x, y, z* coordinates (centers of gravity) are determined for every site and can be displayed in three dimensions. This enables analysis of specific sites of replication among the hundreds to thousands that are present.

The segmentation algorithm was further improved by using a spot detection method which can better discriminate closely spaced replication sites. Visualization software is also improved by modeling the replication sites as solid spheres and using ray-tracing algorithms to render these spheres in three-dimensional space. Higher levels of analysis that are being developed in our laboratory include forming domains of replication sites based on the distance to the nearest neighbor and quantifying various aspects of collocalization of two-channel images which shows DNA replication at different times (Berezney *et al.,* 1995; Samarabandu *et al.,* 1995).

In one application of these approaches, we are combining dual-color three-dimensional laser scanning microscopy and computer image analysis with fluorescence *in situ* hybridization (FISH) to identify where specific genes are replicated and ultimately to "map" the DNA sequences at individual replication sites (Berezney, 1995). This represents a direct test of the "single site–multiple replicon" model and will enable us to examine the specificity of the DNA sequences at individual sites of replication. Initial experiments have studied the replication sites of specific chromosomes in RPMI human lymphoblasts. We have detected

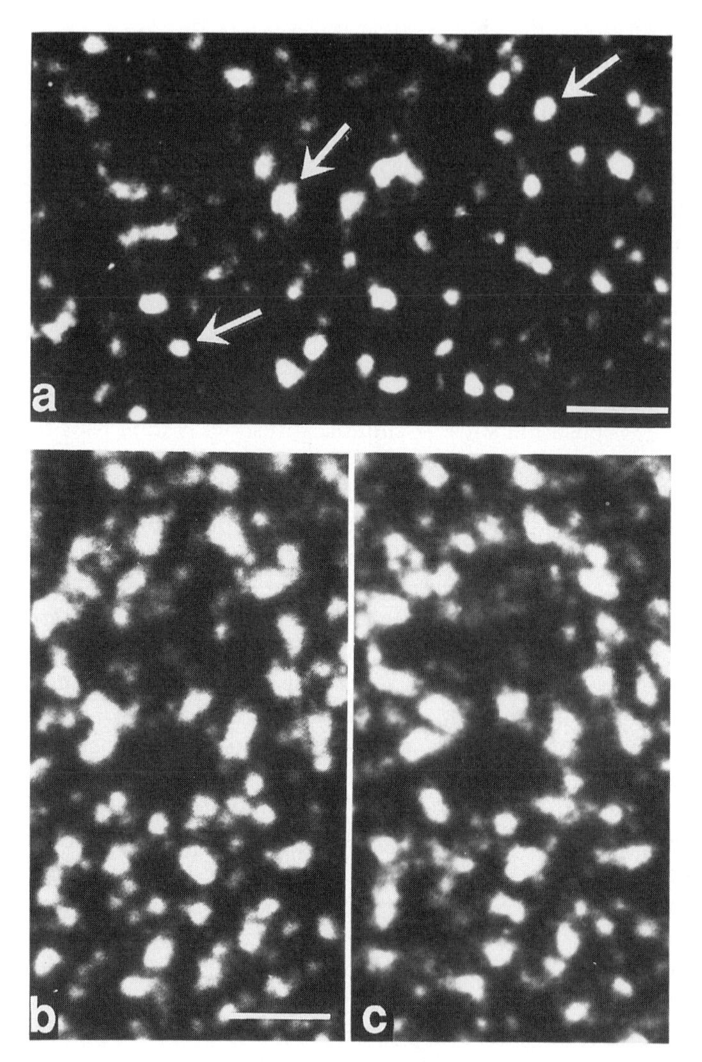

Fig. 13. Higher magnification of replication granules by confocal microscopy. (a) Portion of a typical optical section near the midplane of the nucleus of a 3T3 cell that was pulsed *in vivo* for 60 min with BrdU and processed for immunofluorescence staining with anti-BrdU antibodies and rhodamine-conjugated secondary antibodies. Arrows point to type I granules. (b,c) Stereo pair of a portion of the three-dimensionally reconstructed image derived from seven sections (0.5-μm intervals), including the section shown in (a). Bar: 2.2 μm.

low levels of chromosome 11 replication sites among the multitudes of nonchromosome 11 DNA replication sites (H. Ma and R. Berezney, unpublished results, 1996).

In summary, initial steps have been taken toward defining the macromolecular assembly of individual replication sites in three dimensions. Further refinements of the multidimensional analysis programs, along with high-resolution FISH analysis, should lead to an increasingly clear picture of gene replication inside the cell nucleus.

VII. CONCLUDING REMARKS

The evidence is compelling that genomic organization, function, and regulation are linked through nuclear architecture. First are the studies with isolated nuclear matrices demonstrating the association of a vast array of functional and regulatory properties. Second are the fluorescence and electron microscopic observations of nuclei in whole cells and tissues showing that (1) the chromatin corresponding to individual chromosomes is localized in discrete regions of the nucleus, and (2) the genomic functions of replication, transcription, and RNA splicing occupy discrete spatial sites inside the cell nucleus.

Most excitingly, these two lines of evidence are beginning to merge. Results indicate that the overall organization of sites of replication, transcription, and posttranscriptional RNA transcripts and processing is strikingly maintained in cells extracted for nuclear matrix. These findings corroborate the biochemical results demonstrating functional properties associated with isolated nuclear matrix. They also support the view that *in vitro* nuclear matrix systems represent a valuable new approach for elucidating the relationships of genomic organization, function, and regulation in the mammalian cell nucleus.

In one strategy, proteins or other components known to be involved in genomic processes can be studied in association with the nuclear matrix. In another approach, the nuclear matrix can be used as a starting point to identify novel proteins important for genomic function and regulation. Finally, both previously known proteins and novel nuclear matrix-associated proteins can be probes for studying the spatial organization of these functional sites in intact cells and nuclear matrix preparations.

Another important direction is computer image analysis. The first steps have already been taken in applying multidimensional computer imaging to the three-dimensional organization of functional sites in the cell nucleus. Realizing the explosive development of computer technology in the past decade, it is awesome to consider what will be possible as we approach the 21st century.

ACKNOWLEDGMENTS

We are extremely grateful to Dr. P. C. Cheng (Department of Electrical and Computer Engineering, SUNY at Buffalo) for the use of the confocal microscopy facilities at the Advanced Microscopy and Imaging Laboratory (AMIL) and for assistance and collaboration on experiments involving this instrumentation. Jim Stamos kindly provided the illustrations and photography. The experiments reported from our laboratory were funded by NIH Grant GM 23922.

REFERENCES

Aaronson, R. P., and Blobel, G. (1975). Isolation of nuclear pore complexes in association with a lamina. *Proc. Natl. Acad. Sci. U.S.A.* **72,** 1007–1011.

Adolph, K. W., Cheng, S. M., Paulson, J. R., and Laemmli, U. K. (1977). Isolation of protein scafford from mitotic HeLa cell chromosomes. *Proc. Natl. Acad. Sci. U.S.A.* **74,** 4937–4941.

Allen, S. L., Berezney, R., and Coffey, D. S. (1977). Phosphorylation of nuclear matrix proteins in isolated regenerating rat liver nuclei. *Biochem. Biophys. Res. Commun.* **75,** 111–116.

Altieri, F., Maras, B., Eufemi, M., Ferraro, A., and Turano, C. (1993). Purification of a 57 kDa nuclear matrix protein associated with thiol: Protein-disulfide oxidoreductase and phospholipase C activities. *Biochem. Biophys. Res. Commun.* **194,** 992–1000.

Bachs, O., and Carafoli, E. (1987). Calmodulin and calmodulin binding proteins in liver cell nuclei. *J. Biol. Chem.* **262,** 10786–10790.

Ballard, D. H., and Brown, C. M. (1982). "Computer Vision." Prentice Hall, Englewood Cliffs, NJ.

Belgrader, P., Siegel, A. J., and Berezney, R. (1991a). A comprehensive study on the isolation and characterization of the HeLa S3 nuclear matrix. *J. Cell Sci.* **98,** 281–291.

Belgrader, P., Dey, R., and Berezney, R. (1991b). Molecular cloning of matrin 3. A 125-kilodalton protein of the nuclear matrix contains an extensive acidic domain. *J. Biol. Chem.* **266,** 9893–9899.

Berezney, R. (1974). Large scale isolation of nuclear membranes from bovine liver. *Methods Cell Biol.* **7,** 205–228.

Berezney, R. (1979). Dynamic properties of the nuclear matrix. *In* "The Cell Nucleus" (H. Busch, ed.), Vol. 7, pp. 413–456. Academic Press, New York.
Berezney, R. (1980). Fractionation of the nuclear matrix. *J. Cell Biol.* **85,** 641–650.
Berezney, R. (1984). Organization and functions of the nuclear matrix. *In* "Chromosomal Nonhistone Proteins" (L. S. Hnilica, ed.), Vol. 4, pp. 119–180. CRC Press, Boca Raton, FL.
Berezney, R. (1995). Replicating the mammalian genome in 3D. *J. Cell. Biochem., Suppl.* **21B,** 118.
Berezney, R., and Buchholtz, L. B. (1981a). Isolation and characterization of rat liver nuclear matrices containing high molecular weight deoxyribonucleic acid. *Biochemistry* **20,** 4995–5002.
Berezney, R., and Buchholtz, L. B. (1981b). Dynamic association of replicating DNA fragments with the nuclear matrix of regenerating liver. *Exp. Cell Res.* **132,** 1–13.
Berezney, R., and Coffey, D. S. (1974). Identification of a nuclear protein matrix. *Biochem. Biophys. Res. Common.* **60,** 1410–1417.
Berezney, R., and Coffey, D. S. (1975). The nuclear protein matrix: association with newly synthesized DNA. *Science* **189,** 291–293.
Berezney, R., and Coffey, D. S. (1976). The nuclear protein matrix: Isolation, structure, and functions. *Adv. Enzyme Regul.* **14,** 63–100.
Berezney, R., and Coffey, D. S. (1977). Nuclear matrix: Isolation and characterization of a framework structure from rat liver nuclei. *J. Cell Biol.* **73,** 616–637.
Berezney, R., Macaulay, L. K., and Crane, F. L. (1972). The purification and biochemical characterization of bovine nuclear membranes. *J. Biol. Chem.* **247,** 5549–5561.
Berezney, R., Ma, H., Wei, X., Meng, C., Samarabandu, J. K., and Cheng, P. C. (1995). Elucidating the higher order assembly of replicating sites in mouse 3T3 cells. *J. Cell. Biochem.* **21B,** 137.
Bernhard, W. (1969). A new staining procedure for electron microscopical cytology. *J. Ultrastruct. Res.* **27,** 250–265.
Bernhard, W., and Granboulan, N. (1963). The fine structure of the cancer cell nucleus. *Exp. Cell Res.* **9,** Suppl., 19–53.
Bidwell, J. P., Fey, E. G., van Wijnen, A. J., Penman, S., Stein, J. L., Lian, J. B., and Stein, G. S. (1994). Nuclear matrix proteins distinguish normal diploid osteoblasts from osteosarcoma cells. *Cancer Res.* **54,** 28–32.
Blencowe, B., Nickerson, J. A., Issner, R., Penman, S., and Sharp, P. A. (1994). Association of nuclear matrix antigens with exon-containing splicing complexes. *J. Cell Biol.* **127,** 593–607.
Blobel, G. (1985). Gene gating: A hypothesis. *Proc. Natl. Acad. Sci. U.S.A.* **83,** 8527–8529.
Blobel, G., and Potter, V. R. (1966). Nuclei from rat liver: Isolation method that combines purity with high yield. *Science* **154,** 1662–1664.
Blow, J. J., and Laskey, R. A. (1986). Initiation of DNA replication in nuclei and purified DNA by a cell-free extract of Xenopus eggs. *Cell (Cambridge, Mass.)* **47,** 577–587.
Bridger, J. M., Kill, I. R., O'Farrel, M., and Hutchison, C. J. (1993). Internal lamin structures within G1 nuclei of human fibroblasts. *J. Cell Sci.* **104,** 297–306.
Capco, D. G., Wan, K. M., and Penman, S. (1982). The nuclear matrix: Three-dimensional architecture and protein composition. *Cell (Cambridge, Mass.)* **29,** 847–858.
Capitani, S., Girard, P. R., Mazzei, G. J., Kuo, J. F., Berezney, R., and Manzoli, F. A. (1987). Immunochemical characterization of protein kinase C in rat liver nuclei and subnuclear fractions. *Biochem. Biophys. Res. Commun.* **142,** 367–375.

Cardenas-Corona, M. E., Jacobson, E. L., and Jacobson, M. K. (1987). Endogenous polymers of ADP-ribose are associated with the nuclear matrix. *J. Biol. Chem.* **262,** 14863–14866.

Chaly, N., Bladon, T., Setterfield, G., Little, J. E., Kaplan, J. G., and Brown, D. L. (1984). Changes in distribution of nuclear matrix antigens during the miotic cell cycle. *J. Cell Biol.* **99,** 661–671.

Ciejek, E. M., Nordstrom, J. L., Tsai, M. J., and O'Malley, B. W. (1982). Ribonucleic acid precursors are associated with the chick oviduct nuclear matrix. *Biochemistry* **21,** 4945–4953.

Coghlan, V. M., Langeberg, L. K., Fernandez, A., Lamb, N. J. C., and Scott, J. D. (1994). Cloning and characterization of AKAP 95, a nuclear protein that associates with the regulatory subunit of type II cAMP-dependent protein kinase. *J. Biol. Chem.* **269,** 7658–7665.

Collins, J. M., and Chu, A. K. (1987). Binding of the DNA polymerase α-DNA primase complex to the nuclear matrix in HeLa cells. *Biochemistry* **26,** 5600–5607.

Comings, D. E., and Okada, T. A. (1976). Nuclear proteins. III. The fibrillar nature of the nuclear matrix. *Exp. Cell Res.* **103,** 341–360.

Compton, D. A., Szilak, I., and Cleveland, D. W. (1992). Primary structure of NuMA, an intranuclear protein that defines a novel pathway for segregation of proteins at mitosis. *J. Cell Biol.* **116,** 1395–1408.

Covey, L., Choi, Y., and Prives, C. (1984). Association of simian virus 40 T antigen with the nuclear matrix of infected and transformed monkeys cells *Mol. Cell. Biol.* **4,** 1384–1392.

Cremer, T., Kurz, A., Zirbel, R., Dietzel, S., Rinke, B., Schröck, E., Speicher, M. R., Mathieu, U., Jauch, A., Emmerich, P., Scherthan, H., Ried, T., Cremer, C., and Lichter, P. (1993). Role of chromosome territories in the functional compartmentalization of cell nucleus. *Cold Spring Harbor Symp. Quant. Biol.* **53,** 777–792.

Derenzini, M., Lorenzoni, E., Marinozzi, V., and Barsotti, P. (1977). Ultrastructural cytochemistry of active chromatin in regenerating rat hepatocytes. *J. Ultrastruct. Res.* **59,** 250–262.

Derenzini, M., Novello, F., and Pession-Brizzi, A. (1978). Perichromatin fibrils and chromatin ultrastructural pattern. *Exp. Cell Res.* **112,** 443–454.

Dijkwel, P. A., Mullenders, L. H. F., and Wanka, F. (1979). Analysis of the attachment of replicating DNA to a nuclear matrix in mammalian interphase nuclei. *Nucleic Acids Res.* **6,** 219–230.

Dijkwel, P. A., Wenink, P. W., and Poddighe, J. (1986). Permanent attachment of replication origins to the nuclear matrix in BHK-cells. *Nucleic Acids. Res.* **14,** 3241–3249.

Duda, R. O., and Hart, P. E. (1973). "Pattern Classification and Scene Analysis." Wiley, New York.

Dworetzky, S. I., Fey, E. G., Penman, S., Lian, J. B., Stein, J. L., and Stein, C. S. (1990). Progressive changes in the protein composition of the nuclear matrix during rat osteoblast differentiation. *Proc. Nat. Acad. Sci. U.S.A.* **87,** 4605–4609.

Eisenman, R. N., Tachibana, C. Y., Abrams, H. D., and Hann, S. R. (1985) *v-myc* and *c-myc-* encoded proteins are associated with the nuclear matrix. *Mol. Cell. Biol.* **5,** 114–126.

Fakan, S., and Bernhard, W. (1971). Localization of rapidly and slowly labeled nuclear RNA as visualized by high resolution autoradiography. *Exp. Cell Res.* **67,** 129–141.

Fakan, S., and Bernhard, W. (1973). Nuclear labeling after prolonged ^{3}H-uridine incorporation as visualized by high resolution autoradiography. *Exp. Cell Res.* **79,** 431–444.
Fakan, S., Puvion, E., and Sphor, G. (1976). Localization and characterization of newly synthesized nuclear RNA in isolated rat hepatocytes. *Exp. Cell Res.* **99,** 155–164.
Fawcett, D. W. (1966). An atlas of fine structure. *In* "The Cell, Its Organelles and Inclusions," p. 3 Saunders, Philadelphia.
Feldman, L. T., and Nevins, J. R. (1983). Localization of the adenovirus El A protein, a positive-acting transcriptional factor, in infected cells. *Mol. Cell. Biol.* **3,** 829–838.
Fernandes, D. J., and Catapano, C. V. (1991). Nuclear matrix targets for anticancer agents. *Cancer Cells* **3,** 134–140.
Fey, E., and Penman, S. (1988). Nuclear matrix proteins reflect cell type of origin in cultured human cells. *Proc. Natl. Acad. Sci. U.S.A.* **85,** 121–125.
Fey, E., Krochmalnic, C., and Penman, S. (1986). The nonchromatin substructures of the nucleus: The ribonucleoprotein(RNP) containing and RNP-depleted matrices analyzed by sequential fractionation and resin-less section electron microscopy. *J. Cell Biol.* **102,** 1654–1665.
Fischer, G., Wittmann-Liebold, B., Lang, K., Kiefhaber, T., and Schmid, F. X. (1989). Cyclophilin and peptidyl-prolyl cis-trans isomerase are probably identical proteins. *Nature (London)* **337,** 476–478.
Foster, K. A., and Collins, J. M. (1988). The interrelation between DNA synthesis rates and DNA polymerases bound to the nuclear matrix in synchronized HeLa cells. *J. Biol. Chem.* **260,** 4229–4235.
Fox, M. H., Arndt-Jovin, D. J., Jovin, T. M., Baumann, P. H., and Robert-Nicaud, M. (1991). Spatial and temporal distribution of DNA replication sites localized by immunofluorescence and confocal microscopy. *J. Cell Sci.* **99,** 247–255.
Franke, W. W., and Falk, H. (1970). Appearance of nuclear pore complexes after Bernhard's staining procedure. *Histochemie* **24,** 266–278.
Gasser, S. M., and Laemmli, U. K. (1987). A glimpse at chromosomal order. *Trends Genet.* **3,** 16–22.
Gerace, L., and Blobel, G. (1980). The nuclear envelope lamina is reversibly depolymerized during mitosis. *Cell (Cambridge, Mass.)* **19,** 277–287.
Gething, M.-J., and Sambrook, J. (1992). Protein folding in the cell. *Nature (London)* **355,** 33–45.
Getzenberg, R. H., and Coffey, D. S. (1990). Tissue specificity of the hormonal response in sex accessory tissue in associated with nuclear matrix protein patterns. *Mol. Endocrinol.* **4,** 1336–1342.
Getzenberg, R. H., Pienta, K. J., Huang, E. Y., and Coffey, D. S. (1991). Identification of nuclear matrix proteins in the cancer and normal rat prostate. *Cancer Res.* **51,** 6514–6520.
Goldman, A. E., Moir, R. D., Montag-Lowy, M., and Goldman, R. D. (1992). Pathway of incorporation of microinjected lamin A into the nuclear envelope. *J. Cell Biol.* **119,** 725–735.
Goldman, M. A., Holmquist, G. P., Gray, M. C., Caston, L., and Nag, A. (1984). Replication timing of genes and middle repetitive sequences. *Science* **224,** 686–692.
Graham G. F., Arms, K., and Gurdon, J. B. (1966). The induction of DNA synthesis by egg cytoplasm. *Dev. Biol.* **14,** 349–359.
Gupta, R. C., Dighe, N. R., Randerath, K., and Smith, H. C. (1985). Distribution of initial and persistent 2-acetylaminofluorene-induced DNA adducts within DNA loops. *Proc. Natl. Acad. Sci. U.S.A.* **82,** 6605–6608.

Hadlaczky, G., Went, M., and Ringertz, N. R. (1986). Direct evidence for the non-random localization of mammalian chromosomes in the interphase nucleus. *Exp. Cell Res.* **167,** 1–15.

Hakes, D. J., and Berezney, R. (1991). DNA-binding properties of the nuclear matrix and individual nuclear matrix proteins: Evidence for salt resistant DNA binding sites. *J. Biol. Chem.* **266,** 11131–11140.

Halikowski, M. J., and Liew, C. (1987). Identification of a phosphoprotein in the nuclear matrix by monoclonal antibodies. *Biochem. J.* **241,** 693–697.

Hand, R. (1975). Regulation of DNA replication on subchromosomal units of mammalian cells. *J. Cell Biol.* **64,** 89–97.

Hand, R. (1978). Eucaryotic DNA: Organization of the genome for replication. *Cell (Cambridge, Mass.)* **15,** 315–325.

Harris, H. (1968). "Nucleus and Cytoplasm." Clarendon Press, Oxford.

Harris, H. (1970). "Cell Fusion." Clarendon Press, Oxford.

Harris, S. G., and Smith, H. C. (1988). SnRNP core protein enrichment in the nuclear matrix. *Biochem. Biophys. Res. Commun.* **152,** 1383–1387.

Hassan, A. B., and Cook, P. R. (1993). Visualization of replication sites in unfixed human cells. *J. Cell Sci.* **105,** 541–550.

Hatton, K. S., Dhar, V., Gahn, T. A., Brown, E. H., Mager, D., and Schildkraut, C. L. (1988). Temporal order of replication of multigene families reflects chromosomal location and transcriptional activity. *Cancer Cells* **6,** 335–340.

He, D., Nickerson, J. A., and Penman, S. (1990). Core filaments of the nuclear matrix. *J. Cell Biol.* **110,** 569–580.

He, D., Zheng, C., and Brinkley, B. R. (1995). Nuclear matrix proteins as structural and functional components of the mitotic apparatus. *Int. Rev. Cytol.* **162B,** 1–74.

Hendzel, M. J., Delcuve, G. P., and Davie, J. R. (1991). Histone deacetylase is a component of the internal matrix. *J. Biol. Chem.* **266,** 21936–21942.

Hendzel, M. J., Sun, J.-M., Chen, H. Y., Rattner, J. B., and Davie, J. R. (1994). Histone acetyltransferase is associated with the nuclear matrix. *J. Biol. Chem.* **269,** 22894–22901.

Henry, S. M., and Hodge, L. D. (1983). Nuclear matrix: A cell-cycle-dependent site of increased intranuclear protein phosphorylation. *Eur. J. Biochem.* **133,** 23–29.

Hozák, P., Hassan, A. B., Jackson, D. A., and Cook, P. R. (1993). Visualization of replication factories attached to a nucleoskeleton. *Cell (Cambridge, Mass.)* **73,** 361–373.

Huang, S., and Spector, D. L. (1991). Nascent pre-mRNA transcripts are associated with nuclear regions enriched in splicing factors. *Genes Dev.* **5,** 2288–2302.

Ingber, D. E., Dike, L., Hansen, L., Karp, S., Liley, H., Maniotis, A., McNamee, H., Mooney, D., Plopper, G., and Sims, J. (1994). Cellular tensegrity: Exploring how mechanical charges in the cytoskeleton regulate cell growth, migration, and tissue pattern during morphogenesis. *Int. Rev. Cytol.* **150,** 173–224.

Jackson, D. A., and Cook, P. R. (1985). A general method for preparing chromatin containing intact DNA. *EMBO J.* **14,** 913–918.

Jackson, D. A., and Cook, P. R. (1986a). Replication occurs at a nucleoskeleton. *EMBO J.* **5,** 1403–1410.

Jackson, D. A., and Cook, P. R. (1986b). A cell cycle-dependent DNA polymerase activity that replicates intact DNA in chromatin. *J. Mol. Biol.* **192,** 65–67.

Jackson, D. A., and Cook, P. R. (1986c). Different populations of DNA polymerase α in HeLa cells. *J. Mol. Biol.* **192,** 77–86.

Jackson, D. A., and Cook, P. R. (1988). Visualization of a filamentous nucleoskeleton with a 23 nm axial repeat. *EMBO J.* **7,** 3667–3677.

Jackson, D. A., Hassan, A. B., Errington, R. J., and Cook, P. R. (1993). Visualization of focal sites of transcription within human nuclei. *EMBO J.* **12,** 1059–1065.

Jiménez-García, L. F., and Spector, D. L. (1993). In vivo evidence that transcription and splicing are coordinated by a recruiting mechanism. *Cell (Cambridge, Mass.)* **73,** 47–59.

Jones, C., and Su, R. T. (1982). DNA polymerase α from the nuclear matrix of cells infected with simian virus 40. *Nucleic Acids Res.* **10,** 5517– 5582.

Kaufmann, S. H., Coffey, D. S., and Shaper, J. H. (1981). Considerations in the isolation of rat liver nuclear matrix, nuclear envelope, and pore complex lamina. *Exp. Cell Res.* **132,** 105–123.

Keesee, S. K., Meneghini, M. D., Szaro, R. P., and Wu, Y. J. (1994). Nuclear matrix proteins in human colon cancer. *Proc. Natl. Acad. Sci. U.S.A.* **91,** 1913–1916.

Khanuja, P. S., Lehr, J. E., Soule, H. D., Gehani, S. K., Noto, A. C., Choudhury, S., Chen, R., and Pienta, K. J. (1993). Nuclear matrix proteins in normal and breast cancer cells. *Cancer Res.* **53,** 3394–3398.

Kill, I. R., Bridger, J. M., Campbell, K. H. S., Maldonado-Codina, G., and Hutchison, C. J. (1991). The timing of the formation and usage of replicase cluster in S-phase nuclei of human dipoid fibroblasts. *J. Cell Sci.* **100,** 869–876.

Lao, Y. F., and Arrighi, F. F. (1981). Studies of mammalian chromosome replication. II. Evidence for the existence of defined chromosome replicating units. *Chromosoma* **83,** 721–741.

Lawrence, J. B., Singer, R. H., and Marselle, L. M. (1989). Highly localized tracks of specific transcripts within interphase nuclei visualized by in situ hybridization. *Cell (Cambridge, Mass.)* **57,** 493–502.

Lehner, C. F., Eppenberger, H. M., Fakan, S., and Nigg, E. A. (1986). Nuclear substructure antigens monoclonal antibodies against components of nuclear matrix preparations. *Exp. Cell Res.* **162,** 205–219.

Lewis, C. D., Lebkowski, J. S., Daly, A. K., and Laemmli, U. K. (1984). Interphase nuclear matrix and metaphase scaffolding structures. *J. Cell Sci.* **1,** Suppl., 103–122.

Li, H., and Bingham, P. M. (1991). Arginine/serine-rich domains of the su(wa) and tra RNA processing regulators target proteins to a subnuclear compartment implicated in splicing. *Cell (Cambridge, Mass.)* **67,** 335–342.

Mancini, M. A., Shan, B., Nickerson, J. A., Penman, S., and Lee, W. (1994). The retinoblastoma gene product is a cell cycle-dependent, nuclear matrix-associated protein. *Proc. Natl. Acad. Sci. U.S.A.* **91,** 418–422.

Manuelidis, L. (1985). Individual interphase chromosome domains revealed by in situ hybridization. *Hum. Genet.* **71,** 288–293.

Manuelidis, L. (1990). Individual interphase chromosomes. *Science* **250,** 1533–1540.

Mariman, E. C. M., van Eekelen, C. A. G., Reinders, R. J., Berns, A. J. M., and van Venrooij, W. J. (1982). Adenoviral heterogeneous nuclear RNA is associated with the host nuclear matrix during splicing. *J. Mol. Biol.* **154,** 103–119.

Maul, G. G., Maul, H. M., Scogna, J. E., Liebermann, M. W., Stein, G. S., Hsu, B. Y. I., and Borun, T.W. (1972). Time sequence of nuclear pore formation in phytohemagglutinin-stimulated lympocytes and HeLa cells during the cell cycle. *J. Cell Biol.* **5,** 433–422.

McCready, S. J., Godwin, J., Mason, D. W., Brazell, I. A. and Cook, P. R. (1980. DNA is replicated at the nuclear cage. *J. Cell Sci.* **46,** 365–386.

Meier, U. T., and Blobel, G. (1992). Noppl40 shuttles on tracks between nucleolus and cytoplasm. *Cell (Cambridge, Mass.)* **70,** 127–138.

Milavetz, B. I., and Edwards, D. R. (1986). Synthesis and stability of nuclear matrix proteins in resting and serum-stimulated Swiss 3T3 cells. *J. Cell. Physiol.* **127,** 388–396.

Mirkovitch, J., Mirault, M. F., and Laemmli, U. K. (1984). Organization of the higher-order chromatin loop: specific attachment sites on nuclear scaffold. *Cell (Cambridge, Mass.)* **39,** 223–232.

Moir, R. D., Montag-Lowy, M., and Goldman, R. D. (1994). Dynamic properties of nuclear lamins: lamin B is associated with sites of DNA replication. *J. Cell Biol.* **125,** 1201–1212.

Moir, R. D., Spann, T. P., and Goldman, R. D. (1995). The dynamic properties and possible functions of nuclear lamins. *Int. Rev. Cytol.* **162B,** 141–182.

Monneron, A., and Bernhard, W. (1969). Fine structural organization of the interphase nucleus in some mammalian cells. *J. Ultrastruct. Res.* **27,** 266–288.

Mortillaro, M. J., and Berezney, R. (1995). Association of a nuclear matrix cyclophilin with splicing factors. *J. Cell. Biochem., Suppl.* **21B,** 142.

Mortillaro, M. J., Vijayaraghavan, S., and Berezney, R. (1993). Evidence that matrin 3 is a member of a complex gene family. *Mol. Biol. Cell* **4,** 81S.

Nakamura, H., Morita, T., and Sato, C. (1986). Structural organization of replication domains during DNA synthetic phase in the mammalian nucleus. *Exp. Cell Res.* **165,** 291–297.

Nakayasu, H., and Berezney, R. (1989). Mapping replicational sites in the eucaryotic cell nucleus. *J. Cell Biol.* **108,** 1–11.

Nakayasu, H., and Berezney, R. (1991). Nuclear matrins: Identification of the major nuclear matrix proteins. *Proc. Natl. Acad. Sci. U.S.A.* **88,** 10312–10316.

Nash, R. E., Puvion, E., and Bernhard, W. (1975). Perichromatin fibrils as components of rapidly labeled extranucleolar RNA. *J. Ultrastruct. Res.* **53,** 395–405.

Neri, L. M., Mazzotti, G., Capitani, S., Maraldi, N. M., Cinti, C., Baldini, N., Rana, R., and Martelli, A. M. (1992). Nuclear matrix-bound replicational sites detected in situ by 5-bromodeoxyuridine. *Histochemistry* **98,** 19–32.

Nickerson, J. A., He, D., Fey, F. G., and Penman, S. (1990). The nuclear matrix. *In* "The Eukaryotic Nucleus, Molecular Biochemistry and Macromolecular Assemblies" (P. R. Strauss and S. H. Wilson, eds.), Vol. 2, pp. 763–782. Telford Press, Caldwell, NJ.

Nickerson, J. A., Krockmalnic, G., Wan, K. M., Turner, C. D., and Penman, S. (1992). A normally masked nuclear matrix antigen that appears at mitosis on cytoskeleton filaments adjoining chromosomes, centrioles and midbodies. *J. Cell Biol.* **116,** 977–987.

Noaillac-Depeyre, J., Azum, M., Geraud, M., Mathieu, C., and Gas, N. (1987). Distribution of nuclear matrix proteins in interphase CHO cells and rearrangements during the cell cycle. An ultrastructural study. *Biol. Cell* **61,** 23–32.

O'Keefe, R. T., Henderson, S. C., and Spector, D. L. (1992). Dynamic organization of DNA replication in mammalian cell nuclei: Spatially and temporally defined replication of chromosome-specific α-satellite DNA sequences. *J. Cell Biol.* **116,** 1095–1110.

Paff, M. T., and Fernandes, D. J. (1990). Sythesis and distribution of primer RNA in nuclei of CCRF-CEM leukemia cells *Biochemistry* **29,** 3442–3450.

Painter, R. B., and Young, B. R. (1976). Formation of nascent DNA molecules during inhibition of replicon initiation in mammalian cells. *Biochim. Biophys. Acta* **418,** 146–153.

Pardoll, D. M., Vogelstein, B., and Coffey D. S. (1980). A fixed site of DNA replication in eucaryotic cells. *Cell* **19,** 527–536.

Partin, A. W., Getzenberg, R. H., Carmichael, M. J., Vindivich, D., Yoo, J., Epstein, J. I., and Coffey, D. S. (1993). Nuclear matrix protein patterns in human benign prostate hyperplasia and prostate cancer. *Cancer Res.* **53,** 744–746.

Paulson, J. R., and Laemmli, U. K. (1977). The structure of histone-depleted metaphase chromosomes. *Cell (Cambridge, Mass.)* **12,** 817–828.

Payrastre, B., Nievers, M., Boonstra, J., Breton, M., Verkleij, A. J., and Van Bergen en Henegouwen, P. M. P. (1992). A differential location of phosphoinositide kinases, diacylglycerol kinase, and phospholipase C in the nuclear matrix. *J. Biol. Chem.* **267,** 5078–5084.

Peters, K. E., and Comings, D. E. (1980). Two-dimensional gel electrophoresis of rat liver nuclear washes, nuclear matrix and hnRNA proteins. *J. Cell Biol.* **86,** 135–155.

Peters, K. E., Okada, T. A., and Comings, D. E. (1982). Chinese hamster nuclear proteins. An electrophoretic analysis of interphase, metaphase and nuclear matrix preparations. *Eur. J. Biochem.* **129,** 221–233.

Pienta, K. J., and Lehr, J. E. (1993a). A common set of nuclear matrix proteins in prostate cancer cells. *Prostate* **23,** 61–67.

Pienta, K. J., and Lehr, J. E. (1993b). Inhibition of prostate cancer growth by estramustine and etoposide: Evidence for interaction at the nuclear matrix. *J. Urol.* **149,** 1662–1625.

Puvion, E., and Moyne, G. (1981). In situ localization of RNA structures. *In* "The Cell Nucleus" (H. Busch, ed.), Vol. 8, pp 59–115. Academic Press, New York.

Reeves, R., and Chang, D. (1983). Investigations of the possible functions for glycosylation in the high mobility group proteins. Evidence for a role in nuclear matrix association. *J. Biol. Chem.* **258,** 679–687.

Reiter, T., and Penman, S. (1983). 'Prompt' heat shock proteins: Translationally regulated synthesis of new proteins associated with the nuclear matrix-intermediate filaments as an early response to heat shock. *Proc. Natl. Acad. Sci. U.S.A.* **80,** 4737–474.

Rennie, P. S., Bruchovsky, N., and Cheng, H. (1983). Isolation of 3 S androgen receptors from salt resistant fractions and nuclear matrices of prostatic nuclei after mild trypsin digestion. *J. Biol. Chem.* **258,** 7623–7630.

Sahyoun, N., LeVine, H., Bronson, D., and Cuatrecasas, P. (1984). Ca(2+)-calmodulin-dependent protein kinase in neuronal nuclei. *J. Biol. Chem.* 9341–9344.

Samarabandu, J. K., Cheng, P. C., and Berezney, R. (1994). Application of three-dimensional image analysis to the mammalian cell nucleus. *Proc. IEEE Southeastcon '94,* pp. 98–100.

Samarabandu, J. K., Ma, H., Cheng, P. C., and Berezney, R. (1995). Computer aided analysis of DNA replications sites in the mammalian cell. *J. Cell. Biochem., Suppl.* **21B,** 137.

Shaper, J. H., Pardoll, D. M., Kaufmann, S. H., Barrack, F. R., Vogelstein, B., and Coffey, D. S. (1978). The relationship of nuclear matrix to cellular structure and function. *Adv. Enzyme Regul.* **17,** 213–248.

Sikorska, M., Whitfield, J. F., and Walker, P. R. (1988). The regulatory and catalytic subunits of cAMP-dependent protein kinases are associated with transcriptional active chromatin during changes in gene expression. *J. Biol. Chem.* **263,** 3005–3011.

Smetana, K., Lejnar, J., Vlastiborova, A., and Busch, H. (1971). On interchromatinic dense granules of mature human neutrophil granulocytes. *Exp. Cell Res.* **64,** 105–112.

Smith, H. C. and Berezney, R. (1980). DNA polymerase α is tightly bound to the nuclear matrix of actively replicating liver. *Biochem. Biophys. Res. Commun.* **97,** 1541–1547.

Smith, H. C., and Berezney, R. (1982). Nuclear matrix-bound DNA synthesis: An in-vitro system. *Biochemistry* **21,** 6751–6761.
Smith, H. C., and Berezney, R. (1983). Dynamic domains of DNA polymerase α in regenerating rat liver. *Biochemistry* **22,** 3042–3046.
Smith, H. C., Puvion, F., Buchholtz, L. A., and Berezney, R. (1984). Spatial distribution of DNA loop attachment and replication sites in the nuclear matrix. *J. Cell Biol.* **99,** 1794–1802.
Smith, H. C., Berezney, R., Brewster, J. M., and Rekosh, D. (1985). Properties of adenoviral DNA bound to the nuclear matrix. *Biochemistry* **24,** 1197–1202.
Smith, H. C., Harris, S. G., Zillmann, M., and Berget, S. M. (1989). Evidence that a nuclear matrix protein participates in pre-mRNA splicing. *Exp. Cell Res.* **182,** 521–533.
Somanathan, S., Mortillaro, M. J., and Berezney, R. (1995). Identifying members of the matrin 3 gene family. *J. Cell. Biochem., Suppl.* **21B,** 146.
Spector, D. (1984). Colocalization of U1 and U2 small nuclear RNPs by immunocytochemistry. *Biol. Cell* **51,** 109–112.
Spector, D. L. (1993). Macromolecular domains within the cell nucleus. *Annu. Rev. Cell Biol.* **9,** 265–315.
Spector, D. L., Schrier, W. H., and Busch, H. (1983). Immunoelectron microscopic localization of SnRNPs. *Biol. Cell* **49,** 1–10.
Staufenbiel, M., and Deppert, W. (1983). Nuclear matrix preparations from liver tissue and from cultured vertebrate cells: Differences in major polypeptides. *Eur. J. Cell Biol.* **31,** 341–348.
Staufenbiel, M., and Deppert, W. (1984). Preparation of nuclear matrices from cultured cells: Subfractionation of nuclei in situ. *J. Cell Biol.* **98,** 1886–1894.
Steele, W. J., and Busch, H. (1963). Studies on the acidic nuclear proteins of the Walker tumor and the liver. *Cancer Res.* **23,** 1153–1163.
Stuurman, N., Meijne, A. M. L., van der Pol, A. J., de Jong, L., van Driel, R., and van Renswoude, J. (1990). The nuclear matrix from cells of different origin. Evidence for a common set of matrix proteins. *J. Biol. Chem.* **265,** 5460–5465.
Tawfic, S., and Ahmed, K. (1994). Association of casien kinase 2 with nuclear matrix. *J. Biol. Chem.* **269,** 7489–7493.
Tubo, R. A., and Berezney, R. (1987a). Pre-replicative association of multiple replicative enzyme activities with the nuclear matrix during rat liver regeneration. *J. Biol. Chem.* **262,** 1148–1154.
Tubo, R. A., and Berezney, R. (1987b). Identification of 100 and 150S DNA polymerase α-primase megacomplexes solubilized from the nuclear matrix of regenerating rat liver. *J. Biol. Chem.* **262,** 5857–5865.
Tubo, R. A., and Berezney, R. (1987c). Nuclear matrix-bound DNA primase: Elucidation of an RNA primase system in nuclear matrix isolation from regenerating rat liver. *J. Biol. Chem.* **262,** 6637–6642.
Tubo, R. A., Smith, H. C., and Berezney, R. (1985). The nuclear matrix continues DNA synthesis at in vivo replicational forks. *Biochim. Biophys. Acta* 825, 326–334.
Tubo, R. A., Martelli, A. M., and Berezney, R. (1987). Enhanced processivity of nuclear matrix bound DNA polymerase α from regenerating rat liver. *Biochemistry* **26,** 5710–5718.
Turner, B. M., and Franchi, (1987). Identification of protein antigens associated with the nuclear matrix and with clusters of interchromatin granules in both interphase and miotic cells. *J. Cell Sci.* **87,** 269–282.

van der Velden, H. M. W., and Wanka, F. (1987). The nuclear matrix-its role in the spatial organization and replication of eukaryotic DNA. *Mol. Biol. Rep.* **12,** 69–77.

van Eekelen, C. A. G., Salden, M. H. L., Habets, W. J. A., van de Putte, L. B. A., and van Venrjooij, W. J. (1982). On the existence of an internal nuclear protein structure in HeLa cells. *Exp. Cell Res.* **141,** 181–190.

van Wijnen, A. J., Bidwell, J. P., Fey, E. G., Penman, S., Lian, J. B., Stein, J. L., and Stein, G. S. (1993). Nuclear matrix association of multiple sequence specific DNA binding activities related to SP-1, ATF, CCAAT, C/EBP, Oct-1 and AP-1. *Biochemistry* **32,** 8397–8402.

Verheijen, R., Kuijpers, H., Vooijs, P., van Venrooij, W., and Ramaekers, F. (1986). Protein composition of nuclear matrix preparations from HeLa cells: an immunochemical approach. *J. Cell Sci.* **80,** 103–122.

Verheijen, R., van Venrooij, W. J., and Ramaekers, F. (1988). The nuclear matrix: Structure and composition. *J. Cell Sci.* **90,** 11–36.

Wan, K., Nickerson, J. A., Krockmalnic, G., and Penman, S. (1994). The B1C8 protein is in the dense assemblies of the nuclear matrix and relocates to the spindle and pericentriolar filaments at mitosis. *Proc. Natl. Acad. Sci. U.S.A.* **91,** 594–598.

Wansink, D. G., Schul, W., van der Kraan, I., van Steenel, B., van Driel, R., and de Long, L. (1993). Fluorescent labeling of nascent RNA reveals transcription by RNA polymerase II in domains scattered throughout the nucleus. *J. Cell Biol.* **122,** 283–293.

Wunderlich, F., and Herlan, G. (1977). A reversibly contractile nuclear matrix. *J. Cell Biol.* **73,** 271–278.

Xing, Y., and Lawrence, J. B. (1991). Preservation of specific RNA distribution within the chromatin-depleted nuclear substructure demonstrated by in situ hybrizidation coupled with biochemical fractionation. *J. Cell Biol.* **112,** 1055–1063.

Xing, Y., Johnson, C. V., Dobner, P. R., and Lawrence, J. B. (1993). Higher level organization of individual gene transcription and RNA splicing. *Science* **259,** 1326–1330.

Yang, C. H., Lambie, E. J., and Snyder, M. (1992). NuMA: An unusually long coiled-coil related protein in the mammalian nucleus. *J. Cell Biol.* **116,** 1303–1317.

Zahler, A. M., Neugebauer, K. M., Stolk, J. A., and Roth, M. B. (1993). Human SR proteins and isolation of a cDNA encoding SRp75. *Mol. Cell. Biol.* **13,** 4023–4028.

Zehnbauer, B., and Vogelstein, B. (1985). Supercoiled loops in the organization of replication and transcription in eukaryotes. *BioEssays* **2,** 52–54.

Zeitlin, S., Parent, A., Silverstein, S., and Efstratiades, A. (1987). Pre-mRNA splicing and the nuclear matrix. *Mol. Cell. Biol.* **7,** 111–120.

Zeng, C., He, D., and Brinkley, B. R. (1994a). Localization of NuMA proteins isoforms in the nuclear matrix of mammalian cells. *Cell Motil. Cytoskel.* **29,** 167–176.

Zeng, C., He, D., Berget, S. M., and Brinkley, B. R. (1994b). Nuclear mitotic aparatus proteins: A structural protein interphase between the nucleoskeletons and RNA splicing. *Proc. Natl. Acad. Sci. U.S.A.* **91,** 1505–1509.

Zhelev, N. Z., Todorov, I. T., Philipova, R. N., and Hadjiolov, A. A. (1990). Phosphorylation-related accumulation of the 125K nuclear matrix protein mitotin in human mitotic cells. *J. Cell Sci.* **95,** 59–64.

3

Composition and Structure of the Internal Nuclear Matrix

KARIN A. MATTERN, ROEL VAN DRIEL, AND LUITZEN DE JONG

E.C. Slater Institute
University of Amsterdam
Amsterdam, The Netherlands

I. SCOPE AND INTRODUCTION

The concept of an intranuclear scaffolding structure was proposed for the first time by Berezney and Coffey (1974). This structure has been named nuclear matrix, nuclear scaffold, nucleoskeleton, or karyoskeleton, all referring to essentially the same structure. We will use the term nuclear matrix throughout this chapter. In the past 20 years, more than 500 papers have been published about aspects of the nuclear matrix. Fifty percent of these papers appeared in the last three years. Clearly, the nuclear matrix is enjoying increasing attention. Nuclear matrix structures are ubiquitous in the eukaryotic world and have, for instance, been identified in *Saccharomyces* (Amati and Gasser, 1988), plants (Hall *et al.*, 1991), *Drosophila* (Fisher *et al.*, 1982), and mammals (Berezney and Coffey, 1974).

Thinking about the nuclear matrix has been guided by the concept that this structure has an important role in the spatial and functional organization of the interphase nucleus (de Jong *et al.*, 1990; van Driel *et al.*, 1991; Stuurman *et al.*, 1992a). This idea is fostered by the observation that the machinery for replication, transcription by RNA polymerase II, and RNA processing is associated with the nuclear matrix (see the beginning of Section IV). Also, chromatin is bound in a specific way to the nuclear matrix (see Sections III,A and IV,A). These observations indicate that the nuclear matrix has an important role in handling the genetic information in the interphase nucleus.

After 20 years, the basic concept of a continuous, skeleton-like structure in the nucleus is still valid. However, a number of important questions regarding the nuclear matrix have not been answered, particularly about its composition, molecular structure, dynamics, spatial organization, and interactions with DNA and enzymes of nucleic acid metabolism. In this chapter, we present a concise overview of what is known about the structure and composition of the nuclear matrix and what major problems exist concerning this structure.

II. THE NUCLEAR MATRIX: DEFINITION AND CONFUSION

A. Operational and Conceptual Definition

To analyze and discuss the organization and composition of the nuclear matrix, it is essential to have an adequate definition of this structure. The definition is operational because the *in vivo* structure at the level of molecular interactions is not well understood. The nuclear matrix is the structure that remains after digestion of the DNA in the interphase nucleus with nucleases and subsequent extraction of chromatin and all loosely bound proteins. The resulting structure still has essentially the size and shape of the original nucleus. Three grossly different substructures can be discriminated on electron micrographs: (1) the lamina–pore complex, located at the nuclear periphery, (2) an intranuclear three-dimensional network of fibers or filaments with associated granular material, and (3) residual nucleolar structures. A great deal is known about the structure and composition of the lamina–pore complex (reviewed by, e.g., Gerace and Burke, 1988; Nigg, 1989; Dessev, 1990). In contrast, the molecular structure of the intranuclear fibrogranular network and of the residual nucleolar structure is largely unknown.

For a variety of reasons that will be discussed, results obtained by different investigators are contradictory. This situation has resulted in considerable concern about artifactual structures that may be formed during matrix isolation (e.g., Cook, 1988)—for example, in precipitation of nuclear components during extraction of chromatin with buffers containing a high salt concentration. Although several arguments support the idea that the nuclear matrix exists *in vivo,* definite proof can be offered only if the nuclear matrix is understood at the molecular level. In addition to the operational definition just given, the conceptual definition of the nuclear matrix will also be used: the putative, *in vivo,* three-dimensional network of filaments in the nuclear interior. This will sometimes be indicated by the term *in vivo* nuclear matrix.

B. Different Isolation Protocols

The procedures for the isolation of nuclear matrices involve digestion of DNA and extraction of the fragmented chromatin. However, details of the procedures (e.g., the order of various treatments) may differ considerably (for a comparison, see Belgrader *et al.,* 1991). As may be expected for an operationally defined structure, this results in nuclear matrix preparations that differ in protein composition (Fig. 1), associated nuclear functions, and ultrastructure as analyzed by electron microscopy.

An additional problem is the intrinsic instability of the intranuclear fibrogranular network of the nuclear matrix in many cell types. Several procedures are used to stabilize this structure. In isolation procedures in which a stabilization step is omitted, the internal fibrogranular struc-

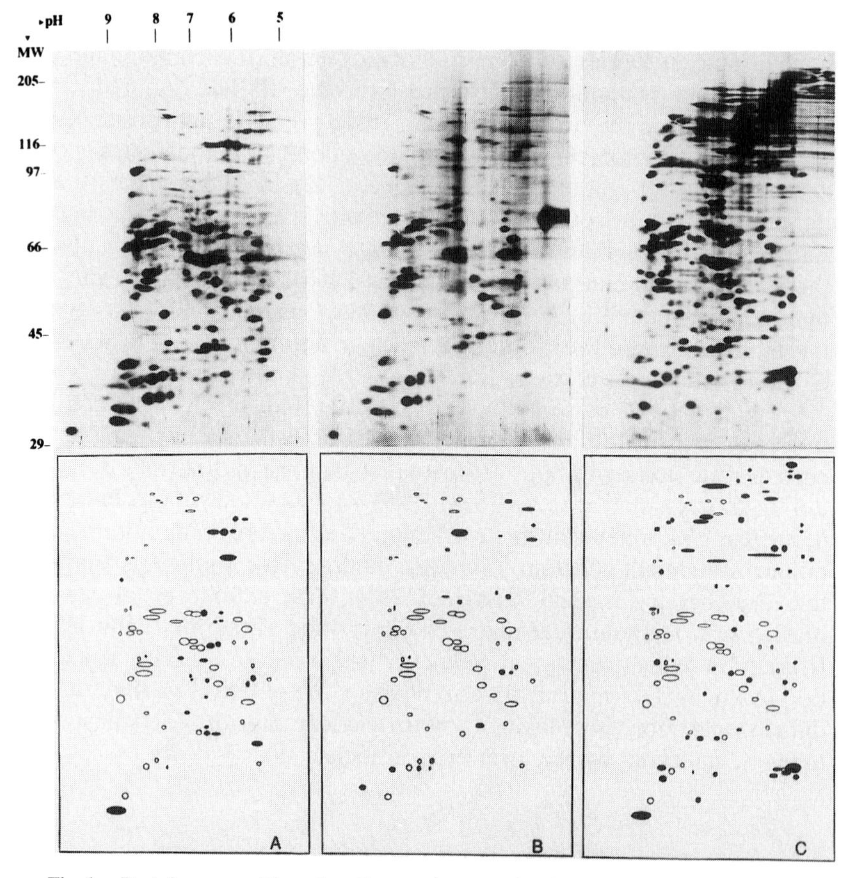

Fig. 1. Protein composition of rat liver nuclear matrices isolated by different protocols. Proteins were separated by nonequilibrium pH gradient electrophoresis followed by sodium dodecyl sulfate–polyacrylamide gel electrophoresis (SDS–PAGE). Molecular weights (kDa) and pH values are indicated in the upper left panel. The upper panels show silver-stained two-dimensional gels; the lower panels show schematic representation of the most abundant proteins, that is, about 35% of all spots discernible in the corresponding gels. The schemes identify proteins that are specific (●) or common (○) to the isolation protocols. Nuclear matrices were isolated essentially according to the protocol of (A) Fey *et al.* (1986), characterized by the use of a cytoskeleton-preserving buffer, VRC, as RNase inhibitor, and 0.25 *M* ammonium sulfate to extract the digested chromatin; (B) Mirkovitch *et al.* (1984), characterized by the use of heat stabilization, LIS to extract the chromatin, and restriction enzymes to digest the DNA; and (C) Kaufmann and Shaper (1984), characterized by the use of NaTT to stabilize the nuclear matrix, DNase and RNase for digestion, and 1.6 *M* NaCl to extract the chromatin. Courtesy of N. Stuurman.

ture may be partially or completely lost. To complicate the situation further, there are strong claims that the integrity of the nuclear matrix depends on the presence of intact RNA, possibly heterogeneous nuclear RNA (hnRNA) (see Section III,E). Therefore, the activity of contaminating ribonuclease (RNase) activity in nuclear matrix preparations may partially destroy its structure. If one adds to these potential causes for differences among nuclear matrix preparations the effects of the use of different cell types and tissues as starting material and the different preparation methods used for electron microscopic analysis of nuclear matrix structures, confusion about the structure and composition of the nuclear matrix is complete. Following is a set of critical parameters that should be considered carefully in comparing different protocols for nuclear matrix isolation.

1. *Nucleases used for digestion of chromatin.* Many investigators use pancreatic deoxyribonuclease (DNase) I (e.g., Berezney and Coffey, 1974; Capco *et al.*, 1982), whereas others employ a cocktail of restriction enzymes (Mirkovitch *et al.*, 1984). In the latter case, more chromatin will remain associated with the nuclear matrix. Sometimes RNase A is used in addition to DNase I (e.g., Kaufmann *et al.*, 1981).
2. *Presence of RNase inhibitors.* Often vanadyl ribonucleoside complex (VRC) is used to inhibit RNase activity (e.g., Fey *et al.*, 1986). VRC also affects a variety of other proteins (Puskas *et al.*, 1982; Ohara-Nemoto *et al.*, 1988). Therefore, it may have additional stabilizing effects.
3. *Ionic strength of buffers used for isolation of nuclei and for extraction of chromatin.* Buffers of low ionic strength in combination with buffers containing 1 to 2 *M* NaCl or KCl have been used (e.g., Kaufmann *et al.*, 1981). Penman and co-workers introduced a buffer optimized to preserve cytoskeletal structures in combination with the same buffer containing 0.25 *M* ammonium sulfate (Capco *et al.*, 1982). Cook and co-workers developed an elegant method to remove digested chromatin at near-physiological ionic strength by electrophoresis after embedding cells in agarose beads (Jackson *et al.*, 1988), whereas Laemmli and co-workers employ lithium 3,5-diiodosalicylate (LIS) in the millimolar concentration range (Mirkovitch *et al.*, 1984).
4. *Stabilization of the fibrogranular network of the nuclear matrix.* The mild oxidizing substance sodium tetrathionate (NaTT) reversibly

stabilizes the fibrogranular structure, presumably by introducing disulfide bonds between proteins (Kaufmann and Shaper, 1984; Belgrader *et al.*, 1991; Stuurman *et al.*, 1992b). Alternative procedures involve incubation with millimolar concentrations of Ca^{2+} or Cu^{2+} (Lebkowski and Laemmli, 1982a,b) or incubation at 37°C to 40°C (Mirkovitch *et al.*, 1984). It is unclear how most of these treatments act at the molecular level.

Fey and Penman (1988) introduced a method to separate nuclear matrix proteins from intermediate filament polypeptides. In this method the matrix–intermediate filaments structure is dissolved in 8 *M* urea under reducing conditions. Subsequently, intermediate filament proteins are allowed to reaggregate on removal of the urea, after which they can be precipitated by centrifugation. The resulting preparation of soluble proteins is enriched in nuclear matrix proteins. However, some nuclear matrix proteins may coaggregate with the intermediate filament proteins, and some cytoskeletal intermediate filament-associated polypeptides may still be present in the mixture of soluble proteins.

Clearly, different protocols result in nuclear matrix preparations with somewhat different protein compositions (Fig. 1) and with different components of the nuclear machinery still attached. It is very difficult to define which proteins are structurally part of the nuclear matrix and which are not. We and others are attempting to identify the major structural components of this intranuclear network. What is required is an understanding of the molecular interactions of the intranuclear fibrogranular structure. Only if this problem is solved will we understand the molecular basis of nuclear organization and higher-order regulation of gene expression.

III. PUTATIVE STRUCTURAL COMPONENTS OF THE INTERNAL NUCLEAR MATRIX

Although the nuclear matrix was first described 20 years ago (Berezney and Coffey, 1974), the molecular basis of the internal nuclear matrix is not known. It is assumed that it consists of one or more (hetero)polymeric structures to which other matrix components are bound (Stuurman *et al.*, 1992a). Components that form this putative nuclear framework are expected to have certain properties: (1) they must be able to form continuous but dynamic structures, (2) they must be able to interact with other nuclear matrix components, (3) they are probably abundant in matrix preparations, and (4) they may be evolutionarily conserved. We will discuss some components of the nuclear matrix which may have a structural function.

None of these components have proved to be the structural basis of the internal nuclear matrix. A systematic biochemical analysis of the components in combination with high-resolution ultrastructural techniques is needed to elucidate the organization of the nuclear matrix (Section V).

A. Matrix-Associated Region (MAR)-Binding Proteins

It is widely believed that the DNA in interphase (Vogelstein *et al.*, 1980) and mitotic chromosomes (Paulson and Laemmli, 1977) is organized in loops of 5–200 kb, the bases of which are attached to the nuclear matrix (interphase) or chromosome scaffold (metaphase). The nuclear matrix has binding sites for certain A + T-rich DNA sequences, called scaffold-associated regions (SARs) (Mirkovitch *et al.*, 1984) or matrix-associated regions (MARs) (Cockerill and Garrard, 1986), which are thought to form the bases of the DNA loops. We have shown that the structural proteins of the nuclear lamina, the A- and B-type lamins, bind MARs *in vitro* (Ludérus *et al.*, 1992, 1994). MAR binding seems to be a general property of intermediate filament and related proteins because desmin and NuMA (Section III,D) are also able to bind MARs *in vitro* (Ludérus *et al.*, 1994). Because desmin is probably not a nuclear protein (Section III,C), it is unlikely that it binds MARs *in vivo.* Whether lamins and NuMA bind MARs *in vivo* is not known.

The nuclear matrix protein scaffold attachment factor A (SAF-A), also called p120 (von Kries *et al.*, 1994) or sp120 (Tsutsui *et al.*, 1993), is identical to heterogeneous nuclear ribonucleoprotein (hnRNP) U (Fackelmayer *et al.*, 1994). *In vitro,* this protein preferentially binds to multiple MAR elements. Ultraviolet (UV)-cross-linking experiments show that it is bound to chromosomal DNA *in vivo* as well. As shown by Romig *et al.* (1992), SAF-A can form large aggregates and mediates the formation of looped DNA structures *in vitro.* Therefore, hnRNP–U/SAF-A may have a function in the organization of chromosomal DNA in addition to its suggested role in hnRNA metabolism (Section III,E).

MARs contain sequences similar to the DNA topoisomerase II consensus sequence (Gasser and Laemmli, 1986). Topoisomerase II preferentially and cooperatively binds MAR-containing DNA, resulting in insoluble protein–DNA aggregates (Adachi *et al.,* 1989). It is a major component in the nuclear matrix of growing cells and chromosome scaffold preparations (Berrios *et al.,* 1985; Earnshaw *et al.,* 1985; Gasser *et al.,* 1986). Topoisomerase II is bound to the bases of the radial loop domains of mitotic chromosomes (Earnshaw and Heck, 1985; Gasser *et al.,* 1986). This implies that topoisomerase II, in addition to its enzymatic function, also has a structural role in the organization of chromosomes. Such a structural role has been questioned by Hirano and Mitchison (1993). These authors found that topoisomerase II activity is necessary for *in vitro* mitotic chromosome assembly and condensation using *Xenopus* egg extracts and sperm chromatin. However, it was easily extracted from these chromosomes under mild conditions (100 m*M* NaCl) without disrupting the shape of the chromosomes. Also, the extremely low concentration of topoisomerase II in nonproliferating cells (Heck and Earnshaw, 1986) and its absence from lampbrush chromosomes (Fischer *et al.,* 1993) do not indicate a structural role for topoisomerase II in the organization of chromosomes.

Another abundant nuclear matrix protein is found to bind MARs. Von Kries *et al.* (1991) identified a 95-kDa chicken protein, named attachment region binding protein (ARBP), that selectively and cooperatively binds several MARs. This protein is soluble only at salt concentrations above 200 m*M* NaCl and in the presence of 2-mercaptoethanol. On dilution or dialysis, purified ARBP forms aggregates that do not redissolve by the addition of NaCl to 2 *M*.

To conclude, in addition to having a structural role in the lamina, the lamins can bind MARs and may be involved in the organization of interphase chromosomes at the nuclear periphery. Of the intranuclear MAR-binding proteins, both hnRNP U and ARBP are the most serious candidates to play both a structural role in the nuclear matrix and a role in the loop organization of DNA. Doubts have been raised about a structural function for topoisomerase II in a putative chromosome scaffold in metaphase and interphase.

B. Actin

Several authors have demonstrated that actin is not only part of the cytoskeleton, but is also a constituent of interphase nuclei and nuclear matrices (Clark and Rosenbaum, 1979; Capco *et al.,* 1982; Nakayasu and Ueda, 1983, 1985a, 1986; Valkov *et al.,* 1989; Verheijen *et al.,* 1986; reviewed by Verheijen *et al.,* 1988). It is not clear in all cases whether this finding is due to the residual cytoskeleton that cofractionates with the nuclear matrix (Staufenbiel and Deppert, 1984). Another reason to be suspicious about the nuclear localization of actin is that several types of stimuli, such as heat shock and dimethyl sulfoxide, can lead to translocation of actin into nuclei (Ono *et al.,* 1993, and references therein). However, the existence of an additional acidic form of actin in nuclei of mouse L-cells (Nakayasu and Ueda, 1986) and the presence of actin in hand-isolated *Xenopus* oocyte nuclei, which are likely to be free of cytoskeleton (Clark and Rosenbaum, 1979), may indicate that actin is present in the nucleus. Also, the last-named authors demonstrated the presence of nuclear actin *in situ* using ultrastructural techniques.

It was long unclear whether the putative nuclear actin was present in the form of filaments. Filamentous actin has been found in neuronal interphase nuclei (Amankwah and De Boni, 1994). These authors used heavy meromyosin (HMM) to decorate the actin filaments. Double labeling with HMM-Au and an antibody to sarcomeric α-actin showed that only a small amount of actin in the nucleoplasm is in filamentous form. Previously, Clark and Rosenbaum (1979), using DNase I inhibition assays, found that in the nuclei of *Xenopus* oocytes approximately 63% of the total nuclear actin exists in a globular state. Ankebauer *et al.* (1989) showed that nuclei of different species contain a heterodimeric, actin-binding protein. This protein increases the critical concentration for actin assembly and, therefore, could be responsible for the relatively high concentration of G-actin in nuclei.

Evidence for a possible role of nuclear actin in RNA metabolism has been documented. Scheer *et al.* (1984) observed that microinjection of actin-binding proteins or antibodies to actin into nuclei of living amphibian oocytes resulted in the inhibition of transcription by RNA polymerase II of the lampbrush chromosomes. Nakayasu and Ueda (1984, 1985b) showed that actin filaments are closely associated with small nuclear ribonucleoproteins (snRNPs) and pre-mRNAs. They also reported that rapidly labeled RNA was released from the nuclear matrix under conditions that cause depolymerization of actin filaments. Later, Sahlas *et al.* (1994) showed colocalization of actin with snRNP and the splicing factor SC-35 in interphase PC12 nuclei, depending on the differentiation state of the cell.

We conclude that F-actin may be a structural part of the nuclear matrix. Some observations suggest a function for actin in transcription,

RNA processing, or RNA transport, but the precise role of actin in these processes is unclear.

C. Intermediate Filament Proteins

In many cases, the quantitatively major proteins are those of the adhering cytoplasmic intermediate filament system rather than those of the nuclear matrix itself (e.g., Capco *et al.*, 1982; Staufenbiel and Deppert, 1984; Verheijen *et al.*, 1986). Evidently, the nuclear matrix is tightly associated with cytoskeletal elements, particularly the intermediate filament cytoskeleton. Extending this observation, it has been suggested that the nuclear matrix, the cytoskeleton, and the extracellular matrix between cells in a tissue constitute one large, continuous structure through which cells may communicate (Getzenberg *et al.*, 1990; Pienta and Coffey, 1992; Ingber, 1993a,b).

Publications by Jackson and Cook (1988), He *et al.* (1990), and Wang and Traub (1991) show nuclear matrices consisting of filaments bearing a strong resemblance to intermediate filaments. However, additional evidence that intermediate filaments are present inside mammalian nuclei is scarce. During the last few years, it has been claimed that intermediate filament-like proteins are present in the nuclear matrix, but mainly in plants (Beven *et al.*, 1990; Frederick *et al.*, 1992; McNulty and Saunders, 1992; Minguez and Moreno Diaz de la Espina, 1993). The evidence is based on cross-reaction of antibodies against intermediate filaments or lamins with proteins in the internal nuclear matrix but not in the residual nucleoli. Aligue *et al.* (1990) found a cytokeratin-like protein in nuclear matrix preparations of regenerating rat liver. Bridger *et al.* (1993) detected internal lamin structures in nuclei of human dermal fibroblast, but only during the G_1 phase. However, Belgrader *et al.* (1991) showed that a monoclonal antibody that recognizes an epitope common to all known intermediate filaments does not cross-react with proteins other than lamins of HeLa nuclear matrices which are free of cytoskeleton. Given the lack of evidence for the presence of known intermediate filaments in the nucleus, it is possible that the intermediate filament-resembling structures mentioned earlier (Jackson and Cook, 1988; He *et al.*, 1990; Wang and Traub, 1991) represent new members of the intermediate filament family.

Because of the structural function of intermediate filaments in the cytoskeleton and lamina, proteins of or closely related to the class of intermediate filament proteins are interesting candidates for structural proteins of the internal matrix. However, such proteins have not been identified unambiguously.

D. Nuclear–Mitotic Apparatus Protein and Other Coiled-Coil Proteins

The nuclear–mitotic apparatus (NuMA) protein was first identified by Lydersen and Pettijohn (1980). In interphase cells it is restricted to the nucleus, but in mitotic cells it localizes to the spindle poles. Although Lydersen and Pettijohn found the protein to be specific to humans, others found it later in other mammalian nuclei as well (Tang *et al.*, 1993; Yang *et al.*, 1992). NuMA is also known as centrophilin (Tousson *et al.*, 1991), spindle pole-nucleus (SPN) antigen (Kallajoki *et al.*, 1993) and SP-H (Maekawa *et al.*, 1991). DNA sequence analysis predicts a protein with two globular domains separated by a large, discontinuous α-helical domain similar to coiled-coil regions in structural proteins such as intermediate filaments (Compton *et al.*, 1992; Yang *et al.*, 1992). Analysis of various cDNA clones has suggested the existence of NuMA isoforms generated by alternative splicing: in addition to the predominant 230-kDa form, a 195-kDa isoform has also been observed (Tang *et al.*, 1993).

During mitosis NuMA appears to play an essential role in spindle microtubule formation. During recovery from microtubule inhibition, NuMA foci act as microtubule organizing centers (Kallajoki *et al.*, 1991; Tousson *et al.*, 1991). Kallajoki *et al.* (1991, 1993) also showed that microinjection of antibodies raised against NuMA blocks mitosis, resulting in daughter cells with micronuclei. Little is known about the function of NuMA in the interphase nucleus. NuMA has been shown to be a component of the nuclear matrix (Lydersen and Pettijohn, 1980; Kallajoki *et al.*, 1991). Zeng *et al.* (1994) showed that NuMA colocalizes with splicing factors (Sm proteins) in interphase nuclei and nuclear matrices of HeLa cells. NuMA is also associated with snRNP in immunoprecipitated complexes from HeLa cells but associates *in vitro* only with splicing complexes that are reconstituted with pre-mRNA. These results have led to the suggestion that NuMA may provide a bridge between RNA processing and the nucleoskeleton (Zeng *et al.*, 1994).

In yeast a related coiled-coil protein (110 kDa) is found called nuclear filament-related protein (NUF1) (Mirzayan *et al.,* 1992), which is also a constituent of the nuclear matrix. Also, in mammalian cells, a related protein of the same size is found. The localization differs during mitosis from NuMA: it is dispersed in the cytoplasm. Nickerson *et al.* (1990) reported a nuclear matrix protein (W511) with stretches of homology to the α-helical regions of intermediate filaments.

Given their primary structure and subnuclear distribution, the coiled-coil proteins NuMA, NUF1, and W511 are candidates for a structural role in the nuclear matrix. In addition, NuMA binds MARs *in vitro* and possibly RNA as well. However, it is not known whether NuMA binds to nucleic acids *in vivo.* Likewise, more knowledge is required about the ability of NuMA, NUF1, and W511 to form homo- or heteropolymers.

E. RNA and Heterogeneous Nuclear Ribonucleoproteins

The nuclear matrix, isolated without RNase, contains about 70% of the nuclear RNA (Long *et al.,* 1979; van Eekelen and van Venrooij, 1981; He *et al.,* 1990). In addition, several agents that inhibit or stimulate transcription induce extensive structural rearrangements in nuclei and nuclear matrices (reviewed by Brasch, 1990). The observation that the nuclear matrix is sensitive to RNase digestion (Kaufmann *et al.,* 1981; Long and Schrier, 1983; Bouvier *et al.,* 1985a,b; Fey *et al.,* 1986; He *et al.,* 1990; Belgrader *et al.,* 1991) may indicate that RNA has a structural role, in combination with certain RNA binding proteins.

Proteins known to be associated with RNA are present in the nuclear matrix, like the hnRNP proteins (Dreyfuss *et al.,* 1984; Fey *et al.,* 1986; Verheijen *et al.,* 1986; He *et al.,* 1991; reviewed by Verheijen *et al.,* 1988). Van Eekelen and van Venrooij (1981) showed that hnRNP C1/C2 could be cross-linked to hnRNA in nuclear matrices from HeLa cells. These proteins could not be released by RNase, which suggests that they are involved in the binding of hnRNA to the nuclear matrix. In this study, the internal matrix seemed to resist RNase treatment. In contrast, Dreyfuss *et al.* (1984) found the hnRNP C proteins to be extracted with RNase from nuclear matrices. Interestingly, some of the hnRNP proteins form filaments under certain conditions (Kloetzel *et al.,* 1982; Lohstein *et al.,* 1985). These helical filaments (diameter 18 nm, pitch 60 nm) are formed by hnRNP $(A2)_3(B1)$ tetramers in low salt concentrations and are insoluble in 2 *M* NaCl. It is not known if these proteins also occur in the matrix in a filamentous form. Gallinaro *et al.* (1983) compared the composition of nuclear matrices with those of salt-resistant hnRNP complex preparations, both derived from adenovirus 2-infected HeLa cells. The results confirmed the similarity between the two structures. Both preparations contained rapidly labeled hnRNA, and they shared the same proteins, snRNAs, and the same adenovirus 2 transcripts. He *et al.* (1991) showed hnRNPs proteins to be associated with nuclear matrix filaments. These results together are compatible with the idea that salt-resistant hnRNP complexes associated with pre-mRNA are an integral part of the filaments of the internal matrix.

In Section III,A we mentioned that the abundant hnRNP U is a MAR-binding protein and binds DNA *in vivo.* Other hnRNP proteins may also have a function in the organization of chromosomes, as hnRNP A1, A2–B1, D, and E can bind to single-stranded d(TTAGGG)$_n$, the human telomeric DNA repeat (Ishikawa *et al.,* 1993).

It is tempting to speculate that (some) hnRNP proteins, especially hnRNP U, play a structural role in nuclear matrix organization because (1) they are abundant in the nucleus and in nuclear matrix preparations, (2) they aggregate under certain conditions, and (3) they bind RNA and DNA *in vitro* and *in vivo.*

F. Concluding Remarks

The most simple concept of the nuclear matrix is a system of one or more three-dimensional frameworks of homo- or heteropolymers. On examination of several potential structural proteins of such polymers, only actin, ARBP, and some hnRNP proteins, in particular hnRNP U, seem to fulfill the criteria summarized at the beginning of Section III. It is also possible that other known nuclear matrix proteins, like NuMA, as well as proteins that have not yet been identified, can form polymers. The protein composition of the nuclear matrix is complex (Fig. 1), and the properties of many proteins that fractionate with the nuclear matrix, independent of the isolation procedure, are not known. We should also seriously consider the possibility that the nuclear matrix is the result of the cooperative interaction of many different components, including RNA. So, despite many years of investigation, the structure of the nuclear matrix remains unclear. Section V briefly indicates how this structure can be determined.

IV. NUCLEAR MATRIX AND NUCLEAR FUNCTION

Certain conditions of DNA digestion and extraction of nuclei allow the isolation of structures which are almost as active as intact nuclei in replication (Jackson and Cook, 1986) and transcription (Jackson and Cook, 1985; Razin *et al.,* 1985), although most of the DNA has been extracted. In other words, replication and transcription are associated with the nuclear matrix. Nuclear matrices have also been prepared that contain preassembled spliceosomes which can splice exogenous RNA substrates rapidly, provided that the matrices have been reconstituted with certain soluble factors (Zeitlin *et al.,* 1989). These results imply that the nuclear matrix has binding sites for spliceosomes and for one or more members of transcription and replication complexes, that is, the DNA template, the polymerase and auxiliary proteins, and the nascent RNA or DNA. These binding sites have not yet been identified.

Although the molecular interactions involved in the association of transcription, replication, and splicing with the nuclear matrix remain elusive, much has been learned about the spatial distribution of these matrix-associated functions in the last two decades. The nuclear matrix is also thought to play a role in the organization of interphase chromosomes. The spatial distribution of individual chromosomes and matrix-associated functions is briefly discussed in Section IV,A because of their importance for our understanding of the *in vivo* structure of the nuclear matrix.

A. Interphase Chromosome Organization and Spatial Distribution of Replication, Transcription, and Splicing

With the aid of chromosome-specific DNA probes, it has become clear that the space occupied by individual interphase chromosomes is restricted to a limited domain within the nucleus. Moreover, individual chromosome domains do not overlap. Apparently, interphase chromosomes remain compact and highly organized (see Haaf and Schmid, 1991, for a review).

The spatial distribution of sites of DNA and RNA synthesis in the cell nucleus has been studied with two methods. In the first method, cells are pulse-labeled with [^{3}H]thymidine, to visualize sites of replication (Fakan and Hancock, 1974), or with [^{3}H]uridine, to visualize sites of transcription (for a review, see Fakan and Puvion, 1980), followed by electron microscopic autoradiography (EMARG). More recently, nucleotide analogs have been used to label nascent DNA (Gratzner *et al.*, 1975; Nakamura *et al.*, 1986; Nakayasu and Berezney, 1989; Wansink *et al.*, 1994) or RNA (Jackson *et al.*, 1993; Wansink *et al.*, 1993). Labeled nucleic acids are then detected by immunofluorescence microscopy or immunogold EM using nucleotide analog-specific antibodies. This method is more rapid and has a higher sensitivity and resolution in immunogold EM than in EMARG. Results from both methods are mutually compatible (D. G. Wansink *et al.*, this volume). Transcription by RNA polymerase II, which produces predominantly pre-mRNA, takes place at the border of clumps or strands of condensed chromatin distributed throughout the nucleoplasm. After a chase following a pulse of [^{3}H]uridine, the labeled pre-mRNA is found mainly in the interchromatin space (Fakan and Puvion, 1980). Also, snRNPs and other factors responsible for splicing are present in the interchromatin space in association with newly synthesized RNA (Fakan *et al.*, 1984; Spector *et al.*, 1991). In addition, splicing factors are present in relatively high concentrations in clusters of interchromatin granules. These irregularly shaped structures, with a diameter between 0.2 and 1 μm, are also present exclusively in the interchromatin space. It is not clear to what extent splicing occurs within or outside these clusters of interchromatin granules. Like transcription, DNA synthesis is predominantly localized at the borders of dense chromatin (Fakan and Hancock, 1974). Importantly, the spatial distribution of nascent DNA (Nakayasu and Berezney, 1989), nascent RNA, and splicing factors (D. G. Wansink *et al.*, this volume) remains unaltered after digestion and extraction of most of the chromatin from sodium tetrathionate-stabilized cells. This indicates that the structural principles underlying the spatial organization of nucleic acid metabolism are reflected in the nuclear matrix.

In conclusion, nucleic acid metabolism is associated with the nuclear matrix, and takes place in the interchromatin space and at the borders of dense chromatin.

B. Implications for the Nuclear Matrix Concept

If we take into consideration the spatial organization of interphase chromosomes and the spatial distribution of nuclear matrix-associated functions, the following picture of the overall nuclear organization emerges. Two main compartments can be envisaged: the chromatin compartment and the interchromatin space. The chromatin compartment is formed by the individual chromosomes, each of which has its own territory. The chromatin compartment can be subdivided in the transcriptionally silent dense chromatin and in the transcriptionally active, presumably more decondensed chromatin in direct contact with the interchromatin space. The interchromatin space forms a channeled network between the individual chromosomes. Also within the interchromatin space, subcompartments can be recognized: clusters of interchromatin granules. Then the question arises of how the nuclear matrix is spatially and functionally related to the chromatin compartment and the interchromatin space. Given the fact that active genes, hnRNA, and snRNPs are major components of the nuclear matrix, we may expect that the *in vivo* nuclear matrix must at least be part of the interchromatin space and its borders with the chromatin compartment. The nuclear matrix is also thought to be involved in the organization of DNA in loops. Could it be that the bases of DNA loops are also attached to this putative framework of the interchromatin space and to the borders of the chromatin compartment? Or are the presumptive filaments of the *in vivo* nuclear matrix also present within the central parts of the chromatin compartment *in vivo?* In other words, is there an interphase pendant of the metaphase chromosome scaffold that organizes the DNA in loops around the central chromosome axis? This question deserves further investigation. However, as noted in Section III,A, *in vitro* experiments indicate that loop organization in interphase may occur at least partially at the periphery of the chromatin compartment, both at the nuclear lamina, mediated by the lamins, and in the nuclear interior, mediated by hnRNP U/SAF-A. If lamins and hnRNP U/SAF-A are involved in the organization of loops *in vivo,* which has yet to be demonstrated, then we can envisage that at least part of the bases of DNA loops are attached to structures present at the periphery of the chromatin compartment.

In conclusion, a model is presented for the *in vivo* spatial organization of the nuclear matrix with regard to the chromatin compartment and the interchromatin space. We propose that the nuclear matrix is a framework in the interchromatin space in contact with the periphery of the chromatin compartment. It is involved in the organization of nucleic acid metabolism. The same perichromosomal framework, together with the nuclear lamina, contains bindig sites for A + T-rich DNA sequences that may form the bases of DNA loops.

V. CHALLENGES AND OUTLOOK

The nucleus is a highly organized organelle containing several functional compartments. The molecular interactions underlying the formation and maintenance of nuclear compartments are not known, but the operationally defined nuclear matrix is expected to provide a clue to determine the structural basis of nuclear organization. This expectation is based on the observation that the spatial distribution of compartmentalized components remains unaltered in nuclear matrices compared with the parent nuclei (Wansink *et al.,* 1993; Nakayasu and Berezney, 1989; Zeng *et al.,* 1994).

How can we determine the structure of the internal nuclear matrix? The most important criterion that should be fulfilled by the structural component(s) is the formation of continuous structures inside the nucleus. This has to be shown by labeling of internal filaments in intact nuclei with antibodies against nuclear proteins, using EM and limited *in vivo* cross-linking, followed by immunoprecipitation and analysis of coprecipitating components. To select such proteins, a thorough biochemical analysis of matrix proteins is needed, preferably using high-resolution two-dimensional gel electrophoresis and computerized analysis systems. The first step is to distinguish peripheral proteins (including contaminating cytoskeleton proteins) and internal matrix proteins (Mattern *et al.,* 1996). This can be done by using a special property of the internal structure: it easily dissociates using RNase and reducing agents (e.g., Belgrader *et al.,* 1991). The next step is to distinguish between adherent proteins and proteins that are necessary for the integrity of the internal nuclear matrix. Probably the structural proteins will be abundant in matrix preparations, independent of the cell type or the isolation procedure used. Stuurman *et al.* (1990) showed that a defined set of matrix proteins is present in many different cell types, referred to as minimal matrix proteins. Such a set of common proteins could also be obtained by comparing differently isolated nuclear matrices (see Fig.

1). These isolation procedures can include those used by several authors, as well as additional procedures aimed at stripping the adhering components from the nuclear matrix, as proposed by Stuurman *et al.* (1992a).

This combined biochemical and ultrastructural approach will ultimately answer the question of how the nuclear matrix is built up. The elucidation of the structure of the internal nuclear matrix is the first step in understanding nuclear organization in relation to the regulation of gene expression.

REFERENCES

Adachi, Y., Käs, E., and Laemmli, U. K. (1989). Preferential, cooperative binding of DNA topoisomerase II to scaffold-associated regions. *EMBO J.* **8,** 3997–4006.

Aligue, R., Bastos, R., Serratosa, J., Enrich, C., James, P., Pujades, C., and Bachs, O. (1990). Increase in a 55-kDA keratin-like protein in the nuclear matrix of rat liver cells during proliferative activation. *Exp. Cell Res.* **186,** 346–353.

Amankwah, K. S., and De Boni, U. (1994). Ultrastructural localization of filamentous actin within neuronal interphase nuclei in situ. *Exp. Cell Res.* **210,** 315–325.

Amati, B. B., and Gasser, S. M. (1988). Chromosomal ARS and CEN elements bind specifically to the yeast nuclear scaffold. *Cell (Cambridge, Mass.)* **54,** 967–978.

Ankebauer, T., Kleinschmidt, J. A., Walsh, M. J., Weiner, O. H., and Franke, W. W. (1989). Identification of a widespread nuclear actin binding protein. *Nature (London)* **342,** 822–825.

Belgrader, P., Siegel, A. J., and Berezney, R. (1991). A comprehensive study on the isolation and characterization of the HeLa S3 nuclear matrix. *J. Cell Sci.* **98,** 281–291.

Berezney, R., and Coffey, D. S. (1974). Identification of a nuclear protein matrix. *Biochem. Biophys. Res. Commun.* **60,** 1410–1417.

Berrios, M., Osheroff, N., and Fisher, P. A. (1985). In situ localization of DNA topoisomerase II, a major polypeptide component of the *Drosophila* nuclear matrix fraction. *Proc. Natl. Acad. Sci. U.S.A.* **82,** 4141–4146.

Beven, A., Guan, Y., Peart, J., Cooper, C., and Shaw, P. (1990). Monoclonal antibodies to plant nuclear matrix reveal intermediate filament-related components within the nucleus. *J. Cell Sci.* **98,** 293–302.

Bouvier, D., Hubert, J., Sève, A. P., and Bouteille, M. (1985a). Nuclear RNA-associated proteins and their relationship to the nuclear matrix and related structures in HeLa cells. *Can. J. Biochem. Cell Biol.* **63,** 631–643.

Bouvier, D., Hubert, J., Sève, A. P., and Bouteille, M. (1985b). Structural aspects of intranuclear matrix disintegration upon RNase digestion of HeLa cell nuclei. *Eur. J. Cell Biol.* **36,** 323–333.

Brasch, K. (1990). Drug- and metabolite-induced perturbations in nuclear structure and function: A review. *Biochem. Cell Biol.* **68,** 408–426.

Bridger, J. M., Kill, I. R., O'Farrell, M. O., and Hutchison, C. J. (1993). Internal lamin structures within G_1 nuclei of human dermal fibroblast. *J. Cell Sci.* **104,** 297–306.

Capco, D. G., Wan, K. M., and Penman, S. (1982). The nuclear matrix: Three-dimensional architecture and protein composition. *Cell (Cambridge, Mass.)* **29,** 847–858.

Clark, T. G., and Rosenbaum, J. L. (1979). An actin filament matrix in hand-isolated nuclei of *X. laevis* oocytes. *Cell (Cambridge, Mass.)* **18,** 1101–1108.

Cockerill, P. N., and Garrard, W. T. (1986). Chromosomal loop anchorage of the kappa immunoglobin gene occurs next to the enhancer in a region containing topoisomerase II sites. *Cell (Cambridge, Mass.)* **44,** 273–282.

Compton, D. A., Szilak, I., and Cleveland, D. W. (1992). Primary structure of NuMA, an intranuclear protein that defines a novel pathway for segregation of proteins at mitosis. *J. Cell Biol.* **116,** 1395–1408.

Cook, P. R. (1988). The nucleoskeleton: Artefact, passive framework or active site? *J. Cell Sci.* **90,** 1–6.

de Jong L., van Driel, R., Stuurman, N., Meijne, A. M. L., and van Renswoude, J. (1990). Principles of nuclear organization. *Cell Biol. Int. Rep.* **14,** 1051–1074.

Dessev, G. N. (1990). The nuclear lamina. An intermediate filament protein structure of the cell nucleus. *In* "Cellular and Molecular Biology of Intermediate Filaments" (R. D. Goldman and P. M. Steinert, eds.), pp. 129–145. Plenum, New York.

Dreyfuss, G., Choi, Y. D., and Adam, S. A. (1984). Characterization of heterogeneous nuclear RNA-protein complexes in vivo with monoclonal antibodies. *Mol. Cell. Biol.* **4,** 1104–1114.

Earnshaw, W. C., and Heck, M. M. S. (1985). Localization of topoisomerase II in mitotic chromosomes. *J. Cell Biol.* **100,** 1716–1725.

Earnshaw, W. C., Halligan, B., Cooke, C. A., Heck, M. M. S. and Liu, L. F. (1985). Topoisomerase II is a structural component of mitotic chromosome scaffolds. *J. Cell Biol.* **100,** 1706–1715.

Fackelmayer, F. O., Dahm, K., Renz, A., Ramsperger, U., and Richter, A. (1994). Nucleic-acid-binding properties of hnRNP-U/SAF-A, a nuclear-matrix protein which binds DNA and RNA in vivo and in vitro. *Eur. J. Biochem.* **221,** 749–757.

Fakan, S., and Hancock, R. (1974). Localization of newly-synthesized DNA in a mammalian cell as visualized by high resolution autoradiography. *Exp. Cell Res.* **83,** 95–102.

Fakan, S., and Puvion, E. (1980). The ultrastructural visualization of nuclear and extranucleolar RNA synthesis and distribution. *Int. Rev. Cytol.* **65,** 255–299.

Fakan, S., Leser, G., and Martin, T. E. (1984). Ultrastructural distribution of nuclear ribonucleoproteins as visualized by immunocytochemistry on thin sections. *J. Cell Biol.* **98,** 358–363.

Fey, E. G., and Penman, S. (1988). Nuclear matrix proteins reflect cell type of origin in cultured human cells. *Proc. Natl. Acad. Sci. U.S.A.* **85,** 121–125.

Fey, E. G., Krochmalnic, G., and Penman, S. (1986). The nonchromatin substructures of the nucleus: The ribonucleoprotein (RNP)-containing and RNP-depleted matrices analyzed by sequential fractionation and resinless section electron microscopy. *J. Cell Biol.* **102,** 1654–1665.

Fischer, D., Hock, R., and Scheer, U. (1993). DNA topoisomerase II is not detectable on lampbrush chromosomes but enriched in the amplified nucleoli of *Xenopus* oocytes. *Exp. Cell Res.* **209,** 255–260.

Fisher, P. A., Berrios, M., and Blobel, G. (1982). Isolation and characterization of a proteinaceous subnuclear fraction composed of nuclear matrix, peripheral lamina, and nuclear pore complexes from embryos of *Drosophila melanogaster. J. Cell Biol.* **92,** 674–686.

Frederick, S. E., Mangan, M. E., Carey, J. B., and Gruber, P. J. (1992). Intermediate filament antigens of 60 and 65 kDa in the nuclear matrix of plants: Their detection and localization. *Exp. Cell Res.* **199,** 213–222.

Gallinaro, H., Puvion, E., Kister, L., and Jacob, M. (1983). Nuclear matrix and hnRNP share a common structural constituent associated with premessenger RNA. *EMBO J.* **2,** 953–960.

Gasser, S. M., and Laemmli, U. K. (1986). Cohabitation of scafford binding regions with upstream/enhancer elements of three developmentally regulated genes of *D. melanogaster. Cell* (*Cambridge, Mass.*) **46,** 521–530.

Gasser, S. M., Laroche, T., Falquet, J., Boy de la Tour, E., and Laemmli, U. K. (1986). Metaphase chromosome structure involvement of topoisomerase II. *J. Mol. Biol.* **188,** 613–629.

Gerace. L., and Burke, B. (1988). Functional organization of the nuclear envelope. *Annu. Rev. Cell Biol.* **4,** 335–374.

Getzenberg, R. H., Pienta, K. J., and Coffey, D. S. (1990). The tissue matrix: Cell dynamics and hormone action. *Endocr. Rev.* **11,** 399–417.

Gratzner, H. G., Leif, R. C., Ingram, D. J., and Castro, A. (1975). The use of antibody specific for bromodeoxyuridine for the immunofluorescent determination of DNA replication in single cells and chromosomes. *Exp. Cell Res.* **95,** 88–94.

Haaf, T., and Schmid, M. (1991). Chromosome topology in mammalian interphase nuclei. *Exp. Cell Res.* **192,** 325–332.

Hall, G., Allen, G. C., Loer, D. S., Thompson, W. F., and Spiker, S. (1991). Nuclear scaffolds and scaffold-attachment regions in higher plants. *Proc. Natl. Acad. Sci. U.S.A.* **88,** 9320–9324.

He, D., Nickerson, J. A., and Penman, S. (1990). Core filaments of the nuclear matrix. *J. Cell Biol.* **110,** 569–580.

He, D., Martin, T., and Penman, S. (1991). Localization of heterogeneous nuclear ribonucleoprotein in the interphase nuclear matrix core filaments and on perichromosomal filaments at mitosis. *Proc. Natl. Acad. Sci. U.S.A.* **88,** 7469–7473.

Heck, M. M. S., and Earnshaw, W. C. (1986). Topoisomerase II: A specific marker for cell proliferation. *J. Cell Biol.* **103,** 2569–2581.

Hirano, T., and Mitchison, T. J. (1993). Topoisomerase II does not play a scaffolding role in the organization of mitotic chromosomes assembled in *Xenopus* egg extracts. *J. Cell Biol.* **120,** 601–612.

Ingber, D. E. (1993a). Cellular tensegrity: Defining new rules of biological design that govern the cytoskeleton. *J. Cell. Sci.* **104,** 613–627.

Ingber, D. E. (1993b). The riddle of morphogenesis: A question of solution chemistry or molecular cell engineering? *Cell* (*Cambridge, Mass.*) **75,** 1249–1252.

Ishikawa, F., Matunis, M. J., Dreyfuss, G., and Cech, T. R. (1993). Nuclear proteins that bind the pre-mRNA 3′ splice site sequence (UUAG/G) and the human telomeric DNA sequence $d(TTAGGG)_n$. *Mol. Cell. Biol.* **13,** 4301–4310.

Jackson, D. A., and Cook, P. R. (1985). Transcription occurs at a nucleoskeleton. *EMBO J.* **4,** 919–925.

Jackson, D. A., and Cook, P. R. (1986). Replication occurs at a nucleoskeleton. *EMBO J.* **5,** 1403–1410.

Jackson, D. A., and Cook, P. R. (1988). Visualization of a filamentous nucleoskeleton with a 23 nm acial repeat. *EMBO J.* **7,** 3667–3677.

Jackson, D. A., Yuan, J., and Cook, P. R. (1988). A gentle method for preparing cyto- and nucleo-skeletons and associated chromatin. *J. Cell Sci.* **90,** 365–378.
Jackson, D. A., Hassan, A. B., Errington, R. J., and Cook, P. R. (1993). Visualization of focal sites of transcription within human nuclei. *EMBO J.* **12,** 1059–1065.
Kallajoki, M., Weber, K., and Osborn, M. (1991). A 210 kDa nuclear matrix protein is a functional part of the mitotic spindle: A microinjection study using SPN monoclonal antibodies. *EMBO J.* **10,** 3351–3362.
Kallajoki, M., Harborth, J., Weber, K., and Osborn, M. (1993). Microinjection of a monoclonal antibody against SPN antigen, now identified by peptide sequences as the NuMA protein, induces micronuclei in PtK2 cells. *J. Cell Sci.* **104,** 139–150.
Kaufmann, S. H., and Shaper, J. H. (1984). A subset of non-histone nuclear proteins reversibly stabilized by the sulfhydryl cross-linking reagent tetrathionate. Polypeptides of the internal nuclear matrix. *Exp. Cell Res.* **155,** 477–495.
Kaufmann, S. H., Coffey, D. S., and Shaper, J. H. (1981). Considerations in the isolation of rat liver nuclear matrix, nuclear envelope, and pore complex lamina. *Exp. Cell Res.* **132,** 105–123.
Kloetzel, P. M., Johnson, R. M., and Sommerville, J. (1982). Interaction of the hnRNA of amphibian oocytes with fibril-forming proteins. *Eur. J. Biochem.* **127,** 301–308.
Lebkowski, J. S., and Laemmli, U. K. (1982a). Evidence for two levels of DNA folding in histone-depleted HeLa interphase nuclei. *J. Mol. Biol.* **156,** 309–324.
Lebkowski, J. S., and Laemmli, U. K. (1982b). Non-histone proteins and long-range organization of HeLA interphase DNA. *J. Mol. Biol.* **156,** 325–344.
Lohstein, L., Arenstorf, H. P., Chung, S., Walker, B. W., Wooley, J. C., and LeStourgeon, W. M. (1985). General organization of protein in HeLa 40S nuclear ribonucleoprotein particles. *J. Cell Biol.* **100,** 1570–1581.
Long, B. H., and Schrier, W. H. (1983). Isolation from Friend erythroleukemia cells of an RNase-sensitive nuclear matrix fibril fraction containing hnRNA and snRNA. *Biol. Cell* **48,** 99–108.
Long, B. H., Huang, C. Y., and Pogo, A. O. (1979). Isolation and characterization of the nuclear matrix in Friend erythroleukemia cells: Chromatin and hnRNA interactions with the nuclear matrix. *Cell (Cambridge, Mass.)* **18,** 1079–1090.
Ludérus, M. E. E., de Graaf, A., Mattia, E., den Blaauwen, J. L., Grande, M. A., de Jong, L., and van Driel, R. (1992). Binding of matrix attachment regions to lamin B_1. *Cell (Cambridge, Mass.)* **70,** 949–959.
Ludérus, M. E. E., den Blaauwen, J. L., de Smit, O. J. B., Compton, D. A., and van Driel, R. (1994). Binding of matrix attachment region DNA to lamin polymers involves single-stranded regions and the minor groove. *Mol. Cell Biol.* **14,** 6297–6305.
Lydersen, B. K., and Pettijohn, D. E. (1980). Human-specific nuclear protein that associates with the polar region of the mitotic apparatus: Distribution in a human/hamster hybrid cell. *Cell (Cambridge, Mass.)* **22,** 489–499.
Maekawa, T., Leslie, R., and Kuriyama, R. (1991). Identification of a minus end-specific microtubule-associated protein located at the mitotic poles in cultured mammalian cells. *Eur. J. Cell Biol.* **54,** 255–267.
Mattern, K. A., Humbel, B. M., Muijsers, A. O., de Jong, L., and van Driel, R. (1996). hnRNP proteins and B23 are the major proteins of the internal nuclear matrix of HeLa S3 cells. *J. Cell. Biochem.* **62,** 275–289.
McNulty, A. K., and Saunders, M. J. (1992). Purification and immunological detection of pea nuclear intermediate filaments: Evidence for plant nuclear lamins. *J. Cell Sci.* **103,** 407–414.

Minguez, A., and Moreno Diaz de la Espina, S. (1993). Immunological characterization of lamins in the nuclear matrix of onion cells. *J. Cell Sci.* **106,** 431–439.

Mirkovitch, J., Mirault, M. E., and Laemmli, U. K. (1984). Organization of the higher-order chromatin loop: Specific DNA attachment sites on nuclear scaffold. *Cell* (*Cambridge, Mass.*) **39,** 223–232.

Mirzayan, C., Copeland, C. S., and Snyder, M. (1992). The *NUF1* gene encodes an essential coiled-coil related protein that is a potential component of the yeast nucleoskeleton. *J. Cell Biol.* **116,** 1319–1332.

Nakamaru, H., Morita, T., and Sato, C. (1986). Structural organizations of replicon domains during DNA synthetic phase in the mammalian nucleus. *Exp. Cell Res.* **165** 291–297.

Nakayasu, H., and Berezney, R. (1989). Mapping replicational sites in the eucaryotic cell nucleus. *J. Cell Biol.* **108** 1–11.

Nakayasu, H., and Ueda, K. (1983). Association of actin with the nuclear matrix from bovine lymphocytes. *Exp. Cell Res.* **143,** 55–62.

Nakayasu, H., and Ueda, K. (1984). Small nuclear RNA-protein complex anchors on the actin filaments in bovine lymphocyte nuclear matrix. *Cell Struct. Funct.* **9,** 317–325.

Nakayasu, H., and Ueda, K. (1985a). Ultrastructural localization of actin in nuclear matrices from mouse leukemia L5178Y cells. *Cell Struct. Funct.* **10,** 305–309.

Nakayasu, H., and Ueda, K. (1985b). Association of rapidly-labeled RNAs with actin in nuclear matrix from mouse L5178Y cells. *Exp. Cell Res.* **160,** 319–330.

Nakayasu, H., and Ueda, K. (1986). Preferential association of acidic actin with nuclei and nuclear matrix from mouse leukemia, L5178Y cells. *Exp. Cell Res.* **163.** 327–336.

Nickerson, J. A., He, D., Fey, A. G., and Penman, S. (1990). The nuclear matrix. *In* "The Eukaryotic Nucleus: Molecular Biochemistry and Macromolecular Assemblies" (P. R. Strauss and S. H. Wilson, eds.) Vol. 2, pp. 763–782. Telford Press, Caldwell, NJ.

Nigg, E. A. (1989). The nuclear envelope. *Curr Opin. Cell Biol.* **1,** 435–440.

Ohara-Nemoto, Y., Yoshida, M., and Ota, M. (1988). Interaction of vanadyl ribonucleoside complex with the androgen receptor. *Biochem. Int.* **17,** 197–202.

Ono, S., Abe, H., Nagaoka, R., and Ovinata, T. (1993). Colocalization of ADF and cofilin in intranuclear actin rods of cultured muscle cells . *J. Muscle Res. Cell Motil.* **14,** 195–204.

Paulson, J. R., and Laemmli, U. K. (1977). The structure of histon-depleted metaphase chromosomes. *Cell* (*Cambridge, Mass.*) **12,** 817–828.

Pienta, K. J., and Coffey, D. S. (1992). Nuclear-cytoskeletal interactions: Evidence for physical connections between the nucleus and cell periphery and their alteration by transformation. *J. Cell. Biochem.* **49,** 357–365.

Puskas, R. S., Manley, N. R., Wallace, D. M., and Berger, S. L. (1982). Effect of ribonucleoside–vanadyl complexes on enzyme-catalyzed reactions central to recombinant deoxyribonucleic acid technology. *Biochemistry.* **21,** 4602–4608.

Razin, S. V., Yarovaya, O. V., and Georgiev, G. P. (1985). Low ionic strength extraction of nuclease-treated nuclei destroys the attachment of transcriptionally active DNA to the nuclear skeleton. *Nucleic Acids Res.* **13,** 7427–7444.

Romig, H., Fackelmayer, F. O., Renz, A., Ramsperger, U., and Richter, A. (1992). Characterization of SAF-A, a novel nuclear DNA binding protein from HeLa cells with high affinity for nuclear matrix/scaffold attachment DNA elements. *EMBO J.* **11,** 3431–3440.

Sahlas, D. J., Milankov, K., Park, P. C., and De Boni, U. (1994). Distribution of snRNP's, splicing factor SC-35 and actin in interphase nuclei: Immunocytochemical evidence for differential distribution during changes in functional states. *J. Cell Sci.* **105,** 347–357.

Scheer, U., Hinssen, H., Franke, W. W., and Jockusch, B. M. (1984). Microinjection of actin-binding proteins and actin antibodies demonstrates involvement of nuclear actin in transcription of lampbrush chromosomes. *Cell (Cambridge, Mass.)* **39,** 111–122.

Spector, D. L., Fu, X.-D., and Maniatis, T. (1991). Association between distinct pre-mRNA splicing components and the cell nucleus. *EMBO J.* **10,** 3467–3481.

Staufenbiel, M., and Deppert, W. (1984). Preparation of nuclear matrices from cultured cells: Subfractionation of nuclei in situ. *J. Cell. Biol.* **98,** 1886–1894.

Stuurman, N., Meijne, A. M. L., van der Pol, A. J., de Jong, L., van Driel, R., and van Renswoude, J. (1990). The nuclear matrix from cells of different origin. Evidence for a common set of matrix proteins. *J. Biol. Chem.* **265,** 5460–5465.

Stuurman, N., de Jong, L., and van Driel, R. (1992a). Nuclear frameworks: Concepts and operational definitions. *Cell Biol. Int. Rep.* **16,** 837–852.

Stuurman, N., Floore, A., Colen, A., de Jong, L., and van Driel, R. (1992b). Stabilization of the nuclear matrix by disulfide bridges. Identification of matrix polypeptides that form disulfides. *Exp. Cell Res.* **200,** 285–294.

Tang, T. K., Tang, C. C., Chen, Y., and Wu, C. (1993). Nuclear proteins of the bovine esophageal epithelium. II. The *NuMA* gene gives rise to multiple mRNAs and gene products reactive with monoclonal antibody W1. *J. Cell Sci.* **104,** 249–260.

Tousson, A., Zeng, C., Brinkley, B. R., and Valdivia, M. M. (1991). Centrophilin: A novel mitotic spindle protein involved in microtubule nucleation. *J. Cell Biol.* **112,** 427–440.

Tsutsui, K., Tsutsui, K., Okada, S., Watarai, S., Seki, S., Yasuda, T., and Shohmori, T. (1993). Identification and characterization of a nuclear scaffold protein that binds the matrix attachment region DNA. *J. Biol. Chem.* **268,** 12886–12894.

Valkov, N. I., Ivanova, M. I., Uscheva, A. A., and Krachmarov, C. P. (1989). Association of actin with DNA and nuclear matrix from Guerin ascites tumour cells. *Mol. Cell. Biochem.* **87,** 47–56.

van Driel, R., Humbel, B., and de Jong, L. (1991). The nucleus: A black box being opened. *J. Cell. Biochem.* **47,** 311–316.

van Eekelen, C. A. G., and van Venrooij, W. J. (1981). HnRNA and its attachment to a nuclear protein matrix. *J. Cell Biol.* **88,** 554–663.

Verheijen, R., Kuijpers, H., Vooijs, P., van Venrooij, W., and Ramaekers, F. (1986). Protein composition of nuclear matrix preparations from HeLa cells: An immunochemical approach. *J. Cell Sci.* **80,** 103–122.

Verheijen, R., van Venrooij, W., and Ramaekers, F. (1988). The nuclear matrix: Structure and composition. *J. Cell Sci.* **90,** 11–36.

Vogelstein, B., Pardoll, D. M., and Coffey, D. S. (1980). Supercoiled loops and eucaryotic DNA replication. *Cell (Cambridge, Mass.)* **22,** 79–85.

von Kries, J. P., Buhrmester, H., and Strätling, W. H. (1991). A matrix/scaffold attachment region binding protein: Identification, purification and mode of binding. *Cell (Cambridge, Mass.)* **64,** 123–135.

von Kries, J. P., Buck, F., and Strätling, W. H. (1994). Chicken MAR binding protein p120 is identical to human heterogeneous nuclear ribonucleoprotein (hnRNP) U. *Nucleic Acids Res.* **22,** 1215–1220.

Wang, X., and Traub, P. (1991). Resinless section immunogold electron microscopy of karyo-cytoskeletal frameworks of eukaryotic cells cultured in vitro. Absence of a salt-stable nuclear matrix from mouse plasmacytoma MPC-11 cells. *J. Cell Sci.* **98,** 107–122.

Wansink, D. G., Schul, W., van der Kraan, I., van Steensel, B., van Driel, R., and de Jong, L. (1993). Fluorescent labeling of nascent RNA reveals transcription by RNA polymerase II in domains scattered throughout the nucleus. *J. Cell Biol.* **122,** 283–293.

Wansink, D. G., Manders, E. E. M., van der Kraan, I., Aten, J. A., van Driel, R., and de Jong, L. (1994). RNA polymerase II transcription is concentrated outside replication domains throughout S-phase. *J. Cell Sci.* **107,** 1449–1456.
Yang, C. H., Lambie, E. J., and Snyder, M. (1992). NuMA: An unusually long coiled-coil related protein in the mammalian nucleus. *J. Cell Biol.* **116,** 1303–1317.
Zeitlin, S., Wilson, R. C., and Efstratiadis, A. (1989). Autonomous splicing and complementation of in vivo-assembled spliceosomes. *J. Cell Biol.* **108,** 765–777.
Zeng, C., He, D., Berget, S. M., and Brinkley, B. R. (1994). Nuclear-mitotic apparatus protein: A structural protein interface between the nucleoskeleton and RNA splicing. *Proc. Natl. Acad. Sci. U.S.A.* **91,** 1505–1509.

NOTE ADDED IN PROOF

Sections III,E and V: Recently we have shown that the major proteins of the internal, fibrogranular nuclear matrix of HeLa S3 cells are B23 (numatrin) and 16 hnRNP proteins, together making up 75% of the total internal matrix protein mass (Mattern *et al.,* 1996).

4

High Unwinding Capability of Matrix Attachment Regions and ATC-Sequence Context-Specific MAR-Binding Proteins

TERUMI KOHWI-SHIGEMATSU[1] AND YOSHINORI KOHWI[1]

The Burnham Institute
La Jolla Cancer Research Center
La Jolla, California

I. INTRODUCTION

The nuclear matrix is thought to contribute to the structural and functional organization of DNA by providing a framework to which specific

[1] Present address: Department of Cancer Biology, Lawrence Berkeley National Laboratory, University of California, Berkeley, California.

NUCLEAR STRUCTURE
AND GENE EXPRESSION

regions of DNA can attach and form topologically independent loop domains. The DNA segments that have high affinity to the nuclear matrix *in vitro* [matrix attachment regions (MARs)] are postulated to be located at the bases of the chromatin loops. MARs have been identified in various types of cells and are AT-rich without consensus.

Evidence shows that MARs play a role in tissue-specific gene expression. MARs associated with the immunoglobulin μ heavy chain locus are essential for transcription of a rearranged μ gene in transgenic B lymphocytes (Forrester *et al.,* 1994). In addition, a cell type-specific MAR-binding protein, special AT-rich binding protein 1 (SATB1), has been identified; this protein is expressed predominantly in thymocytes (Dickinson *et al.,* 1992).

We examined AT-rich MARs from the DNA structural viewpoint and found that they contain a subset of AT-rich sequences that are capable of readily relieving negative superhelical strain by base unpairing or unwinding. This finding on the MAR–DNA structure led to the first cloning of a MAR-binding protein, SATB1. SATB1 binds to double-stranded MAR sequences that will unwind under superhelical strain; however, it will not bind to mutated MARs which are still AT-rich yet have lost their unwinding capability. This chapter describes the DNA structure of MARs, gene cloning of the MAR-binding protein SATB1, and the biological significance of SATB1 and other MAR-binding proteins with similar binding specificities.

II. MATRIX ATTACHMENT REGIONS CONTINUOUSLY BASE UNPAIRED UNDER SUPERHELICAL STRAIN

We used a chemical probe (chloroacetaldehyde) which specifically reacts with unpaired DNA bases to examine the MARs structure *in vitro* in supercoiled DNA. Chloroacetaldehyde reaction is ideal for detecting such unpaired bases because its reaction specificity is very high and remains unchanged under varying ionic conditions. Even a single unpaired base present in supercoiled plasmid DNA can be detected and mapped in a sequencing gel (reviewed in Kohwi-Shigematsu and Kohwi, 1992). The modified DNA bases can be mapped at a single base resolution in polyacrylamide gels (Kohwi and Kohwi-Shigematsu, 1988; Kohwi-Shigematsu and Kohwi, 1992). Using this chemical probing method, we examined various MARs, including DNA sequences surrounding the immunoglobulin heavy chain (IgH) enhancer (Cockerill *et al.,* 1987), the human interferon β (huIFN-β) gene MARs (Bode and Maass, 1988; Mielke *et al.*, 1990), a MAR 5′ of the human β-globin gene (Jarman and Higgs, 1988), a MAR 5′ of the locus control region (LCR) of the human β-globin locus (Yu *et al.*, 1994), potato light-inducible gene (Stockhaus *et al.*, 1987), and the yeast H4 ARS sequence (Amati and Gasser, 1988).

We identified the chloroacetaldehyde-reactive regions in all the MAR segments examined, indicating that MARs typically contain specific regions that have high base-unpairing propensity (base-unpairing region, BUR). Detection of BURs in MARs surrounding the IgH enhancer is shown in Fig. 1. The length of the unpaired regions depends greatly on the salt concentration: the lower the ionic strength, the more extensive the unpairing region. When BURs of various MARs were examined for their base-unpairing potential under varying salt concentrations, a clear preference was observed for the direction in which base unpairing expanded. In the case of the BUR within the MAR located 3′ of the IgH enhancer, unpairing expanded toward the relatively guanine cytosine (GC)-rich core enhancer region as the salt concentration decreased (Kohwi-Shigematsu and Kohwi, 1990; summarized in Fig. 1D). This base unpairing is not due to a high rate of DNA "breathing" in these AT-rich regions; rather, it reflects stably unwound DNA structures; this was demonstrated by a two-dimensional gel analysis of topoisomers with different linking numbers (Fig. 2). Figure 2 shows that at topoisomer −23 there is a jump in mobility, indicating a structural transition beyond the superhelical density of approximately $\rho = -0.05$. The 13 topoisomers with lower linkage than topoisomer −23 (shown by a bracket in Fig. 2) contain stable unwound DNA structure, as they all migrated more slowly in the first dimension than the topoisomer at the structural transition. Mobility of these topoisomers was about the same in the first dimension, with a very slight increase in mobility as the superhelical density increased. This indicates that after the structural transition, progressive unpairing occurs as superhelical density increases. In summary, the more negative the superhelical strain applied to a MAR sequence, the more unwinding occurs to relieve the negative superhelical strain by continuously base unpairing. By doing so, MARs can store negative supercoil energy.

III. DETERMINATION OF CORE UNWINDING ELEMENTS

When MAR segments are examined under low ionic strength, we can delineate BUR using the chloroacetaldehyde method described in

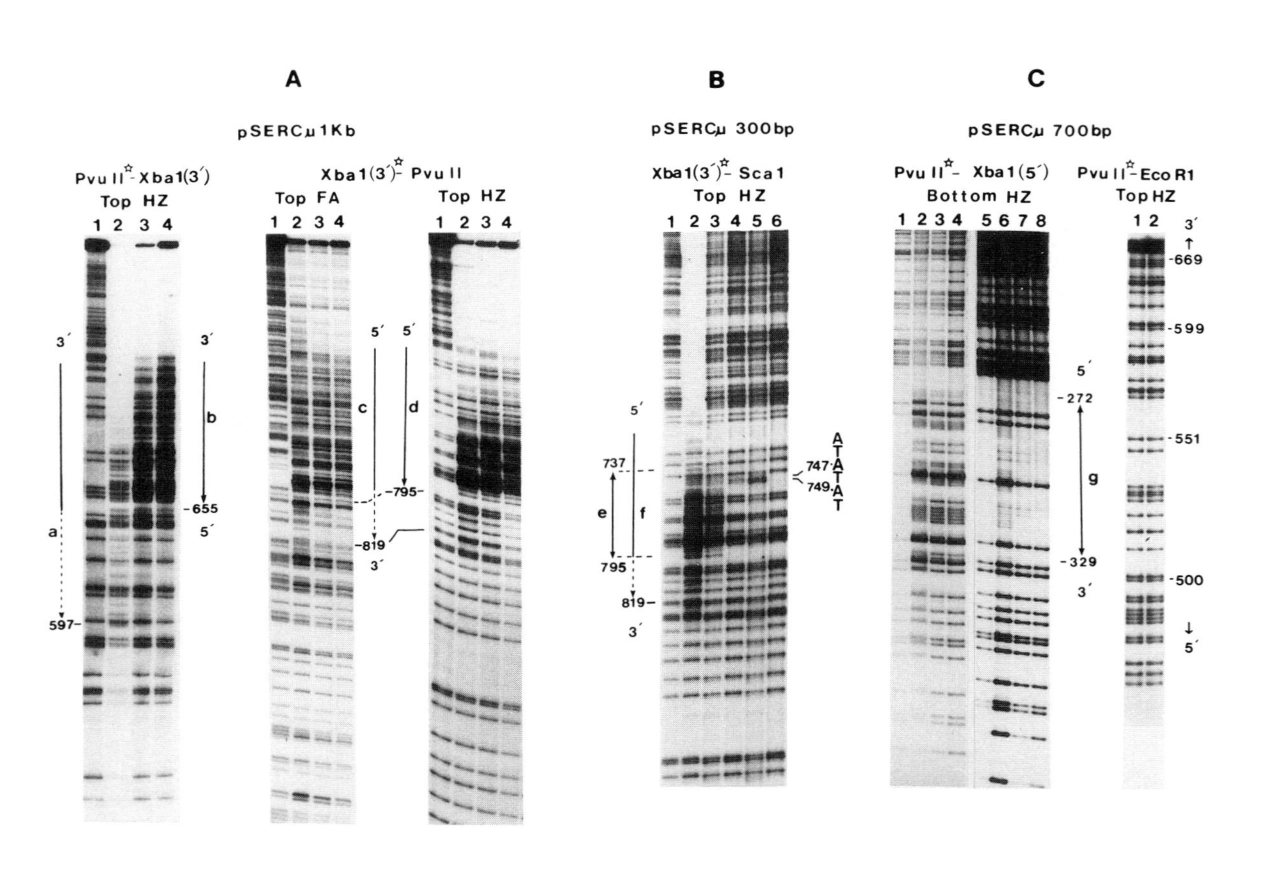
A
pSERCμ1Kb
Pvu II☆- Xba1(3′)
Top HZ
Xba1(3′)☆- Pvu II
Top FA
Top HZ
B
pSERCμ 300bp
Xba1(3′)☆- Sca1
Top HZ
C
pSERCμ 700bp
Pvu II☆- Xba1(5′)
Bottom HZ
Pvu II☆- Eco R1
Top HZ

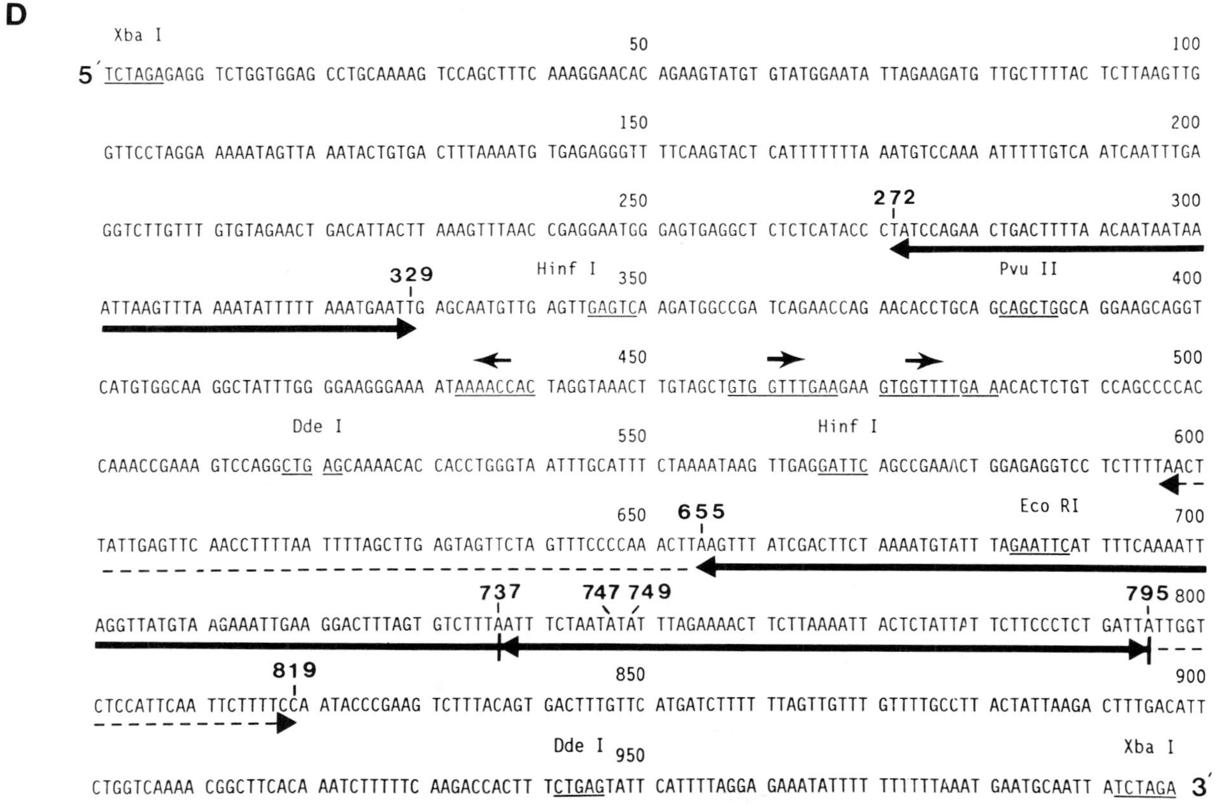

D

```
  Xba I                                                    50                                                         100
5' TCTAGAGAGG TCTGGTGGAG CCTGCAAAAG TCCAGCTTTC AAAGGAACAC AGAAGTATGT GTATGGAATA TTAGAAGATG TTGCTTTTAC TCTTAAGTTG

                                                          150                                                         200
   GTTCCTAGGA AAAATAGTTA AATACTGTGA CTTTAAAATG TGAGAGGGTT TTCAAGTACT CATTTTTTTA AATGTCCAAA ATTTTTGTCA ATCAATTTGA

                                                          250                      272                                300
   GGTCTTGTTT GTGTAGAACT GACATTACTT AAAGTTTAAC CGAGGAATGG GAGTGAGGCT CTCTCATACC CTATCCAGAA CTGACTTTTA ACAATAATAA

                                329           Hinf I      350                                 Pvu II                  400
   ATTAAGTTTA AAATATTTTT AAATGAATTG AGCAATGTTG AGTTGAGTCA AGATGGCCGA TCAGAACCAG AACACCTGCA GCAGCTGGCA GGAAGCAGGT

                                                          450                                                         500
   CATGTGGCAA GGCTATTTGG GGAAGGGAAA ATAAAACCAC TAGGTAAACT TGTAGCTGTG GTTTGAAGAA GTGGTTTTGA AACACTCTGT CCAGCCCCAC

                     Dde I                                550                  Hinf I                                 600
   CAAACCGAAA GTCCAGGCTG AGCAAAACAC CACCTGGGTA ATTTGCATTT CTAAAATAAG TTGAGGATTC AGCCGAAACT GGAGAGGTCC TCTTTTAACT

                                                          650  655                                  Eco RI            700
   TATTGAGTTC AACCTTTTAA TTTTAGCTTG AGTAGTTCTA GTTTCCCCAA ACTTAAGTTT ATCGACTTCT AAAATGTATT TAGAATTCAT TTTCAAAATT

                                     737       747 749                                                         795    800
   AGGTTATGTA AGAAATTGAA GGACTTTAGT GTCTTTAATT TCTAATATAT TTAGAAAACT TCTTAAAATT ACTCTATTAT TCTTCCCTCT GATTATTGGT

                        819                               850                                                         900
   CTCCATTCAA TTCTTTTCCA ATACCCGAAG TCTTTACAGT GACTTTGTTC ATGATCTTTT TTAGTTGTTT GTTTTGCCTT ACTATTAAGA CTTTGACATT

                                                  Dde I   950                                                  Xba I
   CTGGTCAAAA CGGCTTCACA AATCTTTTTC AAGACCACTT TCTGAGTATT CATTTTAGGA GAAATATTTT TTTTTTAAAT GAATGCAATT ATCTAGA 3'
```

Fig. 1. See the following page for legend.

Fig. 1. Fine mapping of the chloroacetaldehyde (CAA)-reactive sites surrounding the IgH enhancer. (A) Supercoiled plasmid pSERCμ1kb was reacted with CAA in sodium acetate buffer at pH 5 containing 25 m*M* Na^+ (lane 2), 37.5 m*M* Na^+ (lane 3), and 50 m*M* Na^+ (lane 4). DNA in lane 1 was not CAA-modified. The CAA-modified and unmodified DNA were 5′-end-labeled at the *PVu*II site and further digested with *Xba*I (left gel) or 3′-end-labeled at the *Xba*I site and digested with *PVu*II (middle and right gels). The labeled DNA fragment containing the CAA-reactive sequence was isolated from a native acrylamide gel and subjected to chemical reaction with hydrazine (HZ) or with formic acid (FA). The DNA was then reacted with piperdine. The DNA sample was loaded on a 6% denaturing urea-polyacrylamide gel. Arrows a–d indicate regions of CAA reactivity. The region shown by broken lines (---) within the arrows indicates additional CAA reactivity resulting under low-salt conditions. The DNA fragments are end-labeled at the restriction site indicated by a star. (B) Supercoiled plasmid pSERCμ300bp containing the *Eco*RI–*Xba*I (3′) 300-bp fragment was reacted with CAA in sodium acetate buffer at pH 5 containing 25 m*M* Na^+ (lane 2), 75 m*M* Na^+ (lane 3), 25 m*M* Na^+ and 2 m*M* Mg^{2+} (lane 4), and 50 m*M* Na^+ and 2 m*M* Mg^{2+} (lane 5). Lanes 1 and 6 represent CAA-unmodified DNA. The DNA was 3′-end-labeled at the *Xba*I site and digested with *Sca*I (at position 157). The *Xba*I (3′)–*Sca*I fragment was isolated from a native acrylamide gel, subjected to HZ–piperdine reaction, and loaded onto a 6% urea-denaturing polyacrylamide gel. The arrow f indicates the region of CAA reactivity under a low-salt condition, and the double-headed arrow e indicates the region of CAA reactivity under a high-salt condition. Positions 749 and 747 are indicated to show CAA reactivity at adenine residues at these positions in the presence of Mg^{2+} (see D for summary). (C) Supercoiled plasmid pSERCμ700bp containing the *Xba*I(5′)–*Eco*R1 700-bp fragment was reacted with CAA in a solution acetate buffer at pH 5 containing 25 m*M* Na^+ (lane 2), 37.5 m*M* Na^+ (lane 3), 50 m*M* Na^+ (lane 4), 75 m*M* Na^+ (lane 6), 25 m*M* Na^+ and 2 m*M* Mg^{2+} (lane 7), and 50 m*M* Na^+ and 2 m*M* Mg^{2+} (lane 8). Lanes 1 and 5 represent CAA-unmodified DNA. The DNA was 5′-end-labeled at the *Pvu*II site and further digested with *Xba*I. The *Pvu*II–*Xba*I(5′) fragment (left and middle gels) and the *Pvu*II–*Eco*RI fragment (right gel) were isolated and subjected to HZ–piperdine reaction. The double-headed arrow g indicates the region of CAA reactivity 5′ of the enhancer. (D) Summary of chemical cleavage sites for the CAA-modified IgH enhancer region. The solid double-headed arrow represents major CAA reactivity observed at 50 m*M* Na^+ or higher. Broken lines represent extended CAA reactivity under 25 and 37.5 m*M* Na^+. Underlined sequences indicated by small arrows are those similar to the core elements common to most viral enhancers (Weiher *et al.*, 1983).

Section II. On the other hand, by increasing the ionic strength and/or adding Mg^{2+} ions [the condition(s) that stabilize double-stranded DNA], it is possible to delineate a small subregion or core unwinding elements within BURs (Kohwi-Shigematsu and Kohwi, 1990). Because these elements persistently remain base-unpaired even under high ionic strength, they have exceptionally high unwinding propensity. For example, in the case of the 300-bp MAR segment 3′ of the IgH enhancer, a 160-bp

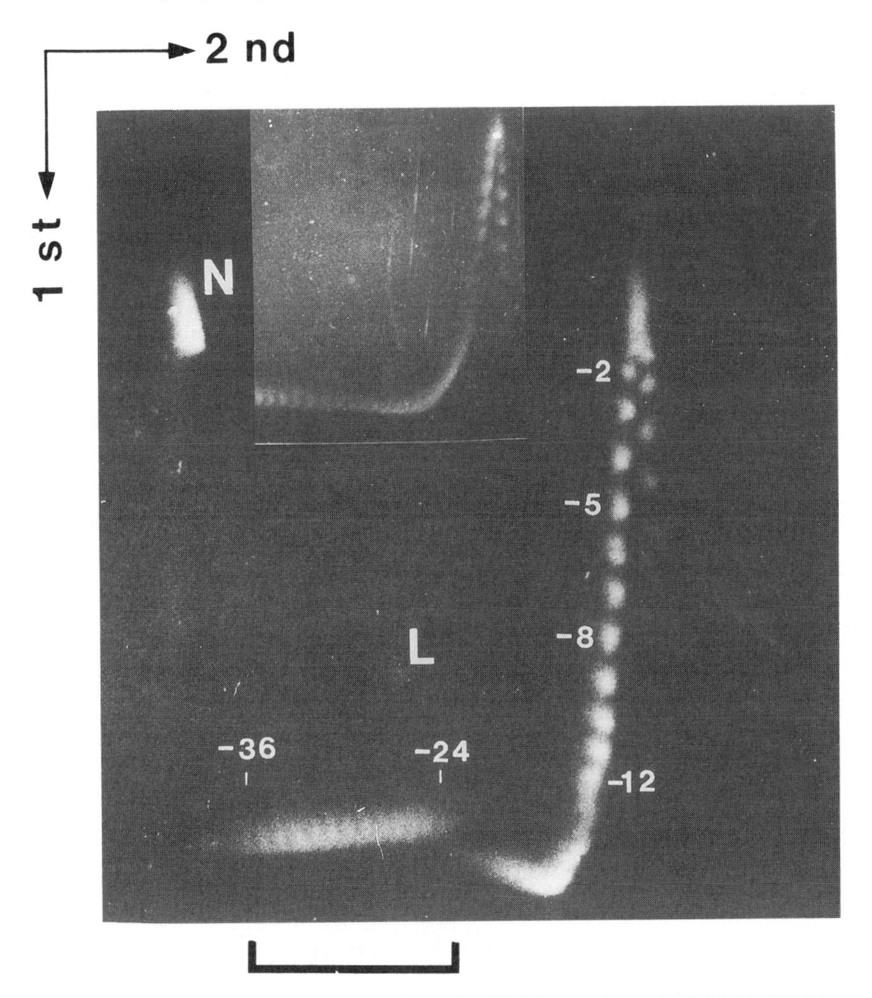

Fig. 2. Two-dimensional gel electrophoresis of DNA containing the IgH *Xba*I 992-bp fragment. A mixture of topoisomers of different mean superhelix density for pTKCATCμ1kb and pBR322 (inset) was prepared using calf thymus topoisomerase 1 (Bethesda Research Laboratories, Gaithersburg, MD) in the presence of various concentrations of ethidium bromide (Kochel and Sinden, 1988) and was electrophoresed from a single well in a 1.2% agarose gel in 80 m*M* Tris-borate and 10 m*M* ethylenediaminetetraacetic acid (EDTA), pH 8.2 at 22°C. The gel was then soaked in the same buffer containing 50 μg/ml chloroquine and rerun in this buffer in a direction of 90° to the original dimension at 22°C. The gel was stained in ethidium bromide and photographed under ultraviolet illumination. The number of topoisomers and the positions of nicked DNA (N) and linear DNA (L) are indicated. The bracket indicates topoisomers that contain stable, unwound DNA structure.

region (positions 655 to 819) was delineated as the BUR under low ionic strength. Under a higher-salt condition, the BUR was confined to a narrower region of 58 bp (positions 737 to 795). Ultimately, an ATATAT motif residing in this 58-bp region remained unpaired even in the presence of Mg^{2+} ions (Fig. 1B, lanes 4 and 5).

The short motif ATATAT in this particular sequence context of the MAR proved to be a core unwinding element by site-directed mutagenesis. Mutagenizing the ATATAT motif to CTGTCT led to the complete abolishment of the unwinding capability of the entire 300-bp MAR region. Therefore, this motif is necessary for the unwinding of the MAR region (Fig. 3; compare lanes 2 and 4).

Are there core unwinding elements in other MARs? We examined the MAR of the huIFN-β gene and found similar results. In the IFN-β MAR, there was also a subregion marked by very high unwinding capability, and a short sequence within this subregion remained unpaired even under high-salt conditions. When this element was mutated, entire MAR sequences became incapable of unwinding under negative superhelical strain (Bode *et al.*, 1992). This particular core unwinding element also contained a sequence A*ATATAT*TT identical to the core element for the MAR 3′ of the IgH enhancer. For certain MARs, the short DNA sequences that remain base-unpaired under high-salt conditions were found to be GC-rich sequences that are capable of forming non-B DNA structures such as an intramolecular triple helix structure under negative superhelical strain (L. A. Dickinson, J. Bode, Y. Kohwi, and T. Kohwi-Shigematsu, unpublished results, 1996). Whether these sequences are core unwinding elements must be studied by site-directed mutagenesis experiments.

IV. UNWINDING CAPABILITY AND MAR ACTIVITY

The observation that MARs share a high unwinding propensity prompted us to investigate the importance of this structural characteristic in MAR activity. MAR activity was measured by its binding to the nuclear matrix and its effect on enhancer/promoter activities *in vivo.* We found a strong correlation between unwinding capability and these biological activities (Bode *et al.*, 1992). We designed an experiment to address this question by employing a short oligonucleotide sequence of 25 bp sequence 5′ TCTTTAATTTCTA*ATATAT*TTAGAA 3′ (wild-type monomer). This was derived from the 3′ MAR of the IgH enhancer

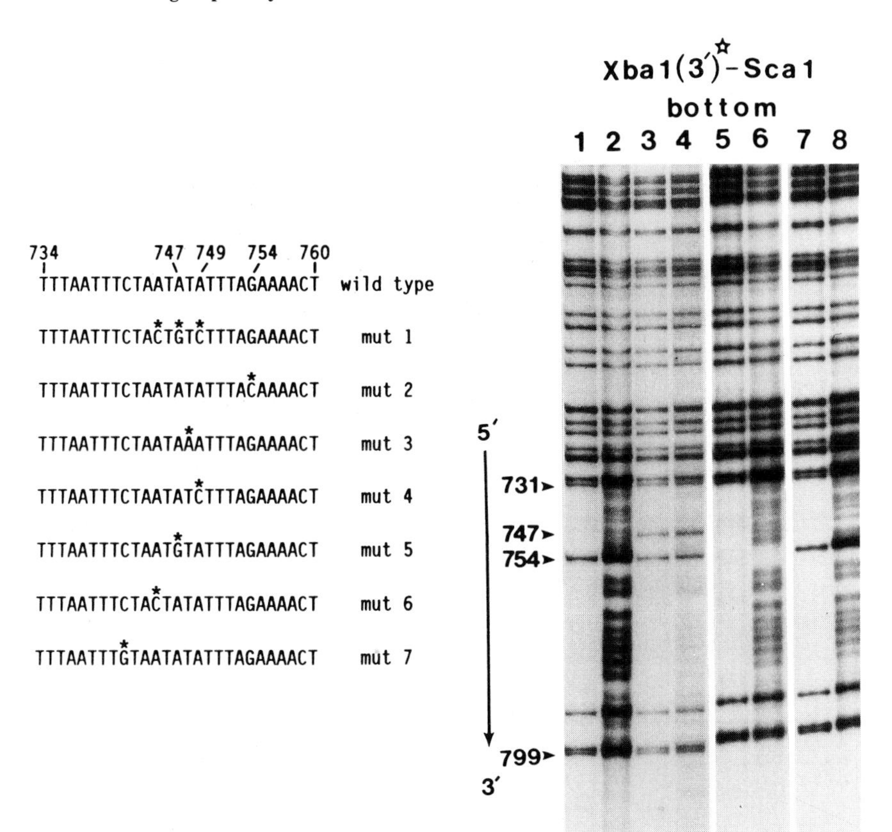

Fig. 3. Effects of site-directed mutagenesis on CAA reactivity in the IgH enhancer region. (Left) List of mutated DNAs employed in this experiment. These mutations were introduced to the pSERCμ1kb plasmid with the use of synthetic oligonucleotides shown here using the kit from Amersham (Arlington Heights, IL). (*) indicates the nucleotide mutated. (Right) Supercoiled plasmid DNA, wild-type (pSERCμ1kb) (lanes 1 and 2), mut 1 (lanes 3 and 4), mut 3 (lanes 5 and 6), and mut 4 (lanes 7 and 8) were either reacted (even-numbered lanes) or not reacted (odd-numbered lanes) with CAA in 25 m*M* sodium acetate buffer at pH 5. The DNA was then digested with *Xba*I and 5′-end-labeled with T4 kinase and [γ-^{32}P]ATP. The DNA was further digested with *Pvu*II, and the *Xba*I (3′)–*Pvu*II fragment was isolated. The labeled DNA was then subjected to HZ–piperdine reaction and electrophoresed on a 6% urea-denaturing polyacrylamide gel.

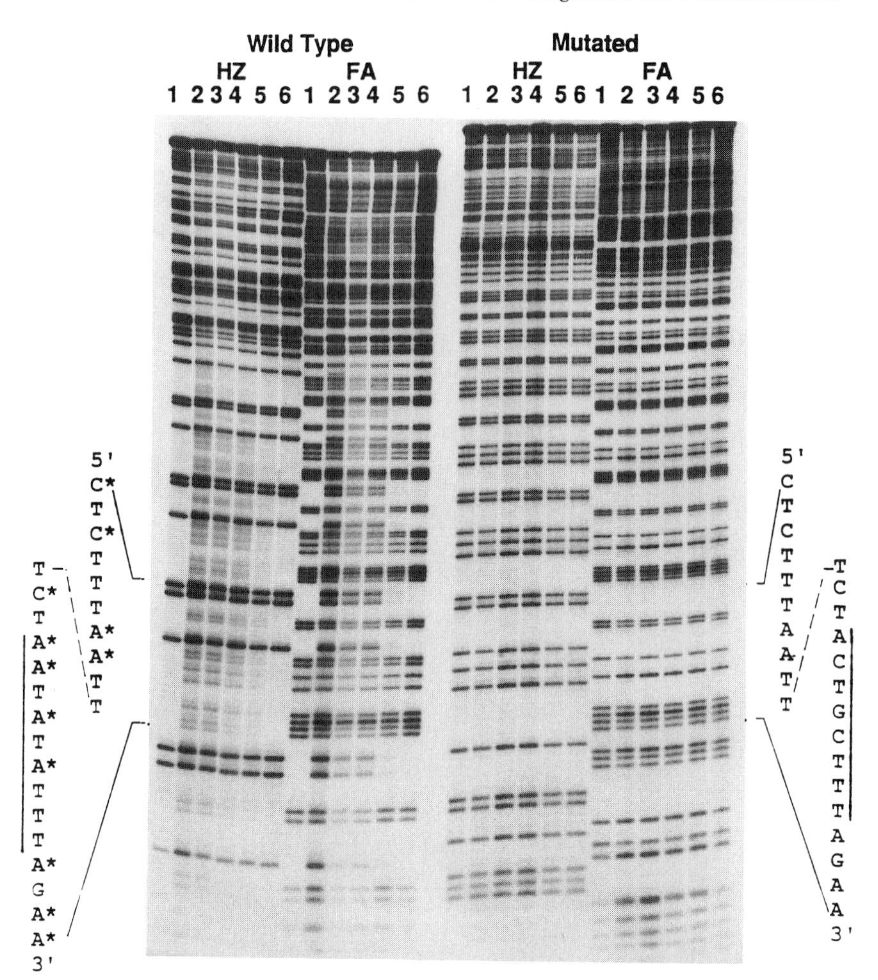

Fig. 4. Base-unpairing property of wild-type and mutated oligonucleotide multimers. Complementary oligonucleotides 5′-TCTTTAATTTCTAATATATTTAGAAttc-3′ and 5′-TTCTAAATATATTAGAAATTAAAGAgaa-3′ that contains the nucleation site of unwinding in the wild-type MAR 3′ of IgH enhancer were hybridized into double-stranded DNA, concatemerized by hybridization through overlapping single-stranded ends, and then digested with mung bean nuclease to remove single-stranded ends. The bases that were added to serve as single-stranded tails are indicated by lowercase letters. Multimers contained one *Eco*RI site per monomer. The orientation of the sequence was the same throughout the multimer. Multimers were separated on an acrylamide gel, and heptamers and octamers were cloned into the *Eco*RV site of a Bluescript vector. For the preparation of mutated multimers, we used 5′-TCTTTAATTTCTACTGCTTTAGAAttc-3′ and 5′-TTCTAAAGCAGTAGAAATTAAAGAAgaa-3′ for hybridization. Supercoiled plasmid

which contains the unwinding core element (underlined) and a homologous 24-bp sequence, 5′-TCTTTAATTTCTA*CTGC*TTTAGAA 3′ (mutated monomer where three adenines are mutated and one thymine is deleted), as underlined. Both sequences were made into a duplex DNA and multimerized in a unidirectional manner to prepare seven repeats for the wild type [wild-type $(25)_7$] and eight repeats for the mutated sequence [mutated $(24)_8$]. The resulting wild-type $(25)_7$ and mutated $(24)_8$ were both AT-rich sequences, yet they conferred contrasting structural properties in supercoiled plasmid DNA. Under negative superhelical strain, the wild-type $(25)_7$ readily unwinds by extensive base unpairing, whereas the mutated $(24)_8$ sequence completely resists unwinding (Fig. 4).

When these two AT-rich sequences were examined for their affinity to the nuclear matrix *in vitro,* the wild-type $(25)_7$ strongly bound to the nuclear matrix, while the mutated $(24)_8$ showed only weak binding (Fig. 5). Usually a long stretch of native MAR segments is required to confer binding to the nuclear matrix *in vitro;* 300 bp is considered the minimum length to detect any binding. The artificial MAR sequence, wild-type $(25)_7$, was less than 200 bp, yet it showed an affinity to the nuclear matrix comparable in strength to that of the native 800-bp MAR segment "IV" of IFN-β gene (Bode *et al.,* 1992). This indicates that the sequence enriched in unwinding core elements serves as a critical site for binding to the nuclear matrix. These results show that not any AT-rich sequence will confer binding to the nuclear matrix and that the unwinding property of an AT-rich sequence is important for the sequence to be classified as a MAR.

The effects of these two types of AT-rich sequences on the transcriptional activity of the linked reporter gene were examined in stable transformants. The strong MAR sequences from the huIFN-β gene, a light-inducible potato gene (ST-LS1), and the chicken lysozyme gene have been known to augment substantially the activity of heterologous promoters after integration of the transfected DNA (Phi-Van *et al.,* 1990;

DNA (Bluescript DNA) with either (A) a wild-type $(25bp)_7$ or (B) a mutated $(24bp)_8$ sequence was reacted with CAA. CAA modification was performed in sodium acetate buffer (pH 5) that contained 25 m*M* Na^+ (lane 2), 50 m*M* Na^+ (lane 3), 75 m*M* Na^+ (lane 4), or 25 m*M* Na^+ plus 2 m*M* $MgCl_2$ (lane 5). Lanes 1 and 6 show unmodified control DNA. The CAA-modified sites in the *Hind*III–*Bgl*I fragment, 3′ end-labeled at the *Hind*III site, were determined as in Fig. 1. CAA-modified DNA bases are indicated with (*). FA, formic acid.

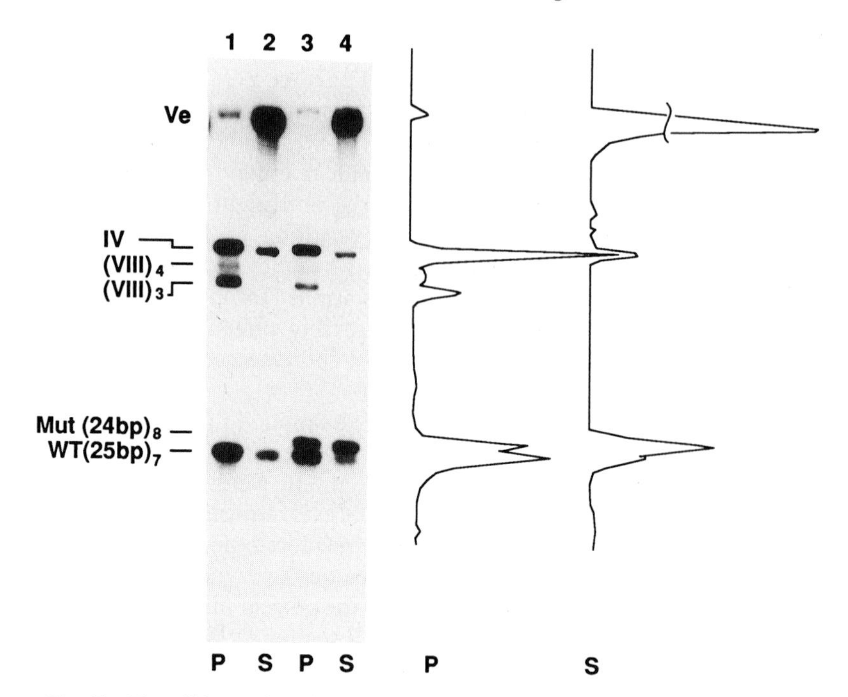

Fig. 5. The wild-type $(25bp)_7$ bound to the nuclear scaffold. Wild-type $(25bp)_7$ and mutated $(24bp)_8$ were analyzed *in vitro* for their affinity for the nuclear scaffold in comparison with known MAR and non-MAR fragments: Ve, vector part (*Cla*I–*Xba*I fragment of pBluescript) as a non-MAR control; IV, 800-bp core fragment from the MAR upstream of huIFN-β; $(VIII)_4$ and $(VIII)_3$, multimers of a 150-bp "sub-MAR" fragment that resides in IV [designations for DNA fragments as in Mielke *et al.* (1990)]. All fragments were labeled with Klenow polymerase and [α-^{32}P]dATP (2′-deoxy-5′-triphosphate). To determine the affinity of these DNA sequences to scaffolds, we used an *in vitro* reconstitution protocol that selects for DNA fragments with an affinity for the extraction-resistant nuclear scaffold (Mielke *et al.*, 1990). A mixture of all fragments without (lanes 1 and 2) or with (lanes 3 and 4) mutated $(24bp)_8$ was incubated at 37°C overnight with nuclear scaffold prepared from more L cells, then separated into a pellet (P) and a supernate (S) fraction by centrifugation (4°C, 14,000*g*, 45 min). DNA was then purified and dissolved in 200 μl of 10 mM Tris-HCl (pH 7.5) and 1 m*M* EDTA buffer. DNA was applied (2000 cpm per slot) to a 3% agarose gel. After electrophoresis, the gel was dried and autoradiographed overnight.

Stief *et al.*, 1989; Mielke *et al.*, 1990; Klehr *et al.*, 1991). Wild-type $(25)_7$ and mutated $(24)_8$ were placed 5′ of the SV40 early promoter, which is fused with the luciferase indicator gene (Fig. 6, top). This DNA construct was stably integrated into the mouse L cell genome. For the construct containing wild-type $(25)_7$, the activity of the luciferase gene was aug-

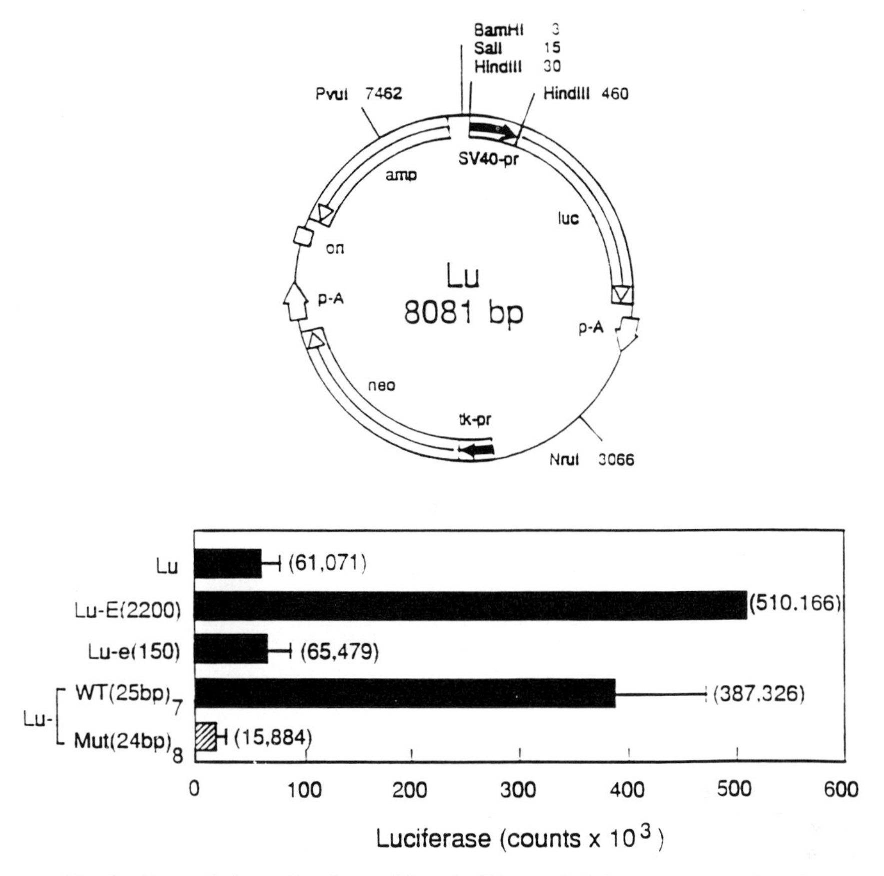

Fig. 6. Transcription-enhancing activity of wild-type $(25bp)_7$ and mutated $(24bp)_8$ *in vivo*. The wild-type $(25gp)_7$ or mutated $(24bp)_8$ sequence was excised as a *Bam*HI–*Sal*I fragment from a Bluescript vector and cloned into plasmid Lu (Klehr *et al.*, 1991) at the *Bam*HI and *Sal*I restriction sites in a position upstream of the SV40 promoter to generate Lu-$(25bp)_7$ and Lu-$(24bp)_8$ plasmids. Plasmid Lu-E(2200) contains the 2.2-kb upstream MAR element of huIFN-β, and Lu-e(150) contains a 150-bp sub-MAR fragment; both were inserted upstream of the SV40 promoter. One microgram of each of the vector plasmids Lu, Lu-E(2200), Lu-$(25bp)_7$, and Lu-$(24bp)_8$ was linearized at the *Pvu*I site and transfected into mouse L cells, as described (Mielke *et al.*, 1990). This method resulted in a small number of integrations per cell and no indication of tandem integration. Stable transformant cells were selected with neomycin analog G-418 (700 mg/ml) for 10 days. The clones were counted on day 12. The cells from 100 to 200 independent colonies were grown close to confluence. Extracts were prepared from a defined number (0.8×10^6 to 1.4×10^6) of cells from the pools resistant to G-418 and were tested for luciferase activity (Williams *et al.*, 1989). Luciferase fluorescence activity is reported per 10^6 cells and is normalized to a single copy per cell. The standard deviations of the luciferase activity, derived from several experiments, are shown by error bars.

mented more than sixfold in comparison to the original vector construct (pLu) without the insert. This level of augmentation is close to the eight-fold increase monitored for the 2.2-kb MAR 5′ of the huIFN-β gene (Fig. 6, bottom), which contains multiple subsites, including an 800-bp core fragment [which, on its own, yields a mere twofold enhancement (data not shown)]. The construct containing mutated $(24)_8$, after stable integration, did not augment the transcription; instead, it reduced luciferase gene activity in reference to the basal level activity of the control vector plasmid (pLu). This demonstrates that a short 190-bp sequence rich in the core unwinding element [wild-type $(25)_7$] is capable of strongly augmenting the transcriptional activity of the SV40 enhancer/promoter-driven reporter gene; this effect correlates well with the nuclear matrix binding affinity, which, in turn, depends on the ease of base unpairing.

MARs unwind only when subjected to negative superhelical strain. Because unwinding is biologically significant for MARs, one must consider whether such supercoiling occurs inside eukaryotic cells. In bacterial and yeast systems, active transcription of a gene induces local negative supercoiling 5′ and positive supercoiling 3′ of the gene (Liu and Wang, 1987; Giaever and Wang, 1988; Wu *et al.,* 1988), as shown in topoisomerase-defective mutant cells. In eukaryotes, the potential also exists to generate superhelical strain by the local removal of histone octamers; this issue, however, is controversial. Previously, we examined whether active transcription also induces local negative superhelical strain in mammalian chromatin, using poly(dG)-poly(dC) sequences of varying length as probes. These sequences form an intramolecular dG · dG · dC triplex only when subjected to negative superhelical strain, and their potential for adopting the dG · dG · dC triplex structure is strikingly length dependent (Kohwi and Kohwi-Shigematsu, 1988, 1991). The results of such experiments strongly suggest that negative superhelical strain can exist, at least transiently, in mammalian chromatin upstream of a transcriptionally active gene. Such superhelical strain can be relieved by specific DNA sequences which adopt altered DNA conformations (Kohwi and Kohwi-Shigematsu, 1991). It is possible that MARs are used to relieve local superhelical strain generated during the processes of replication and transcription.

V. CLONING OF cDNA ENCODING MAR-BINDING PROTEIN SATB1

If the DNA structural property is functionally significant, there may be a protein(s) that can distinguish DNA sequences with specific structural properties from others that lack these properties. To test this hypothesis, a human testis λgt11 expression library was screened with the radiolabeled wild-type $(25)_7$ probe for a protein that binds to this sequence. This led to the isolation of a cDNA that encodes a 763 amino acid protein, which we named special AT-rich binding protein 1 (SATB1) (Dickinson *et al.*, 1992). SATB1 strongly binds to wild-type $(25)_7$, which has a high unwinding propensity; on the other hand, it shows virtually no affinity to mutated $(24)_8$, which lacks unwinding propensity (Fig. 7A). SATB1 is therefore capable of clearly distinguishing similar AT-rich sequences based on their contrasting structural properties (Fig. 7B). Because all MAR sequences tested so far in our laboratory share this strong base-unpairing property, we expected SATB1 to bind to MARs. SATB1 did show a strong affinity to various MARs based on gel-mobility shift assay (Fig. 7C) but not to non-MAR AT-rich sequences. The dissociation constant, K_d, was determined for each of the MARs and ranged between 1×10^{-9} and 1×10^{-10} (Dickinson and Kohwi-Shigematsu, 1995). These K_d values for SATB1 indicate that SATB1 binds to its target sequences as tightly as many sequence-specific transcription factors.

VI. SATB1 RECOGNITION OF SPECIFIC SEQUENCE CONTEXT

Examination of the way SATB1 distinguishes certain AT-rich sequences from others would increase our understanding of the key components of MAR sequences. To study this problem, direct contact sites of SATB1 binding were determined for the MAR located immediately 3′ of the IgH enhancer. The deoxyribonuclease (DNase) 1 footprinting method was inappropriate for this purpose; instead, we employed the missing nucleoside technique to determine which nucleosides are contacted by SATB1 on the DNA (Hayes and Tullius, 1989). This alternative approach involved hydroxyl radical treatment of DNA, resulting in the loss of a deoxyribose residue with its attached base, causing the phophodiester backbone to break at this position. The DNA fragments containing an average of one random nucleoside gap per molecule were used as probes to examine SATB1 binding. Direct contact sites determined by this hydroxyl radical interference experiment are underlined immediately below the nucleotide sequences in Fig. 8A. These direct contact sites were all AT-rich sequences without any consensus motif. Instead

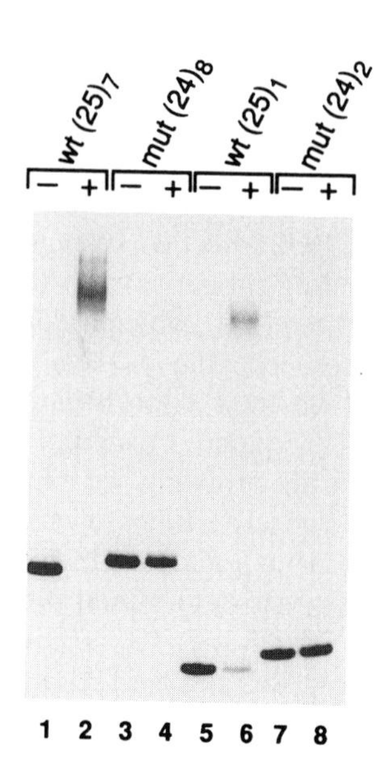

C.

Yeast - MAR
H4ARS
CEN III
Human-MAR IFN-β
Plant-MAR ST-LSI
TATAbox
poly$(dA)_{18}$
$(AT)_{32}$
SATBI
comp.
sp
ns
1 2 3 4 5 6 7 8 9 10 11 12 13 14 15 16 17 18 19 20 21 22 23

of suggesting a loosely defined consensus, we noticed that all of the direct contact sites resided within a specific DNA context (indicated by brackets), in which one strand consists of As, Ts, and Cs and the other strand of well-mixed As, Ts, and Gs (ATC sequence context). From our previous chloroacetaldehyde analysis of base-unpairing sequences, we were able to correlate the sites of ATC sequence context and high base-unpairing propensity. When stretches of sequences with the ATC sequence context were found in tandem, strong unpairing was detected under negative superhelical strain. It is significant that SATB1 direct contact sites are found within a sequence context that correlates with high unwinding potential. As shown in Fig. 8C, major contact sites II, III, and IV colocalize with the base-unpairing sequences, as detected by chloroacetaldehyde under negative superhelical strain; in fact, site IV corresponds exactly to the core unwinding element. These direct contacts sites are located in the region where sequences of ATC context are found in tandem. One remote site, I, confers SATB1 binding, probably because it has a fairly long stretch of ATC sequence context by itself. However, it does not confer base unpairing because it is a single unit.

To confirm that SATB1 recognizes an ATC sequence context rather than the specific DNA sequence with which it makes direct contact, we kept the direct contact sites intact and mutated the surrounding sequences to destroy the ATC sequence context. Such mutation greatly

---

Fig. 7. Sequence requirement for SATB1 binding and mobility shift assay with MAR and non-MAR DNA probes. (A) Autoradiograph of a gel retardation assay with SATB1, synthesized in reticulocyte lysate, and various repeats of wild-type or mutant synthetic oligonucleotide probes. Lanes 1 and 2: [wild-type $(25)_7$], lanes 3 and 4: [mut$(24)_8$], lanes 5 and 6: [wild-type $(25)_1$], lanes 7 and 8: [mut $(24)_2$]. Uneven-numbered lanes (−) are controls without added protein; even-numbered lanes (+) contain SATB1 synthesized in reticulocyte lysate. (B) Sequences of the top strand of the wild-type (wt) and mutant (mut) oligonucleotide monomers. The nucleation site for unpairing, ATATAT in the wt oligonucleotide, was changed to CTG-CT in the mut oligonucleotide, as indicated by boldface letters. (C) SATB1 synthesized in reticulocyte lysate was used in gel-mobility shift assays, with the probes indicated on top of the lanes. Lanes indicated by (−) contain samples with no protein (SATB1) or competitor (comp.) added to the binding reaction; lanes marked by (+) contain samples to which protein was added. A 200-fold molar excess of specific competitor (sp) was present in the samples shown in lanes 3, 7, 10, 14, 17, 20, and 23, and nonspecific competitor (ns) was added to the samples in lanes 4 and 11. The nonspecific competitor was a 445-bp *Pvu*II restriction fragment isolated from Bluescript.

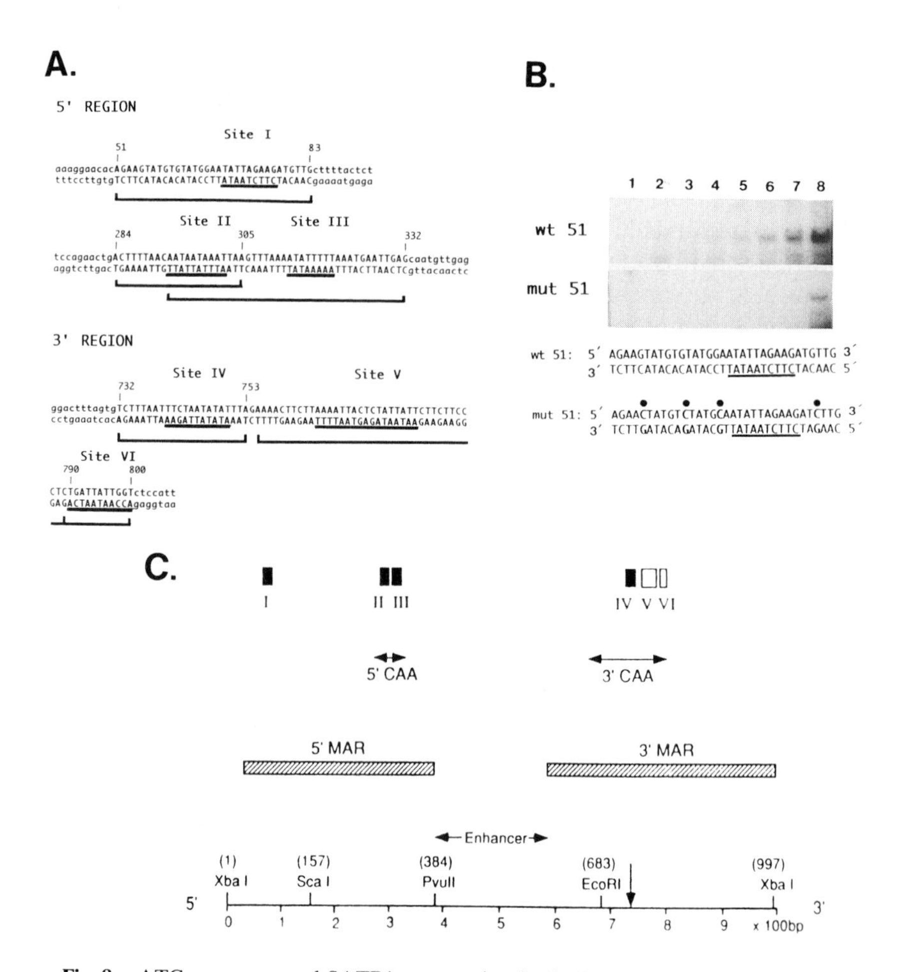

Fig. 8. ATC sequences and SATB1 contact sites in the IgH enhancer fragment. (A) ATC sequences within the 5′ and 3′ regions of the IgH enhancer fragment are indicated by capital letters and brackets, and their positions are shown by numbers. The six SATB1 contact sites are underlined and indicated by roman numerals. (B) Band shift assay with bacterially produced SATB1 and oligomers derived from the ATC sequence 51–83. Only the shifted bands are shown for better comparison between the wild-type (wt 51) and mutant (mut 51) ATC sequence 51–83. Lanes 1–8 contain the following protein concentrations in nanograms: 0 (lane 1), 31 (lane 2), 62.5 (lane 3), 125 (lane 4), 250 (lane 5), 500 (lane 6), 1000 (lane 7), 2000 (lane 8). The sequences of the double-stranded oligomers wt 51 and mut 51 are shown at the bottom; the dots indicate Gs that were replaced by Cs in the mutated DNA. The SATB1 contact site is underlined. (C) Relative location of SATB1-binding sites and sequences with known activities in the IgH enhancer. The 997-bp *Xba*I fragment containing the enhancer, flanked by the 5′ and 3′ regions, is pictured with the major restriction sites. The arrowhead indicates the location of the ATATAT nucleation site for unpairing. The MARs located 5′ and 3′ of the enhancer core are shown by hatched bars (Cockerill *et al.*, 1987). CAA-reactive areas are represented by double-headed arrows (Kohwi-Shigematsu and Kohwi, 1990). The four major and two minor recognition sites for SATB1 are given by solid and open bars, respectively.

reduced the affinity of SATB1 (Fig. 8B). Therefore, it is not the direct contact site that SATB1 recognizes, but rather the specific sequence context. Thus, SATB1 is not a sequence-specific DNA-binding protein, but a sequence context-specific binding protein.

Clustering of DNA sequences with the ATC sequence context is commonly found in MARs. This is summarized for two MARs, the yeast centromere *CENIII* and the potato light-inducible gene (*ST-LS1*) (Fig. 9). These regions containing clustered ATC sequences have a strong propensity to unwind under negative superhelical strain.

In summary, SATB1 recognizes the specific regions (BURs) in MARs that readily become unpaired when subjected to negative superhelical strain. If the unwinding capability of MARs is important for MAR activity and the BUR is a target for the sequence context-specific MAR-binding protein SATB1, the BUR of MARs may be a key region for the biological function of MARs.

VII. UNUSUAL MODE OF BINDING FOR SATB1

Although ATC sequence clusters confer a high propensity for base unpairing under negative superhelical strain, these sequences remain double-stranded in a linear DNA fragment. These double-stranded sequences are the targets for SATB1 binding. How does SATB1 identify ATC sequences with high unpairing potential before they are actually unpaired? To study this recognition mechanism, we examined the binding mode of SATB1 in detail. The results of competition experiments using minor groove binding drugs, distamycin A and berenil, showed that SATB1 binds in the minor groove of DNA (Fig. 10A). SATB1 made very little contact with the bases; modification of the bases at the major groove with dimethyl sulfate and chloroacetaldehyde did not interfere with SATB1 binding. In addition, depurination of bases at the frequency of a single base per DNA template had no adverse effect on SATB1 binding (Fig. 10B,C). These results show that SATB1 binds primarily in the sugar–phosphate backbone. It is surprising that SATB1 binding occurs at specific sites in MARs, making almost no direct contact with the specific bases. It is most likely that SATB1 preferentially binds specific sites indirectly by recognizing an altered sugar–phosphate backbone structure that is dictated by the ATC sequence context. These results may suggest that ATC sequences already have an altered double-

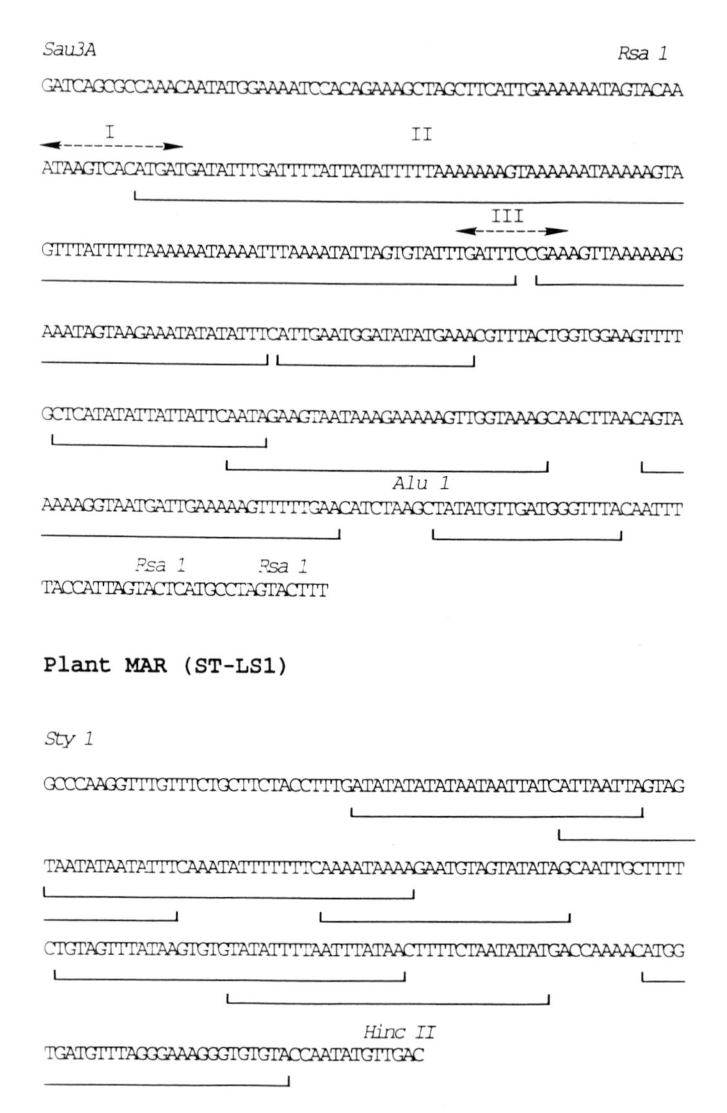

Fig. 9. Clustering of ATC sequences in yeast and plant MARs. (*Top*) The sequence of the 400-bp *Sau*3A–*Rsa*I fragment of the centromere of chromosome III (*CENIII*) is shown. The 320-bp *Rsa*I–*Rsa*I region of this *Sau*3A–*Rsa*I fragment is a MAR (Amati and Gasser, 1988). The minimal-functional *CENIII* region has been mapped to an *Rsa*I–*Alu*I fragment (Carbon and Clarke, 1984) and contains two distinct, short consensus regions (elements I and III, shown by double-headed arrows) flanking an 87-bp A+T-rich region (element II; Fitzgerald-Hayes *et al.*, 1982). (*Bottom*) the *Sty*I–*Hinc*II fragment of a MAR from potato light-inducible gene (*ST-LS1*) is shown. ATC sequences are shown by brackets. Some ATC sequences are superimposed.

stranded DNA structure which can be detected by a protein such as SATB1. Such sequences may unwind readily when subjected to negative superhelical strain.

It is unknown whether SATB1 recognizes ATC sequences when they are unpaired in supercoiled plasmid DNA. This is because it is difficult to fine-map the SATB1 direct contact sites using supercoiled plasmid DNA as the templates. Neither heat-denatured wild-type$(25)_7$ nor a single-stranded synthetic monomer wild-type $(25)_1$ binds SATB1 (L. A. Dickinson and T. Kohwi-Shigematsu, unpublished results, 1992). While SATB1 does not bind to single-stranded DNA, it is possible that ATC sequences in supercoiled plasmid DNA contain an altered but unknown conformation.

VIII. NOVEL MOTIF FOUND IN MAR-BINDING DOMAIN OF SATB1

It is of interest to determine the MAR-binding domain of SATB1 because this protein exhibits an unusual mode of DNA recognition; therefore, the DNA-binding domain may contain a new motif. We determined the MAR-binding domain in a newly cloned SATB1 gene from mouse thymus (Nakagomi *et al.*, 1994). The mouse SATB1 protein is 98% homologous to the human SATB1 originally isolated from testis. Both human and mouse SATB1 contains two sets of amino acid repeats (boxes I and II), as well as a glutamine stretch (Q_{15}) (Fig. 11, top). We delineated a 150 amino acid polypeptide (*Sca*I–*Ase*I) as the binding domain of mouse SATB1. This region confers full DNA-binding activity, recognizes the specific sequence context, and makes direct contact with DNA at the same nucleotides as the whole protein. As expected, this DNA-binding domain was found to contain a novel DNA-binding motif. When fewer than 21 amino acids at either N- or C-terminal sequences of the binding domain were deleted, the majority of its DNA-binding activity was lost (Fig. 11). Concomitant presence of both terminal sequences was mandatory for binding. The two sets of short amino acid sequences (a and b regions) at the N-terminal and (a′ and b′ regions) the C-terminal regions of the 150 amino acid DNA-binding domain share some homology. These N- and C-terminal regions of the MAR-binding domain probably extend toward the DNA sites, as shown in the model (Fig. 11, bottom). All DNA binding analyses and delineation of DNA-binding domains were performed in the presence of dI · dC as a nonspecific DNA competitor.

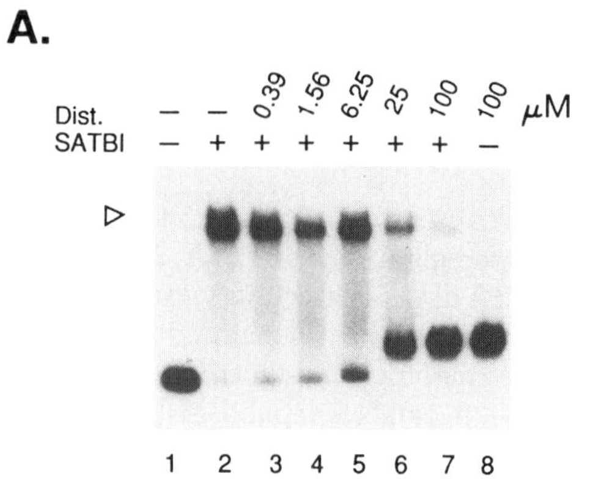
A.
Dist. — — 0.39 1.56 6.25 25 100 100 μM
SATBI — + + + + + + —
1 2 3 4 5 6 7 8

B.

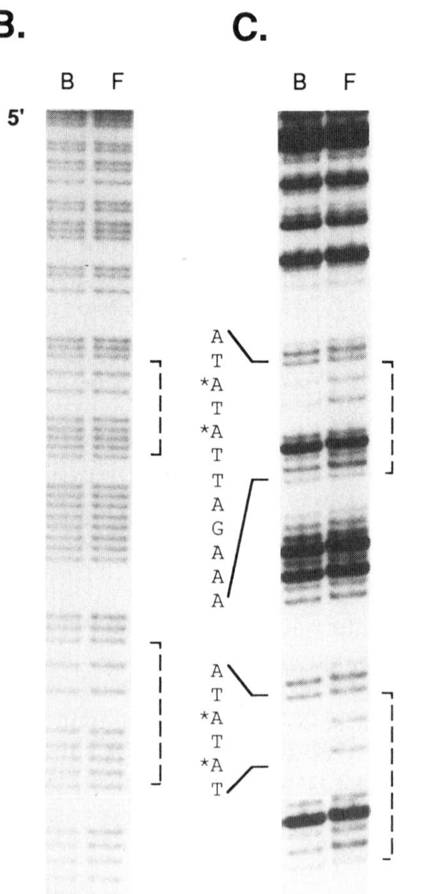
C.
B F
B F
5'
3'
A
T
*A
T
*A
T
T
A
G
A
A
A
A
T
*A
T
*A
T

To study further the MAR-binding motif of SATB1, we employed an independent approach. We used a phage library displaying nonamer random peptides with no built-in structure to identify a MAR-binding motif (Wang *et al.,* 1995). The phage that bound specifically to a MAR affinity column displayed one predominant cyclic peptide motif, CRQNWGLEGC. This peptide exhibited 50% identity with a segment in SATB1 (amino acid 355–363) and could replace the native sequence in a chimeric SATB1 construct without loss of binding affinity or specificity. Mutations of conserved residues greatly reduced MAR binding. The novel MAR-binding motif may help identify other proteins similar to SATB1 that belong to a new class of DNA-binding proteins.

IX. BIOLOGICAL SIGNIFICANCE OF SATB1

The proteins believed to bind MARs contribute to the structural organization of chromatin by anchoring the base of the chromatin to the nuclear matrix. If this is true, these MAR-binding proteins are expected to be more or less ubiquitous. In fact, other MAR-binding proteins, such as topoisomerase II (Berrios *et al.,* 1985; Earnshaw and Heck, 1985; Gasser *et al.,* 1986; Adachi *et al.,* 1989; Sperry *et al.,* 1989), lamin B_1 (Ludérus *et al.,* 1992), and heterogeneous nuclear ribonucleoprotein U

Fig. 10. Distamycin A competition of SATB1 binding and chemical interference experiments. (A) Distamycin A competition: autoradiograph of a gel retardation assay with SATB1 synthesized in rabbit reticulocyte lysate and radiolabeled 3′-En fragment [*Eco*RI(683)–*Xba*I(997)]. Distamycin A (Dist.) was added at the start of the binding reaction in the concentrations indicated in μM at the top of the lanes. The open arrowhead indicates SATB1-bound DNA, and the solid arrowhead indicates distamycin-bound DNA. SATB1 was omitted or present in the samples indicated by (−) or (+), respectively. (B) Depurination interference and (C) methylation interference with end-labeled [wild-type $(25)_2$] bound to thrombin cleaved bacterially produced SATB1. Lanes F and B represent the free and protein-bound probes, respectively, that were separated in a gel mobility shift assay. The stippled bars indicate SATB1 contact sites, as determined by hydroxyl radical interference (missing nucleoside experiment), and the full contact sequence and the location of the ATATAT sequence within the other contact site are given in (C). Two adenine residues in each repeat that interfered with binding when methylated are indicated by stars.

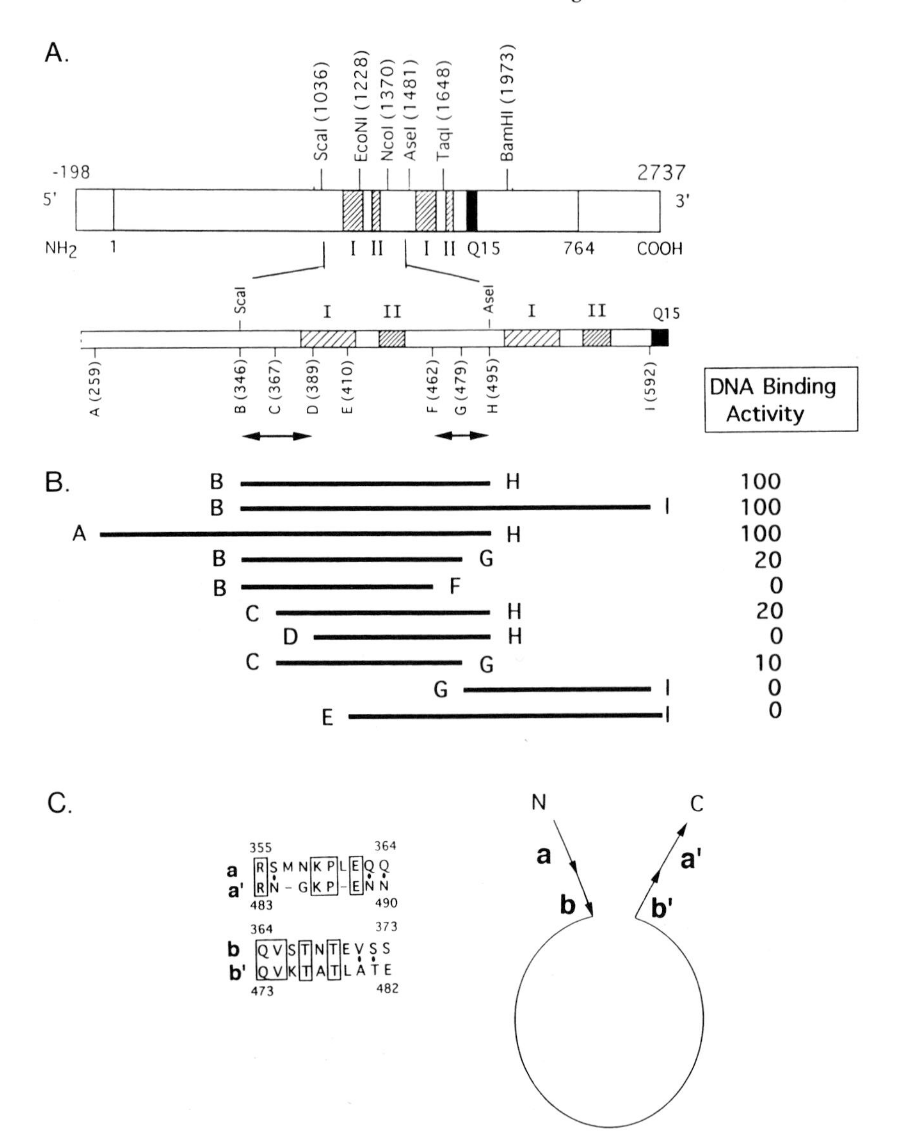

Fig. 11. Fine mapping of the DNA-binding domain and a model for the DNA-binding domain structure of mSATB1. (A) The top panel shows the schematic structure of mSATB1 in the amino acid regions 259–592 containing the DNA-binding region. Amino acid positions used to create deletion clones are indicated in parentheses and by capital letters. Positions of restriction sites *Sca*I and *Ase*I were used to prepare BH peptide. The double-headed arrows indicate essential regions for DNA-binding activity of mSATB1. The

(hnRNP-U) protein [formally called scaffold attachment factor A (SAF-A)] (Romig *et al.,* 1992; Fackelmayer *et al.,* 1994; von Kries *et al.,* 1994; K. Tsutsui and K. Tsutsui, personal communication), are all ubiquitous proteins. Therefore, it was surprising to find that SATB1 was a tissue-specific protein, predominantly expressed in thymocytes (Fig. 12A). At the time of SATB1 cDNA cloning, this tissue-specific expression of SATB1 was not known, and the gene was cloned from a human testis cDNA expression library as a rare gene. Ribonuclease (RNase) protection assay showed that SATB1 is expressed in testis, but the level of expression is substantially lower than that in thymus (Fig. 12B). The thymocyte-specific expression of SATB1 has an important biological implication; MAR-binding proteins may play an important role in gene regulation. Many MAR regions colocalize with or reside in close proximity to regulatory sequences, including enhancers (Cockerill and Garrard, 1986; Gasser and Laemmli, 1986; Cockerill *et al.,* 1987; Jarman and Higgs,

---

positions for the homologous amino acid sequences a and a′ and b and b′ are indicated by short arrows below the double-headed arrows. The specific regions of cDNA from which polypeptide was made are shown by bars. The positions of the fragment ends are indicated by capital letters where the corresponding amino acid positions are indicated on the schematic structure above. These cDNA fragments were prepared with the use of oligonucleotide primers and *Taq* polymerase and were cloned into pRIT2T (Pharmacia, Piscataway, NJ) in frame with Protein A. The DNA-binding activities of protein synthesized from these clones as Protein A fusion proteins are shown on the right column relative to the BH peptide in terms of kilodaltons of a given peptide per kilodalton of BH peptide ×100. N indicates N-terminal and C indicates C-terminal. (B) Gel-shift mobility assay using various concentrations of proteins. The column-purified Protein A fusion proteins were used for the assay. Results for mSATB1 protein, BH peptide, and CG peptide as Protein A fusion proteins are shown. Concentrations: 0 μg (lane 1), 0.05 μg (lane 2), 0.075 μg (lane 3), 0.1 μg (lane 4), 0.2 μg (lane 5), 0.3 μg (lane 6), 0.5 μg (lane 7), 0.75 μg (lane 8), and 1.0 μg (lane 9). (C) Plot showing DNA-binding activity for all SATB1 fragments as Protein A fusion protein based on the gel-mobility shift assay. The densitometric scans of autoradiograms was used to calculate the percentage of DNA fragments that are shifted. The two sets of short amino acid sequences at the N- and C-terminals of BH peptide also share some homology, as indicated. The repeats designated as **a** and **b** are at the N-terminal, and those designated as **a′** and **b′** are at the C-terminal. Identical amino acids are boxed, and chemically similar amino acids are shown by dots. Gaps are indicated by bars. The position of amino acids are indicated. The DNA-binding structure consists of a loop region that contains Box I and Box II, and the two subdomains **a, b** at the N-terminal region (indicated as N) and **b′** and **a′** at the C-terminal region (indicated as C) extending toward the DNA sites.

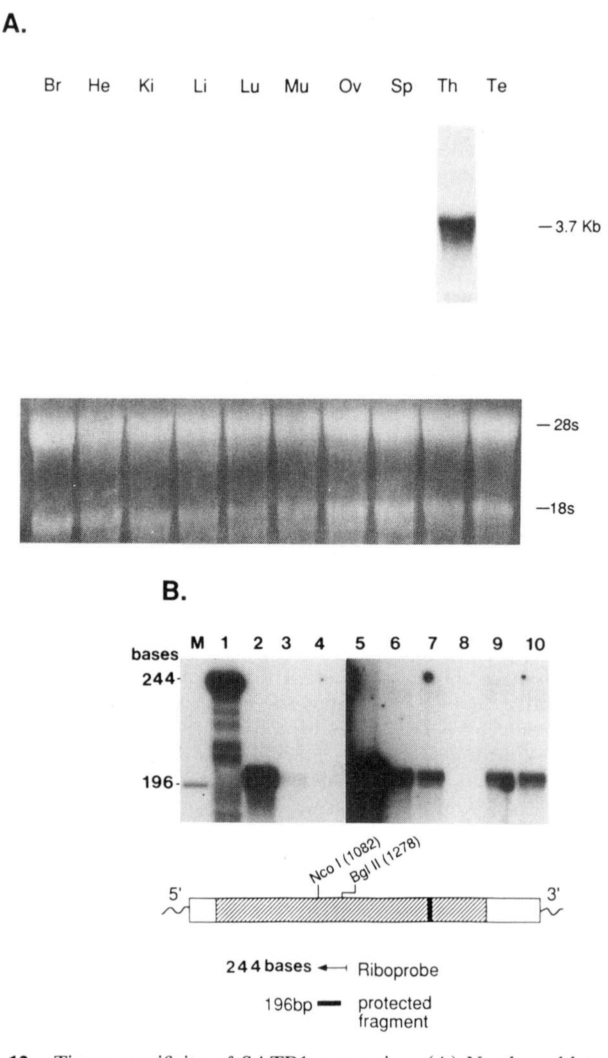

Fig. 12. Tissue specificity of SATB1 expression. (A) Northern blot analysis. In this study, 10 μg each of total cellular RNA from mouse tissues was analyzed using a radiolabeled *Ava*I(392)–*Bgl*II(1278) restriction fragment from pAT1146 (Fig. 1A) as a probe. The major transcript of 3.7 kb in thymus is marked. A transcript of the same size was also detected in human thymus RNA (data not shown). Br, Brain; He, heart; Ki, kidney; Li, liver; Lu, lung; Mu, muscle; Ov, ovaries; Sp, spleen; Th, thymus; Te, testis. The bottom panel is a photograph of the ethidium bromide-stained 28S and 18S rRNA bands on the formaldehyde gel before transfer to the hybridization membrane, showing that equal amounts of RNA were loaded in each track. (B) RNase protection assay. The

1988). The tissue specificity of the MAR-binding protein SATB1 raises the possibility that some of these regulatory sequences are MARs themselves and not regulatory sequences that happen to colocalize with MAR segments. More specifically, regulatory sequences may correspond to BURs or possibly to the core unwinding elements of BURs. Cotransfection experiments with a SATB1-expressing plasmid DNA and a reporter gene linked to wild-type $(25)_7$ showed that SATB1 suppresses the transcriptional enhancement activity of wild-type $(25)_7$ in stable transformants (T. Kohwi-Shigematsu and J. Bode, unpublished results, 1996).

Computer-aided protein analysis of SATB1 revealed an atypical homeodomain and two sets of repeated elements in SATB1 that share homology with the Cut repeats found in the Cut homeodomain protein family. The homeodomain has identified a class of master control genes that play a key role in the specification of the body plan and in the genetic control mechanisms of eukaryotic gene regulation (reviewed in Gehring, 1992). The SATB1 homeodomain by itself binds poorly and with low specificity to DNA. However, the combined action of the homeodomain and the MAR-binding domain confers an even higher level of binding specificity by directing the protein toward a core unwinding element among multiple ATC sequence stretches clustered in a MAR (Dickinson *et al.*, submitted, 1996). For example, the preferential binding of SATB1 to site IV over sites V and VI in the MAR 3′ of the IgH enhancer (Fig. 8C) is achieved by the SATB1 homeodomain together with the MAR-binding domain. This effect of the homeodomain can be detected in the absence of poly[(dI-dC) · (dI · dC)] in the DNA binding assay. The fact that SATB1 is a homeodomain protein and targets a specific element that controls overall MAR structure suggests that developmental regulation could take place at the level of chromatin structure mediated by MARs.

RNA probe used is shown schematically underneath the autoradiograph. The sizes of the undigested probe (244 bases) and of the RNase-resistant fragments (196bp; lane M) are indicated. The control lane (1) contains undigested probe. Total cellular RNA (25 μg each) was from human thymus (lanes 2 and 5), testis (lanes 3, 6, and 9), and the human testicular cell line Germa-1 (Hoffmann *et al.*, 1989) (lanes 4, 7, and 10); lane 8 does not contain RNA. The panel on the right-hand side with lanes 5 to 10 is an overexposure of the autoradiograph shown on the left-hand side to confirm the presence of a 196-bp protected RNA transcript in testis and Germa-1.

SATB1 may play a critical role in T-cell development. We examined whether SATB1 is expressed at specific stages of the T-cell lineage and detected a high level of expression of SATB1 in immature thymocytes for both $CD4^+$ $CD8^+$ double-positive and single-positive cells, but SATB1 is undetectable in mature T cells (T. Kohwi-Shigematsu, I. deBelle, and Y. Kohwi, unpublished results, 1995). In collaboration with Dr. D. Y. Loh and Dr. J. D. Alvarez at the Washington University, St. Louis, Missouri, we are in the process of preparing *SATB1* gene knock-out mice to examine the biological function of SATB1.

SATB1 may also be involved in T-cell proliferation. A report identified the *SATB1* gene as an immediate early gene induced after stimulation of the peripheral T cells by either the lymphokine interleukin-2 (IL-2) or the T-cell antigen receptor–CD3 complex (TCR–CD3) (Beadling *et al.,* 1993). Quiescent lymphocytes are stimulated initially by antigen-presenting cells via the TCR–CD3, together with accessory molecules such as CD28, which enable the cells to produce and respond to IL-2. Subsequently, the interaction of IL-2 with its receptor promotes cell cycle progression and DNA replication (Smith, 1988). Of eight IL-2-induced genes isolated, seven were specific to the IL-2-mediated G_1 "progression" phase of the cell cycle. Only one of these genes is induced during the T-cell receptor-triggered G_0–G_1 "competence" phase and identified as SATB1 (Beadling *et al.,* 1993).

In summary, SATB1 is a very interesting protein that is likely to have an important role in cellular development and proliferation. The functional role of SATB1 is probably mediated by its binding targets, MARs.

X. OTHER MAR-BINDING PROTEINS THAT CAN DISTINGUISH WILD-TYPE $(25)_7$ FROM MUTATED $(24)_8$

SATB1 is a sequence context-specific DNA-binding protein. To search for other proteins similar to SATB1, we constructed a DNA affinity column containing concatemerized, double-stranded wild-type (25) oligonucleotides. We chose human erythroleukemia K562 cells to isolate a MAR-binding protein using this affinity column because whole cell extracts prepared from these cells confer high MAR-binding activity, even though only a small amount of SATB1 is present. Such MAR-binding activity was retained after K562 cell extracts were mixed with anti-SATB1 antibody; the activity was completely lost by preincubation of thymyocyte extracts with the serum. The affinity column purified a protein of 100 kDa to near homogeneity which was not recognized by anti-SATB1 antibody. After antibody production and subsequent cloning of the cDNA, the protein was identified as nucleolin. Nucleolin is the major nucleolar protein or rapidly dividing cells and plays an important role in the regulation of ribosomal RNA gene transcription and assembly (Jordan, 1987). Nucleolin is known to be an RNA-binding protein that also binds single-stranded DNA. When double-stranded wild-type $(25)_5$ and mutated $(24)_8$ were tested for binding to nucleolin, nucleolin was found to bind specifically to sequences of high unwinding potential (Fig. 13). Nucleolin binds wild-type $(25)_5$ with an affinity approximately 50-fold higher than that of mutated $(24)_8$. Furthermore, nucleolin binds a variety of MAR sequences; therefore, the binding

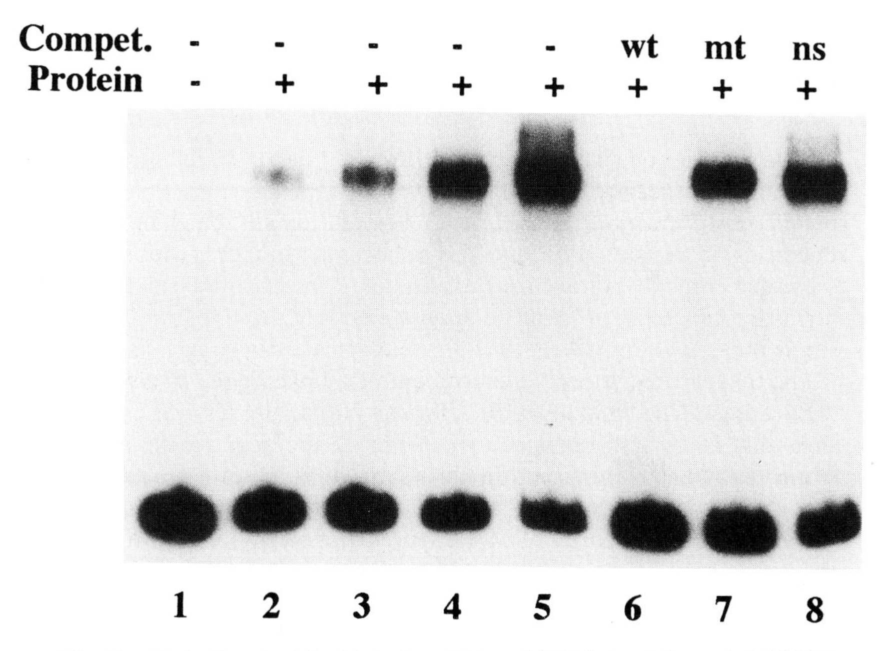

Fig. 13. Nucleolin selectivity binds the wild-type MAR but not the mutated MAR. Gel retardation with wild-type $(25)_5$ probe. Nucleolin concentrations were 0, 1.5, 3, and 12 ng/20 μl in lanes 1–5, respectively, and 12 ng/20 μl in lanes 6–8. A 200-fold molar excess of unlabeled wild-type $(25)_5$ (wt) (lane 6), mutated $(24)_8$ (mt)(lane 7), and a nonspecific 445-bp *Pvu*II fragment from Bluescript (ns)(lane 8) were added as competitors.

specificity of nucleolin for double-stranded DNA was very similar to that of SATB1. The binding affinity of nucleolin was, in general, considerably lower than that of SATB1 and showed some preference for certain MAR sequences, especially *CENIII* ($K_d = 5 \times 10^{-9}$) (Dickinson and Kohwi-Shigematsu, 1995).

Unlike SATB1, nucleolin binds single-stranded DNA much more preferentially than double-stranded DNA. Nucleolin shows a clear strand preference; the T-rich strand of wild-type $(25)_5$ is strongly favored over its complementary strand. The complementary strand binds with a similar affinity to double-stranded wild-type $(25)_5$. From these findings, it can be concluded that nucleolin is a MAR-binding protein that also binds to single-stranded DNA with clear strand preference. Its tight association with the nucleolar matrix is consistent with the discovery that nucleolin is a MAR-binding protein. The multifunctional roles that nucleolin plays in cells are likely to be mediated by MAR sites in the ribosomal DNAs.

XI. CONCLUSION AND FUTURE PROSPECTS

Characterization of the structural property of MARs in supercoiled plasmid DNA has led to the discovery of MARs that consist of a subset of AT-rich sequences (BURs) that possess an unusually high base-unpairing propensity. BUR contains a cluster of sequences with the ATC sequence context. Mutagenesis experiments specifically destroying the unpairing capability of MARs demonstrated that this unpairing propensity is important for MARs to confer biological activity such as binding to the nuclear matrix and enhancement of a linked gene *in vivo*.

Because MARs contain BURs, they can potentially serve as an energy sink; that is, by unwinding, they can relieve the negative superhelical strain that transiently occurs in chromatin. It is important to reiterate that the concept of the ATC sequence context was derived from both *in vitro* structural analysis of MARs and characterization of the binding site recognition by the cloned MAR-binding protein SATB1. MARs are not merely AT-rich sequences with no consensus; they have a novel structural consensus that has never been characterized for other DNA binding sequences.

Although MARs contain the unique region BUR, this does not necessarily mean that all MAR-binding proteins specifically recognize BURs. There may be some MAR-binding proteins, especially those involved in the structural organization of chromatin, that bind any given MAR segment larger than several hundred base pairs. In fact, lamin B_1 appears to be such an example. For lamin B_1, more than 600–800 bp of AT-rich sequences are required to confer any binding; this binding characteristic resembles that of the purified nuclear matrix itself (Dr. E. Ludérus, personal communication). It remains unknown whether MAR-binding proteins that specifically bind BURs of MAR segments, like SATB1, share regulatory roles. We have identified a MAR-binding protein that associates with human breast carcinomas, but not with benign breast lesions or normal proliferating epithelial cells (Yanagisawa *et al.,* 1996). The breast tumor MAR-binding protein is distinct from SATB1 but has similar binding recognition properties. Once cDNA cloning is achieved, we will learn more about this new class of proteins.

BURs of MARs may have multifunctional roles in transcription and replication, either as targets of regulatory proteins or as replication origins. The latter idea is worth pursuing because there is evidence that MARs may be replication origin sites (Razin *et al.*, 1986; Hozák *et al.*, 1993). In fact, the yeast autonomous replication region (ARS) is a MAR/SAR (Amati and Gassar, 1988) consisting of an ATC sequence cluster and becomes readily unpaired under negative superhelical strain (Umek and Kowalski, 1988; T. Kohwi-Shigematsu and Y. Kohwi, unpublished results, 1994). The possibility that the high unwinding propensity of specific DNA regions facilitates unwinding of DNA replication origins may also apply to mammalian systems.

REFERENCES

Adachi, Y., Käs, E., and Laemmli, U. K. (1989). Preferential, cooperative binding of DNA topoisomerase II to scaffold-associated regions. *EMBO J.* **8,** 3997–4006.

Amati, B. B., and Gasser, S. M. (1988). Chromosomal ARS and CEN elements bind specifically to the yeast nuclear scaffold. *Cell* (*Cambridge, Mass.*) **54,** 967–978.

Beadling, C., Johnson, K. W., and Smith, K. A. (1993). Isolation of interleukin 2-induced immediate-early genes. *Proc. Natl. Acad. Sci. U.S.A.* **90,** 2719–2723.

Berrios, M., Osheroff, N., and Fisher, P. A. (1985). In situ localization of DNA topoisomerase II, a major polypeptide component of the *Drosophila* nuclear matrix fraction. *Proc. Natl. Acad. Sci. U.S.A.* **82,** 4142–4146.

Bode, J., and Maass, K. (1988). Chromatin domain surrounding the human interferon-β gene as defined by scaffold-attached regions. *Biochemistry* **27,** 4706–4711.

Bode, J., Kohwi, Y., Dickinson, L., Joh, T., Klehr, D., Mielke, C., and Kohwi-Shigematsu, T. (1992). Biological significance of unwinding capability of nuclear matrix-associating DNAs. *Science* **255,** 195–197.

Carbon, J., and Clarke, L. (1984). Structural and functional analysis of a yeast centromere (CEN3). *J. Cell Sci., Suppl.* **1,** 43–58.

Cockerill, P. N., and Garrard, W. T. (1986). Chromosomal loop anchorage of the kappa immunoglobulin gene occurs next to the enhancer in a region containing topoisomerase II sites. *Cell (Cambridge, Mass.)* **44,** 273–282.

Cockerill, P. N., Yuen, M.-H., and Garrard, W. T. (1987). The enhancer of the immunoglobulin heavy chain locus is flanked by presumptive chromosomal loop anchorage elements. *J. Biol. Chem.* **262,** 5394–5397.

Dickinson, L. A., and Kohwi-Shigematsu, T. (1995). Nucleolin is a matrix attachment region DNA-binding protein that specifically recognizes a region with high base-unpairing potential. *Mol. Cell. Biol.* **15,** 456–465.

Dickinson, L. A., Joh, T., Kohwi, Y., and Kohwi-Shigematsu, T. (1992). A tissue-specific MAR/SAR DNA binding protein with unusual binding site recognition. *Cell (Cambridge, Mass.)* **70,** 631–645.

Dickinson, L. A., Dickinson, C. D., and Kohwi-Shigematsu, T. (1996). The nuclear matrix attachment region (MAR)-binding protein SATB1 contains a homeodomain that promotes specific recognition of the core unwinding element of a MAR. Submitted for publication.

Earnshaw, W. C., and Heck, M. M. S. (1985). Localization of topoisomerase II in mitotic chromosomes. *J. Cell Biol.* **100,** 1716–1725.

Fackelmayer, F. O., Dahm, K., Renz, A., Ramsperger, U., and Richter, A. (1994). Nucleic-acid-binding properties of hnRNP-U/SAF-A, a nuclear-matrix protein which binds DNA and RNA *in vivo* and *in vitro. Eur. J. Biochem.* **221,** 749–757.

Fitzgerald-Hayes, M., Clarke, L., and Carbon, J. (1982). Nucleotide sequence comparisons and functional analysis of yeast centromere DNAs. *Cell (Cambridge, Mass.)* **28,** 235–244.

Forrester, W. C., van Genderen, C., Jenuwein, T., and Grosschedl, R. (1994). Dependence of enhancer-mediated transcription of the immunoglobulin μ gene on nuclear matrix attachment regions. *Science* **265,** 1221–1225.

Gasser, S. M., and Laemmli, U. K. (1986). Cohabitation of scaffold binding regions with upstream/enhancer elements of three developmentally regulated genes of *D. melanogaster. Cell (Cambridge, Mass.)* **46,** 521–530.

Gasser, S. M., Laroche, T., Falquet, J., Boy de la Tour, E., and Laemmli, U. K. (1986). Metaphase chromosome structure—Involvement of topoisomerase II. *J. Mol. Biol.* **188,** 613–629.

Gehring, W. J. (1992). The homeobox in perspective. *Trends Biochem. Sci.* **17,** 277–280.

Giaver, G. N., and Wang, J. C. (1988). Supercoiling of intracellular DNA can occur in eukaryotic cells. *Cell (Cambridge, Mass.)* **55,** 849–856.

Hayes, J. J., and Tullius, T. D. (1989). The missing nucleoside experiment: A new technique to study the recognition of DNA by protein. *Biochemistry* **28,** 9521–9527.

Hofmann, M.-C., Jeltsch, W., Brecher, J., and Walt, H. (1989). Alkaline phosphatase isozymes in human testicular germ cell tumors, their precancerous stage, and three related cell lines. *Cancer Res.* **49,** 4696–4700.

Hozák, P., Hassan, A. B., Jackson, D. A., and Cook, P. R. (1993). Visualization of replication factories attached to a nucleoskeleton. *Cell (Cambridge, Mass.)* **73,** 361–373.

Jarman, A. P., and Higgs, D. R. (1988). Nuclear scaffold attachment sites in the human globin gene complexes. *EMBO J.* **7,** 3337–3344.

Jordan, G. (1987). At the heart of the nucleolus. *Nature* (*London*) **329,** 489–490.

Klehr, D., Maass, K., and Bode, J. (1991). Scaffold-attached regions from the human interferon beta domain can be used to enhance the stable expression of genes under the control of various promoters. *Biochemistry* **30,** 1264–1270.

Kochel, T. J., and Sinden, R. R. (1988). Analysis of trimethylpsoralen photoreactivity to Z-DNA provides a general in vivo assay for Z-DNA: Analysis of the hypersensitivity of (GT)n B-Z junctions. *Biotechniques* **6,** 532–543.

Kohwi-Shigematsu, T., and Kohwi, Y. (1990). Torsional stress stabilizes extended base unpairing in suppressor sites flanking immunoglobulin heavy chain enhancer. *Biochemistry* **29,** 9551–9560.

Kohwi-Shigematsu, T., and Kohwi, Y. (1992). Detection of non-B DNA structures at specific sites in supercoiled plasmid DNA and chromatin with haloacetaldehyde and diethyl pyrocarbonate. *In* "Methods in Enzymology" (D. M. J. Lilley and J. E. Dahlberg, eds.), Vol. 212, pp. 155–180. Academic Press, San Diego, CA.

Kohwi, Y., and Kohwi-Shigematsu, T. (1988). Magnesium ion-dependent, triple-helix structure formed by homopurine–homopyrimidine sequences in supercoiled plasmid DNA. *Proc. Natl. Acad. Sci. U.S.A.* **85,** 3781–3785.

Kohwi, Y., and Kohwi-Shigematsu, T. (1991). Altered gene expression correlates with DNA structure. *Genes Dev.* **5,** 2547–2554.

Liu, L. F., and Wang, J. C. (1987). Supercoiling of the DNA template during transcription. *Proc. Natl. Acad. Sci. U.S.A.* **84,** 7024–7027.

Ludérus, M. E. E., de Graaf, A., Mattia, E., den Blaauwen, J. L., Grande, M. A., de Jong, L., and van Driel, R. (1992). Binding of matrix attachment regions to lamin B1. *Cell* (*Cambridge, Mass.*) **70,** 949–959.

Mielke, C., Kohwi, Y., Kohwi-Shigematsu, T., and Bode, J. (1990). Hierarchical binding of DNA fragments derived from scaffold-attached regions: Correlation of properties *in vitro* and function *in vivo*. *Biochemistry* **29,** 7475–7485.

Nakagomi, K., Kohwi, Y., Dickinson, L. A., and Kohwi-Shigematsu, T. (1994). A novel DNA-binding motif in the nuclear matrix attachment DNA-binding protein SATB1. *Mol. Cell. Biol.* **14,** 1852–1860.

Phi-Van, L., von Kries, J. P., Ostertag, W., and Strätling, W. H. (1990). The chicken lysozyme 5′ matrix attachment region increases transcription from a heterologous promoter in heterologous cells and dampens position effects on the expression of transfected genes. *Mol. Cell. Biol.* **10,** 2302–2307.

Razin, S. V., Kekelidge, M. G., Lukanidin, E. M., Scherrer, K., and Georgiev, G. O. (1986). Replication origins are attached to the nuclear skeleton. *Nucleic Acids Res.* **14,** 8189–8207.

Romig, H., Fackelmayer, F. O., Renz, A., Ramsperger, U., and Richter, A. (1992). Characterization of SAF-A, a novel nuclear DNA binding protein from HeLa cells with high affinity for nuclear matrix/scaffold attachment elements. *EMBO J.* **11,** 3431–3440.

Smith, K. A. (1988). Interleukin-2: Inception, impact, and implications. *Science* **240,** 1169–1176.

Sperry, A. O., Blasquez, V. C., and Garrard, W. T. (1989). Dysfunction of chromosomal loop attachment sites: Illegitimate recombination linked to matrix association regions and topoisomerase II. *Proc. Natl. Acad. Sci. U.S.A.* **86,** 5497–5501.

Stief, A., Winter, D. M., Strätling, W. H., and Sippel, A. E. (1989). A nuclear DNA attachment element mediates elevated and position-independent gene activity. *Nature* (*London*) **341,** 343–345.

Stockhaus, J., Eckes, P., Blau, A., Schell, J., and Willmitzer, L. (1987). Organ-specific and dosage-dependent expression of a leaf-stem specific gene from potato after tagging and transfer into potato and tobacco plants. *Nucleic Acids Res.* **15,** 3470–3491.

Umek, R. M., and Kowalski, D. (1988). The ease of DNA unwinding as a determinant of initiation at yeast replication origins. *Cell* (*Cambridge, Mass.*) **52,** 559–567.

von Kries, J. P., Buck, F., and Strätling, W. H. (1994). Chicken MAR binding protein p120 is identical to human heterogeneous nuclear ribonucleoprotein (hnRNP) U. *Nucleic Acids Res.* **22,** 1215–1220.

Wang, B., Dickinson, L. A., Koivunen, E., Ruoslahti, E., and Kohwi-Shigematsu, T. (1995). A novel matrix attachment region DNA binding motif identified using a random phage peptide library. *J. Biol. Chem.* **270,** 23239–23242.

Weiher, H., Konig, M., and Gruss, P. (1983). Multiple point mutations affecting the simian virus 40 enhancer. *Science* **219,** 626–631.

Williams, T. M., Burlein, J. E., Ogden, S., Kricka, L. J., and Kant, J. A. (1989). Advantages of firefly luciferase as a reporter gene: Application to the interleukin-2 gene promoter. *Anal. Biochem.* **176,** 28–32.

Wu, H.-Y., Shyy, S., Wang, J. C., and Liu, L. F. (1988). Transcription generates positively and negatively supercoiled domains in the template. *Cell* (*Cambridge, Mass.*) **53,** 433–440.

Yanagisawa, J., Ado, J., Nakayama, J., Kohwi, Y., and Kohwi-Shigematsu, T. (1996). A matrix attachment region (MAR)-binding activity due to a p114 kilodalton protein is found only in human breast carcinomas and not in normal and benign breast disease tissues. *Cancer Res.* **56,** 457–462.

Yu, J., Bock, J. H., Slightom, J. L., and Villeponteau, B. (1994). A 5′ beta-globin matrix attachment region and the polyoma enhancer together confer position-independent transcription. *Gene* **139,** 139–145.

5

The Cyclin/Cyclin-Dependent Kinase (cdk) Complex: Regulation of Cell Cycle Progression and Nuclear Disassembly

R. CURTIS BIRD

Department of Pathobiology
Auburn University
Auburn, Alabama

I. INTRODUCTION

Mitosis and nuclear division in many ways embody the essence of cell cycle regulation and control. Placed as they are, at the end of each cell cycle, mitosis and nuclear division represent the culminating reorganization of the nucleus, including both the DNA and the nuclear matrix, followed by division of the genome and the remaining cellular components. We are just beginning to understand the mechanisms by which control is exerted. Complex biochemical mechanisms regulate these processes, ensuring first that the genome has been completely replicated and, second, that there are no chromosome breaks. Chromosome condensation then ensues as mitosis begins with the breakdown of the nuclear membrane. Unique organelles such as the Golgi apparatus fragment in preparation for division, while membrane traffic as well as synthetic processes such as transcription and translation are transiently inhibited (Moreno and Nurse, 1990; Warren, 1993). Vast cytoskeletal reorganization within the cytoplasm resculpts the cell into a transient mitotic figure including a newly constructed mitotic spindle that divides the chromatids into each new cell. Very rapidly, the cell returns to its premitotic morphology with the reformation of the nuclear membrane, the disassembly of the mitotic spindle, and the decondensation of the chromatin and nuclear matrix. As the cytoskeleton reforms and the cells return to their premitotic morphology, a new cell cycle ensues (Mitchison, 1971; Beach *et al.,* 1988).

Regulation of cell proliferation, including the events just described, is largely the province of three distinct classes of genes: the dominant oncogenes, the tumor suppressor genes, and the cyclin/cdk gene families. The dominant oncogenes encode representatives of many of the components of the signal transduction system and are most closely associated with the transition from quiescence (G_0) to active proliferation. However, evidence demonstrates that expression of at least some of these genes is required throughout the cell cycle. For the oncogenes c-*fos* and c-*myc* at least, proliferation through the continuous cell cycle cannot continue if expression is transiently inhibited with antisense oligonucleotides (Bird *et al.,* 1990; Pai and Bird, 1992, 1994a,b).

The tumor suppressor genes are thought to exert an opposing influence on proliferation compared to the dominant oncogenes (Cobrinik *et al.,* 1992; Hamel *et al.,* 1992). Tumor suppressor genes inhibit cell proliferation, and inactivation or deletion of both copies results in unregulated growth. As principal examples, the normal function of the retinoblastoma (Rb) gene product is thought to consist of pushing cells to retire from the cell cycle into a quiescent or differentiation pathway during G_1 phase. The normal function of the p53 gene product is thought to consist of pushing cells into an apoptotic pathway in the event that unreplicated

DNA persists after S phase (Vogelstein and Kinzler, 1992; Marin *et al.*, 1994). Tumor suppressor gene absence or activation results in a failure to exit the cell cycle in the case of Rb or a failure to enter an apoptotic pathway in the case of p53.

The cyclin/cdk gene families include a large and growing group of paired polypeptides which encode a cdk (cyclin-dependent kinase) and a cell cycle phase-specific cyclin cofactor (Nurse, 1990; Moreno and Nurse, 1990). The different cdks (cdk1–cdk8) are generally present throughout the cell cycle, while the cyclins (cyclin A–cyclin H) are transiently synthesized and then degraded, facilitating activation of the appropriate cdk during the appropriate cell cycle phase (van den Heuvel and Harlow, 1993; Pines, 1993; Hunter and Pines, 1994; Nurse, 1994; Morgan, 1995). These activities, which are potentiated by several specific phosphorylation and dephosphorylation events, are responsible for driving cells from one phase of the cell cycle to the next and thus represent the fundamental internal signals which drive cell cycle progression. The activities of both the tumor suppressors and the dominant oncogenes are thought to be linked through the cyclin/cdk complexes. Dominant oncogenes are linked through the initiation of cell growth through direct initiation of transcription of cyclin D_1, which then moves cells into G_1 phase from quiescence (Pines, 1993). The products of both the Rb and p53 tumor suppressor genes interact directly with cyclin/cdk complexes to down-modulate proliferative activities associated with this complex (Cobrinik *et al.*, 1992; Vogelstein and Kinzler, 1992).

II. MECHANISMS OF CYCLIN/cdk FUNCTION

In proliferating eukaryotic cells, a complex network of regulatory biochemical pathways coordinates the timing and execution of the major transitions of the cell cycle, including the transition from G_1 phase, the initiation of DNA synthesis, and the induction of mitosis. During G_1 phase of the cell division cycle, growth factors regulate the accumulation of critical gene products that are required for entry into S phase (Pardee, 1989). During this restriction point, known as START in yeast, cells are most sensitive to the influence of external cues which lead them either to proliferate or to exit the cell cycle and become quiescent, senescent, or differentiated (Pringle and Hartwell, 1981; Jacobs *et al.*, 1985; Lumpkin *et al.*, 1986; Pardee, 1989; Rittling *et al.*, 1986).

Genetic investigation of the eukaryotic cell cycle has revealed many genes that either are essential for cell division or influence its regulation, including the regulatory subunit cyclin gene family and the catalytic subunit cdk gene family (Beach *et al.*, 1982; Nurse, 1975; Nurse *et al.*, 1976). The prototypical member of the cdk family, $p34^{cdc2}$ (cdk1) in *Schizosaccharomyces pombe* (Beach *et al.*, 1982), forms complexes with cyclin A or B and acts both in G_1, prior to DNA replication, and in G_2 at the initiation of mitosis (Hunt, 1989; Murray and Kirschner, 1989; Nurse, 1990; Nurse and Bissett, 1981). In the budding yeast *Saccharomyces cerevisiae,* G_1 cyclins (CLN1, CLN2, and CLN3) interact with $p34^{CDC28}$ (Reed and Nasmyth, 1981) to ensure passage through START and commitment to S phase (Hadwiger *et al.*, 1989; Nash *et al.*, 1988; Richardson *et al.*, 1989). Regulation of this transition seems to be exerted primarily at the level of transcription of the G_1 cyclins (McKinney and Heintz, 1991), although posttranslational events are also required.

In mammalian cells, regulation of the G_1/S transition is controlled by the activities of cyclin D and cyclin E in complex association with cdk2, cdk4, cdk5, and cdk6. Cyclin E is first expressed just prior to the G_1/S transition (Lew *et al.*, 1991) and associates with cdk2 (Dulic *et al.*, 1992; Koff *et al.*, 1991). Formation and activation of a cyclin E/cdk2 complex occurs during G_1 phase of the cell cycle (Koff *et al.*, 1992). Cyclin D types (cyclin D_1–cyclin D_3) can associate with cdk2, cdk4, and cdk6, activating the kinase complexes during G_1, followed by a decline in S phase. The D-type cyclins initiate phosphorylation of Rb and the Rb-like protein p107 but not histone H1 *in vitro* (Dowdy *et al.*, 1993; Ewen *et al.*, 1993; Matsushime *et al.*, 1992). In contrast to the cyclin B/cdk1 kinase complex, the timing of expression and specificity of the substrate of various cyclin D/cdk kinase complexes (Matsushime *et al.*, 1992) suggests that it may be indirectly involved in controlling gene expression during G_1 phase by releasing the transcription factor E2F via phosphorylation of Rb protein, whose phosphorylation appears to be necessary for the G_1/S transition (Buchkovich *et al.*, 1989; DeCaprio *et al.*, 1989; Ludlow *et al.*, 1990). D-type cyclin synthesis appears to be tied directly to growth factor stimulation required to initiate proliferation of quiescent cells. Thus, growth factor stimulation can be tied directly to transcription activation through the activation of several dominant oncogene products that result in the synthesis of cyclin D and activation of cdk complexes.

It is becoming clear that ubiquitin-dependent degradation plays an important role in several facets of cell cycle regulation. Of primary importance is the rapid degradation of each cyclin subtype which is accompanied by release and inactivation of its bound cdk (Barinaga, 1995). The p27 cdk inhibitor is also degraded by the ubiquitin pathway and so has an impact on cell cycle regulation (Pagano *et al.,* 1995). However, ubiquitin-mediated degradation has been shown to play an important role in nuclear events associated with particular phases of the cell cycle not dependent on cyclin degradation. Anaphase initiation has been shown to be triggered by proteolysis rather than by the inactivation of cyclin B/cdk1 (Holloway *et al.,* 1993). Anaphase and chromosome pair separation are blocked by inhibiting ubiquitin addition but are unaffected by inhibition of cyclin ubiquitination. In addition to imposing a regulatory activity on cyclin/cdk complexes, the ubiquitin system has itself been shown to include substrates of cyclin/cdk. The first enzyme in the ubiquitin–protein complex formation cascade, E1, has been identified as a cdk1 substrate linking ubiquitination generally to cell cycle regulation (Nagai *et al.,* 1995). A component of the third enzyme in this cascade has been shown to play a regulatory role in anaphase initiation (reviewed by Barinaga, 1995). Thus, ubiquitin-mediated protein degradation must also be considered an important regulator in timing the events of the cell cycle, including nuclear events, which is intimately associated with cyclin/cdk regulation/activity.

Both cyclin and cdk proteins have been determined to reside principally, though not exclusively, in the nucleus, at least when they are thought to be active (Riabowol *et al.,* 1989; Maridor *et al.,* 1993). In the case of cyclin A, which is localized to the nucleus throughout the cell cycle, nuclear localization requires the C-terminal 15 amino acids and an intact cyclin box domain. Cyclin B, in contrast, is located largely in the cytoplasm until just prior to mitosis, when it migrates to the nucleus (Pines and Hunter, 1991; Gallant and Nigg, 1992; Bailly *et al.,* 1992; Ookata *et al.,* 1992). Also, cdk1 is largely nuclear throughout interphase and associates with centrosomes, in addition to being located in the nuclear and cytoplasmic regions during mitosis.

Expression of certain members of the cdk/cyclin gene family is subject to phosphorylation-sensitive events. Inhibition of the protein phosphatase 1 and 2A (PP1 and PP2A) enzyme families with okadaic acid has been utilized to study the function of dependent gene expression during the cell cycle (You and Bird, 1995). Experiments were designed to detect mRNA levels of the retinoblastoma tumor suppressor gene (Rb), cdk1, cyclin A, and cyclin B genes in HeLa cells. Data demonstrate that okadaic acid selectively elevates mRNA levels for this group of gene products that have been shown to associate as a complex with the transcription factor E2F and that are needed for successful completion of mitosis. Many of the genes whose promoters contain E2F-binding sites are required for cell cycle progression, particularly DNA synthesis (e.g., dihydrofolate reductase, thymidine kinase, DNA polymerase α, cdk1, and B-Myb) (La Thangue, 1994). Such inducible expression has not been paid much attention until recently. Overexpression of members of the cdk/cyclin gene families, and their regulatory cofactors, have been observed in tumor specimens and derived cell lines from a variety of different types of neoplasms (Okayama *et al.,* 1991; Koff *et al.,* 1993; Serrano *et al.,* 1993; Zhang *et al.,* 1993; Khatib *et al.,* 1993; Jiang *et al.,* 1993; Xiong *et al.,* 1993; Leach *et al.,* 1993; de Boer *et al.,* 1993; Bianchi *et al.,* 1993; Tsuruta *et al.,* 1993; Motokura and Arnold, 1993; Keyomarsi *et al.,* 1993; Kamb *et al.,* 1994; Nobori *et al.,* 1994; Peters, 1994; Michalides *et al.,* 1995). Locations of the members of the cdk gene family have also been mapped to chromosomal regions known to exhibit loss of heterozygosity in human cancers (Bullrich *et al.,* 1995). If there exists a central convergent point regulating cell proliferation in the cause of cancer, then it has been suggested that the cyclin/cdk families may represent such a universal target (Steel, 1994).

The cdk/cyclin complexes are themselves the subject of negative or inhibitory activities which appear to fine-tune the timing of cdk/cyclin kinase activity. The p21 protein is a tyrosine phosphatase which has been variously named the cdk interacting protein (Cip1), the cyclin-dependent kinase interactor (Cdi1), and the product of the wild-type p53-activated fragment 1 (*WAF1*) gene (Harper *et al.,* 1993; Gyuris *et al.,* 1993; El-Deiry *et al.,* 1993). The p21 protein appears to be a downstream effector of p53 activity but can be uncoupled from the apoptotic pathway. The p21 protein can also be subverted in its proliferation suppressive activity by c-*myc* activation (Steinman *et al.,* 1994).

The activity of p21 is differentially regulated, depending on the growth stimulatory/inhibitory environment. TGF-β induces activation of cdk2/cyclin E and suppresses p21 in C3H 10T1/2 fibroblasts, which results in proliferation stimulation (Ravitz *et al.,* 1995). However, in colon cancer cell lines which are growth inhibited by TGF-β, p21 activity is induced (Li *et al.,* 1995). The p21 protein has also been associated with cell cycle arrest in the transition to the terminally differentiated state (Halevy *et al.,* 1995). Cell cycle arrest under these conditions requires active hypophosphorylated Rb protein. Rb is normally inactivated by phosphorylation by cdk/cyclin complexes during G_1 phase. The cdk/cyclins are suppressed, and further cell cycle progression is also suppressed by the action of p21. Expression of p21 was induced following expression of the muscle-specific differentiation factor encoded by the *MyoD* gene. This suggests that there is a mutually exclusive mechanism which precludes further cell cycle progression and proliferation after successful induction of the signal to differentiate in a variety of tissue types (Parker *et al.,* 1995). *MyoD* is normally expressed in myoblasts but does not induce differentiation until the cells retire from the cell cycle to G_0 phase (Skapek *et al.,* 1995). Transfection of cyclin D1, but not other cyclins, results in phosphorylation of *MyoD* and inhibits *MyoD trans*-activation of muscle-specific genes. Transfection of p21, in contrast, enhances muscle-specific gene expression even in the absence of the low-serum mitogen signal to differentiate. This suggests a further complication in which *MyoD* expression is initiated but its activity is suppressed by cdk/cyclin. This is followed by retirement from the cell cycle induced by p21 activity through suppression of cdk/cyclin. A number of other cdk inhibitors have also been identified which inhibit the activity of various members of the cdk family (Pines, 1994). Their function seems to be to inhibit cdk activation at the time of cyclin synthesis and binding until the appropriate point in the cell cycle. While their complexity is intriguing, further efforts will be required to determine their role in cell cycle regulation.

Besides these transcription regulatory activities and associated polypeptides, a variety of other target molecules and structural elements, which are thought to be phosphorylated by cyclin/cdk, have been identified (Nurse, 1990; Moreno and Nurse, 1990; Draetta, 1990). These include other cyclins, histone H1, RNA polymerase II, SV40 T antigen, the tumor suppressor proteins Rb and p53, nucleolin, lamins, translation elongation factors EF-1β and EF-1γ, and the oncogenes c-*abl* and pp60$^{c\text{-}src}$ (reviewed in Draetta, 1990; Moreno and Nurse, 1990). Such targets could account for the breakdown and reorganization of organelles, as well as the inhibition of synthetic processes which are suppressed during mitosis. It has been suggested that the reason for such a large number of diverse cdk1 substrates has to do with the nature of the target sequence (reviewed in Moreno and Nurse, 1990). The target motif is predicted to form a β-turn which could bind the minor groove of DNA. Phosphorylation would alter this conformation, thereby inhibiting binding. Support has been lent to this model by the fact that the target sequence is most frequently observed in putative transcription factors near other recognized active sites. Phosphorylation could inhibit binding and thus inhibit expression, as well as facilitate chromosome condensation, resulting in the complete reorganization of the nucleus, particularly the nuclear matrix, during mitosis (Moreno and Nurse, 1990).

The other events in mitosis which seem to be triggered by cdk1 activity are also amenable to speculation. Membrane vesicle breakdown and organelle fragmentation ensure partition of portions of these organelles to each daughter cell. Massive reorganization of the cytoskeleton provides an abundance of unpolymerized tubulin dimers and accessory proteins for the transient assembly of the mitotic spindle. The details of these events and their regulation remain unclear; however, further details are emerging rapidly. It should not be long before we have a coherent model which explains most of the known aspects of mitotic regulation.

III. TARGET SUBSTRATES OF CYCLIN/cdk

A. Nuclear Lamins and Nuclear Envelope Breakdown

Lamins are intermediate filament proteins which form the subunits of the polymeric nuclear lamina. This is a fibrous cytoskeletal network which underlies the nuclear membrane of interphase nuclei. At the onset of mitosis, as the nuclear membrane breaks down, the nuclear lamina disassembles and monomeric A-type lamins are released free into the cytoplasm, while B-type lamins are released but are associated with membranes. These changes in assembly occur coincident with a general increase in cellular phosphorylation levels and an increase in lamin phosphorylation in particular (Lohka *et al.,* 1987; Maller and Smith, 1985; Moreno and Nurse, 1990; Enoch *et al.,* 1991). Lamin phosphorylation occurs as mitosis begins coincident with a breakdown of the interphase ring structure and its replacement with punctate staining dispersed throughout the cytoplasm (Peter *et al.,* 1990). This has been shown to be the result of a mitosis-specific lamin kinase which coprecipitates from yeast extracts with cdk1 and displays temperature sensitivity in cdc2 mutants. Consistent with a role as a mitotic initiator, cdk1 has been proposed as the lamin kinase which induces nuclear lamina disassembly (Peter *et al.,* 1990; Enoch *et al.,* 1991). A cell-free extract which assembles and then disassembles lamin C in response to cyclin/cdk activity has been described (Ward and Kirschner, 1990), and lamin B has been shown to be phosphorylated by cdk1 at the same sites *in vitro* as *in vivo* (Peter *et al.,* 1990). Cyclin/cdk phosphorylates specific sites on lamin C, which is also a substrate of other kinase activities. Mutation of conserved serine phosphorylation sites in both lamin C and lamin A prevents cyclin/cdk phosphorylation and prevents nuclear lamina disassembly at mitosis (Heald and McKeon, 1990).

B. Cytoskeleton Rearrangement and Mitosis

1. Microtubules

The cytoskeleton undergoes enormous reorganization during mitosis. The cytoskeletal framework is largely disassembled during late G_2 phase and reformed into the mitotic spindle at the initiation of mitosis. Microtubules, which form one of the primary structural elements of the cytoskeleton, are affected in two general ways during this period. First, a general disassembly of microtubules to component tubulin occurs, followed by nucleation of microtubule polymerization at the poles of the new mitotic spindle (Alfa *et al.,* 1990). As a major component of the reorganized mitotic spindle, cytoskeletal components like microtubules represent a likely target of cyclin/cdk activity. Addition of cdk1 to preparations of tubulin and accessory proteins has been shown to alter microtubule polymerization dynamics in a manner which mimics microtubule dynamics during metaphase (Verde *et al.,* 1990; Buendia *et al.,* 1992; Ohta *et al.,* 1993). This has been shown to be due most likely to an indirect effect, as cdk1 does not phosphorylate the tubulin subunits and does not appear to have this effect in purified tubulin preparations. In addition, cdk2 and cdk5 have been shown to phosphorylate microtubule accessory proteins *tau* (τ) (Baumann *et al.,* 1993) and possibly MAP-1B, -2, and -4 (Pelech *et al.,* 1990; Tombes *et al.,* 1991). Also, cdk1/cyclin B has been shown to be directly associated with microtubules during interphase as well as mitotic asters, and may be required to maintain the integrity of the spindle during mitosis and meiosis (Ookata *et al.,* 1993, 1995; Kubiak *et al.,* 1993). This interaction appears to be mediated through direct interaction with the proline-rich carboxyterminal region of MAP-4. In addition, cdk1/cyclin B can phosphorylate MAP-4, inhibiting its microtubule-stabilizing activity. There is also evidence that cdk1/cyclin B may interfere with cyclic adenosine monoyphosphate (cAMP)-dependent protein kinase association with MAP-2 (Keryer *et al.,* 1993).

2. *Vimentin and Desmin Intermediate Filaments*

Intermediate filaments undergo dynamic changes during mitosis which mirror in many ways the disassembly of cytoplasmic microtubules. Both vimentin and desmin intermediate filaments are depolymerized to cytoplasmic aggregates as cells enter mitosis. Study of vimentin intermediate filaments, particularly, has suggested a direct link between the activity of the cyclin/cdk complex family and intermediate filament assembly. The cdk1 complexes have been shown to phosphorylate vimentin at sites unique from other kinases (e.g., serine 55), and cdk1 has been shown to colocalize with phosphorylated vimentin (Tsujimura *et al.,* 1994). Desmin is also phosphorylated at specific sites during mitosis (Chou *et al.,* 1990). Immunofluorescence investigation has revealed that phosphorylated vimentin is present throughout the cytoplasm only during mitosis. Vimentin becomes dephosphorylated as the cells complete cytokinesis. Because the level of cdk1 activity has also been correlated with the disassembly of vimentin intermediate filaments, it has been suggested that cdk1 is directly responsible for disassembly of this intermediate filament class at the initiation of mitosis (Tsujimura *et al.,* 1994). The vimentin kinase has been shown to copurify with cdk1-like activities and to contain bona fide cdk1 in the preparations (Chou *et al.,* 1990). These data strongly suggest that intermediate filament proteins are mitotic substrates of cdk1 and are important mediators of the global reorganization of the cell architecture which occurs during cell division.

C. Regulation of Chromosome Condensation

1. *Histone H1*

Histone H1 was one of the first substrates of cdk1 to be identified. It was used principally as a convenient substrate of cdk1 *in vitro* initially but came to be appreciated for its role as a natural substrate of this enzyme complex (Moreno and Nurse, 1990). Histone H1 protein contains several sites which are phosphorylated by cdk1. The phosphorylation sites are located in two basic arms which interact with DNA between the nucleosomes (Langan *et al.,* 1989). It is believed that cdk1 phosphorylates Ser/Thr residues in Lys-Ser/Thr-Pro-X-Lys sequences in substrate histone H1 polypeptides. It has been suggested that such phosphorylation events could significantly affect chromosome condensation during mitosis by altering nucleosome packing (Bradbury *et al.,* 1974).

2. *RNA Polymerase II*

A general inhibition of transcription accompanies mitosis coincident with chromosome condensation and the onset of mitotic spindle formation. Such observations have led to the investigation of RNA polymerase activity during mitosis. RNA polymerase becomes highly phosphorylated on the carboxy-terminal domain, which consists of a conserved and highly repeated sequence containing a cdk1 phosphorylation site (Moreno and Nurse, 1990; Corden, 1990). This repeated sequences (26–52 copies of Ser-Pro-Thr-Ser-Pro-Ser-Tyr) has been predicted to form a β-turn conformation which binds DNA through intercalation of the tyrosine residues (Suzuki, 1990). It has been proposed that phosphorylation of these residues by cdk family members could alter the affinity of RNA polymerase II for DNA, causing dissociation from the transcribed strand and enhancing chromosome condensation (Moreno and Nurse, 1990). Evidence has shown a direct association between RNA polymerase II and cyclin/cdk, making members of these gene families subunits of the RNA polymerase II protein complex (Liao *et al.,* 1995).

The transcription factor IIH (TFIIH) complex contains a kinase activity which can bind and phosphorylate the carboxy-terminal domain of RNA polymerase II. This domain is essential to RNA polymerase, enabling it to respond to regulation by transcription factors. The kinase is composed of cdk7 and its cofactor cyclin H (Shiekhattar *et al.,* 1995). The TFIIH/cdk7/cyclin H complex contains a cdk-activating kinase which can phosphorylate other cdk/cyclin complexes, including both cdk1 and cdk2. Thus, evidence linking RNA polymerase and the cell cycle exists in at least two areas, suggesting involvement in the regulation of cell cycle progression. It has been suggested that the role of the cyclin/cdks in this interaction seems to be to enhance interaction between general transcription factors associated with RNA polymerase II and other promoter-specific transcription factors (O'Neill and O'Shea, 1995).

3. *RCC1, Repressor of Chromosome Condensation*

Eukaryotic cells possess a mechanism that effectively suppresses mitosis until the end of S phase. This mechanism prevents cell division in the presence of unreplicated DNA, thus ensuring that the genome is completely replicated prior to mitosis (Hartwell and Weinert, 1989; Nishimoto *et al.,* 1992; Murray, 1992). Once cells reach the end of S phase, this inhibitory activity is inactivated and mitosis can proceed once appropriate cdk/cyclin molecules are activated. This mechanism requires an exquisite sensitivity to even small amounts of unreplicated DNA and precise timing such that the mechanism is inactivated at the end of S phase. It has been proposed that such a signal must be part of the replicating DNA itself or of proteins associated with it (Dasso, 1993). Only one mutation, RCC1 (regulator of chromosome condensation), appears to be involved in mitotic suppression and also encodes a DNA-associated protein (Enoch *et al.,* 1992; Roberge, 1992). Because RCC1 mutations also affect a wide variety of other cellular processes, such as nuclear assembly, transcription regulation, posttranscriptional processing of pre-rRNA, and mRNA transport, explanation of the mechanisms describing its specific effects on cell cycle progression has been difficult.

It has been shown, however, that in RCC1 mutants inhibition of DNA replication occurs as a result of defective chromosome condensation (Nishimoto *et al.,* 1978; Eilen *et al.,* 1980). RCC1 encodes a highly conserved DNA-binding protein which is lost from the DNA at the time of mitosis. RCC1 loss also seems to uncouple the completion of DNA replication from the mechanism which activates cdk1 (Nishitani *et al.,* 1991). Analysis of other mutations induced in this gene allowed their recovery during functional mutant screens. Analysis of these mutations also suggests that they are analogous, reflecting a conservation of function between RCC1 homologs from different species (Dasso, 1993). Additionally, RCC1 seems to function as a guanine nucleotide exchange factor for a closely associated Ras-related protein called Ran (Coutavas *et al.,* 1993). Both proteins are found in abundance in cultured cells, with Ran being present at approximately a 25-fold molar excess to RCC1.

Two potential functions have been proposed for RCC1 (Murray, 1992; Dasso, 1993). RCC1 could function as a signal transduction factor which detects unreplicated DNA and alters nucleotide exchange activity in association with Ran in response. If Ran functions as other Ras-like proteins, it would then interact with a GAP cofactor, enabling inhibition of mitosis. Once S phase is complete, RCC1 activity would diminish, allowing inactive Ran to accumulate and replication to cease. Alternatively, RCC1 could function as part of the chromatin structure. Loss of RCC1 from the chromatin might result in structural defects which affect chromatin function. The mechanisms by which this hypothesis might operate, however, are unknown. Because chromatin rearrangements and nuclear matrix rearrangements are such important parts of mitosis, understanding their regulation and cdk control of their activity will prove very important in understanding how mitosis is delayed until S phase is complete. It will also prove important in understanding regulation of reorganization of the chromatin during mitosis.

4. Mitotic Reorganization of Nucleolar Chromatin

Ribosome assembly occurs in the nucleolus during interphase and is tied to the presence of this structure, as the nucleolus is the site of rDNA transcription. During mitosis, when the nucleus has undergone reorganization to form a spindle apparatus, the nucleolus becomes inactive and is disassembled as the multiple chromosomal regions which compose the nucleolar organizer disaggregate. This process is also accompanied by chromosome condensation. There is a subset of nucleolar proteins which remain associated with the nucleolar organizer DNA regions during mitosis and have been suggested as possible regulatory components controlling this dynamic disassembly and assembly during and after mitosis (Belenguer *et al.,* 1990). Included in these proteins is nucleolin, which is involved in nucleolar chromatin condensation and the packaging of pre-rRNA (Belenguer *et al.,* 1990; Bugler *et al.,* 1987; Gas *et al.,* 1985; Herrera and Olson, 1986; Lapeyre *et al.,* 1987; Erard *et al.,* 1988; Olson and Thompson, 1983). Phosphorylation of nucleolin occurs at times when rRNA transcription is active during cell proliferation. When cells become quiescent on attaining confluence, rRNA transcription drops to approximately 5% of proliferating cell levels and nucleolin becomes relatively dephosphorylated (Belenguer *et al.,* 1989; Geahlen and Harrison, 1984; Schneider *et al.,* 1986; Suzuki *et al.,* 1985, 1987; Caizergues-Ferrer *et al.,* 1987). The sites of phosphorylation conform to a repeated TPXK motif which is similar to the phosphorylation motif in histone H1 utilized by cdk1 (Wells and McBride, 1986; Shenoy *et al.,* 1989; Hayes and Nurse, 1989).

The increase in cdk1 activity at the onset of mitosis leads to chromosome condensation and localization of the cdk1 cofactor cyclin in the nucleolus (Bradbury *et al.,* 1973; Westendorf *et al.,* 1989). Due to its location in the nucleolus and because nucleolin contains the target phosphorylation motif for cdk1 as well as its tight association with chromatin, nucleolin was investigated as a potential cdk1 substrate responsible for nuclear rearrangements at mitosis (Belenguer *et al.,* 1990). Nucleolin was found to be specifically phosphorylated during mitosis on threonine by cdk1 both *in vitro* and *in vivo.* In addition, casein kinase II, which also phosphorylates nucleolin, was active during interphase and not during mitosis and phosphorylated serine residues in nucleolin. Thus, cdk1 specifically induces the active form of nucleolin at a period during the cell cycle, during mitosis, which is specifically associated with chromosome condensation and nuclear rearrangement.

D. Caldesmon

Caldesmon is an accessory protein of the cytoskeleton (Bachs *et al.,* 1990; Matsumura and Yamashiro, 1993). It is located in most nuclear compartments, including the nuclear matrix but excluding the nucleolus. Caldesmon is involved in regulation of actomyosin adenosine triphosphatase (ATPase) and cross-link formation, as well as in actin filament polymerization and stability. In nonmuscle cells, caldesmon functions in the regulation of cell motility and cytoskeleton reorganization during mitosis. Caldesmon is a substrate of cdk, becoming phosphorylated during mitosis (Yamashiro and Matsumura, 1991; Yamashiro *et al.,* 1990, 1991; Mak *et al.,* 1991; Yamakita *et al.,* 1992). Late in mitosis, during cytokinesis as actin filaments appear in the cleavage furrows, caldesmon migrates from a diffuse cytoplasmic distribution to colocalize with the actin filaments (Hosoya *et al.,* 1993). Dephosphorylation occurs at the time caldesmon becomes actin associated. Phosphorylation by cdk1 may be required at the initiation of mitosis to allow dissociation of caldesmon from actin filaments, facilitating their assembly and reorganization during spindle assembly (Yamashiro and Matsumura, 1991; Yamashiro *et al.,* 1990, 1991). This function suggests that cdk1 phosphorylation of caldesmon plays a functional role in mitotic cell rounding and cytokinesis.

E. Retinoblastoma Protein Family and Cell Cycle Progression

The retinoblastoma protein (Rb) was first identified as a mutation associated with the early childhood tumor retinoblastoma. The gene encodes a 105,000-Da, nuclear protein which can be phosphorylated at multiple sites with an apparent molecular size of 115,000–117,000 Da (Hensey *et al.,* 1994). The Rb protein is functional when hypophosphorylated and inactive in the hyperphosphorylated state (Buchkovich *et al.,* 1989; DeCaprio *et al.,* 1989). The Rb protein undergoes changes in phosphorylation associated with transitions in cell cycle phase (Ewen *et al.,* 1993; Ludlow *et al.,* 1990; reviewed by Lewin, 1990; Draetta, 1990). Phosphatases 1 (PP1) and 2A (PP2A) are two of the four principal serine/threonine phosphatases involved in reversal of such regulatory posttranslational modifications. Several PP1 homologs are required for exit from mitosis in yeast (Doonan and Morris, 1989; Ohkura *et al.*, 1989; Booher and Beach, 1989), and PP2A has been shown to inhibit cdk1 activity (Félix *et al.*, 1990; Lee *et al.*, 1991; Kinoshita *et al.*, 1990). Both PP1 and PP2A are thought to exert a direct suppressive influence over cell cycle progression through their activation of Rb (Hamel *et al.*, 1992; Cobrinik *et al.*, 1992).

The Rb protein and the Rb-related p107 protein are known to be active during G_1 phase and to bind cdk/cyclin complexes, particularly those containing cyclin D_1, as well as members of the transcription factor E2F family during this phase of the cell cycle (Shan *et al.*, 1992; Peeper *et al.*, 1993; Ewen *et al.*, 1993). This Rb-binding activity sequesters these factors, rendering them inactive during G_1 phase until Rb is phosphorylated by cyclin/cdk. Through this interaction, Rb is known to be an important suppressor of cell cycle progression and transcription, particularly of genes transcribed by RNA polymerase II (Nevins, 1992). However, because Rb has been associated with the upstream binding factor (UBF) which is required for transcription by RNA polymerase I, it appears that Rb can extend its inhibitory activity to these transcripts as well (Cavanaugh *et al.*, 1995). Rb accumulates in the nucleoli of differentiating monocyte-like cells. This Rb accumulation correlates with the inhibition of rDNA transcription and inhibits *in vitro* transcription by RNA polymerase I (see also (Section III,E). Rb is able to suppress transcription from these genes by sequestering UBF, a molecule with which Rb is tightly associated *in vivo,* thus offering an additional pathway by which Rb can inhibit proliferation.

The principal pathway that employs this proliferation-suppressive mechanism appears to be cell differentiation which usually requires permanent withdrawal from the cell cycle. Rb activity is required for skeletal muscle myoblasts to complete terminal differentiation to myotubes (Halevy *et al.*, 1995). This process is inhibited by cdk activity through its inactivation of Rb or by forced induction of cyclin D_1, although not other cyclins (Skapek *et al.*, 1995). *MyoD,* which induces myoblast differentiation (Buckingham, 1992), also induces the activity of the cdk inhibitor p21 (Halevy *et al.*, 1995). This induction is independent of tumor suppressor p53 activity, which is also known to induce p21 (Halevy *et al.*, 1995; Parker *et al.*, 1995). Thus, a simple mechanism coupling cell cycle arrest and differentiation can be constructed. The inducer of differentiation is able to induce withdrawal from the cell cycle by stimulation of p21 prior to induction of differentiation-specific genes associated with terminal differentiation of skeletal muscle.

There is also evidence that Rb-related p107 does not bind equally well to all members of the cyclin/cdk gene families or even to cyclin or cdk alone (Peeper *et al.,* 1993). Binding of cyclin A/cdk complexes (including either cdk1 or cdk2) to p107 requires both polypeptides. Only cyclin A/cdk binds and phosphorylates p107; cyclin B/cdk does not. This provides evidence to support the common assumption that not only are the different cyclin/cdk complexes synthesized and active during different cell cycle phases, but that they differ qualitatively in their associations and in their substrates.

While the mechanisms by which protein phosphorylation influences gene expression are not clearly understood, two models, the protein–protein contact and chromatin accessibility models, have been developed (Chen *et al.,* 1989; Mahadevan and Willis, 1990; Mahadevan *et al.,* 1991). It has been proposed that the activating signal must be delivered to specific regulatory proteins (Prywes *et al.,* 1988), resulting in localized chromatin relaxation (Chen and Allfrey, 1987). Cell cycle regulatory proteins such as Rb may integrate the signal initiated by okadaic acid stimulation; it has been speculated that this could occur by stabilization of existing phosphorylation events. Such events could result in release of inhibition of transcription factors, allowing transcription of cellular genes (Hunter and Karin, 1992; Hamel *et al.,* 1992; Cobrinik *et al.,* 1992; McKinney and Heintz, 1991).

1. Retinoblastoma as Trans-Activator of Transcription

The *Rb* gene plays a central role in the control of cell division cycle and differentiation (Klein, 1987; Weinberg, 1989, 1991). Evidence indicates reversal of tumorigenicity resulting in senescence on introduction of wild-type *Rb* into a variety of tumor cells lacking Rb protein (Bookstein *et al.,* 1990; Uzvolgyi *et al.,* 1991). The role of its DNA-binding domain in transcriptional regulation, the many cell cycle regulatory proteins it is known to bind (including products of members of the cyclin, cdk, and E2F gene families), and its nuclear location suggest a central regulatory role for *Rb* (Lee *et al.,* 1987). The interaction with DNA appears to be sequence specific (Wang *et al.,* 1990). Rb expression levels are essentially constant during the cell cycle, except for some enhancement in accumulation of the hypophosphorylated form which occurs as cells exit the cell cycle to differentiate (Akiyama and Toyoshima, 1990). Rb is primarily regulated by posttranslational modifications, the most important of which is the phosphorylation state (Chen *et al.,* 1989; Mihara *et al.,* 1989; Xu *et al.,* 1989; Furukawa *et al.,* 1990; Ludlow *et al.,* 1990; Stein *et al.,* 1990; Goodrich *et al.,* 1991). The hypophosphorylated form of Rb which is present during G_1 phase is the active form.

Naturally occurring mutations which inactivate the *Rb* gene have been found to be clustered in a region of the protein known as the pocket domain (Weinberg, 1991; La Thangue, 1994). Few mutations occur outside this region, and mutations within the pocket domain interfere with Rb binding to viral transforming antigens/proteins. Pocket mutations also lose the ability to suppress cell cycle progression, suggesting that Rb functions to sequester proliferation-promoting activities. Members of the E2F transcription factor gene family appear to be the principal targets bound by Rb at this site in normal cells. Although differing somewhat in the timing of their interactions, both Rb-related proteins p107 and p130 also appear to function in this manner (Shirodkar *et al.,* 1992; Schwarz *et al.,* 1993; Cobrinik *et al.,* 1993). This association links cyclin/cdk to E2F through mutual association with Rb, thus allowing potential regulation of E2F availability by differential regulation of Rb phosphorylation and therefore E2F binding.

We have observed proportional overexpression of Rb in response to the level of overexpression of c-*fos* in clonal lines of transfected cells. Such evidence indicates that Rb may be functioning as a proportional brake which balances the stimulatory effect of c-*fos* overexpression (Pai and Bird, 1994a,b). Rb may repress c-*fos* transcription through interaction with a discrete element on the c-*fos* promoter, utilizing a common regulatory motif termed the retinoblastoma control element (RCE) (Kim *et al.,* 1991, 1992; Pietenpol *et al.,* 1990; Robbins *et al.,* 1990). The c-*fos* RCE has also been shown to confer transcriptional repression by Rb on a heterologous promoter (Robbins *et al.,* 1990). Deletion analysis has further indicated that the 5′ RCE located between -97 and -86 in the c-*fos* gene is sufficient to allow regulation of c-*fos* by Rb (Kim *et al.,* 1991). Evidence has identified a single binding site (5′-GCGCCACCCC-3′) for the retinoblastoma control proteins (RCPs) within the c-*fos* RCE. RCPs and Rb are thought to be involved in regulatory control mechanisms that govern the expression of growth response genes such as c-*fos,* c-*myc,* and TGF-β_1 (Udvadia *et al.,* 1992).

Both tumor suppressor proteins Rb and p53 become hyperphosphorylated when cells are treated with the PP1/2A inhibitor okadaic acid (Yatsunami *et al.,* 1993). Treatment with okadaic acid also results in a selective upshift in mRNA levels of most of the transcriptions encoding proteins associated with these tumor suppressor proteins. These include cyclins A, B, and D_1, as well as cdk1, cdk2, and Rb, but not cyclins D_2 or D_3 (You and Bird, 1995). This activation occurs in both quiescent G_0 phase cells and exponentially growing cells primarily as a result of transcription activation. This activity has been localized to two discrete regions of the promoter in the case of human cdk1 (H. Liu and R. C. Bird, unpublished results, 1996). While posttranscriptional mechanisms cannot be ruled out entirely, at least for cdk1 the primary regulation appears to be at the level of transcription initiation.

F. Signal Transduction

Stathmin is a highly conserved phosphorylated protein which is thought to function as a relay, allowing integration of many of the most well-described phosphorylation-dependent signal transduction systems. Stathmin encodes consensus sequences for cAMP-dependent protein kinase, cdk1, cdk2, mitogen associated protein (MAP) kinase, and casein kinase II (Sobel, 1991; Doye *et al.,* 1989, 1992; Schubart *et al.,* 1989; Maucuer *et al.,* 1990). It has been proposed that these multiple kinases are responsible for the multiple phosphorylated forms of stathmin (Beretta *et al.,* 1993). That stathmin is a substrate of cdk1 is quite likely; however, it is unclear what mechanisms are involved beyond its potential as a substrate for many of the major signaling pathways in the cytoplasm. This fact makes stathmin a promising molecule at which cross-talk between signaling pathways could occur.

The cellular homolog of the Rous sarcoma transforming virus, $pp60^{c\text{-}src}$, is a cytoplasmic protein tyrosine kinase which is also a substrate of cyclin/cdk activity (Morgan *et al.,* 1989; Shenoy *et al.,* 1989). Thus, substrates like $pp60^{c\text{-}src}$ represent likely targets through which cdk1 initiates a phosporylation cascade resulting in mitosis-specific phosphorylation of proteins which are not direct substrates of cdk1, such as lamin C (Ward and Kirschner, 1990). This represents an important mechanism because cdk1 is unable to phosphorylate directly a number of proteins normally phosphorylated during mitosis (Lohka, 1989; Shenoy *et al.,* 1989).

IV. FUTURE OUTLOOK

Efforts to investigate the regulatory mechanisms involved in cell cycle regulation surrounding cdk/cyclin complexes have focused on p21 as well as other cdk regulators and their role in suppression/activation of this complex. These efforts have been driven by a desire to elucidate the global regulatory strategy which drives the cdk/cyclin complex family during each phase of the cell cycle. While extraordinarily important in understanding how proliferation of cells is controlled, such studies leave open the question of what events occur downstream from the regulatory gateway represented by cdk/cyclin complexes. In reviewing the known and potential substrates of this complex, I have described a large and varied array of cellular components which comprise many aspects of the cell architecture. Yet, that is the nature of the extensive changes which occur during mitosis. The entire cell is reorganized and recruited to effectively partition the genome and other components of the cell with a high level of fidelity.

What is most striking is the vast ignorance which persists in this area. Little is known regarding the mechanisms which drive these changes. Tantalizing details abound, however, and the future holds the promise of unraveling the details of mitotic control mechanisms. Understanding these processes represents discovery of one of the cornerstones of life itself, as the ability to divide successfully is one of the oldest and best-conserved cellular processes. Further efforts to understand how cdk/cyclin activity affects specific changes in nuclear reorganization and initiation of mitosis will reveal regulatory mechanisms which are applicable to all eukaryotic organisms.

ACKNOWLEDGMENTS

The author would like to acknowledge the contributions of Dr. S. Su, Ms, J. You, and Mr. R. Young-White. This work was supported by the NIH, the USDA, and the Morris Animal Foundation.

REFERENCES

Akiyama, T., and Toyoshima, K. (1990). Marked alteration in phosphorylation of the Rb protein during differentiation of human promyelocytic HL60 cells. *Oncogene* **5,** 179–183.

Alfa, C. E., Ducommun, B., Beach, D., and Hyams, J. S. (1990). Distinct nuclear and spindle pole body population of cyclin-cdc2 in fission yeast. *Nature (London)* **347,** 680–682.

Bachs, O., Lanini, L., Serratosa, J., Coll, M. J., Bastos, R., Aligue, R., Rius, E., and Carafoli, E. (1990). Calmodulin-binding proteins in the nuclei of quiescent and proliferatively activated rat liver cells. *J. Biol. Chem.* **265,** 18595–18600.

Bailly, E., Pines, J., Hunter, T., and Bornens, M. (1992). Cytoplasmic accumulation of cyclin B1 in human cells: Association with a detergent-resistant compartment and with the centrosome. *J. Cell Sci.* **101,** 529–545.

Barinaga, M. (1995). A new twist to the cell cycle. *Science* **269,** 631–632.

Baumann, K., Mandelkow, E. M., Biernat, J., Piwnica-Worms, H., and Mandelkow, E. (1993). Abnormal Alzheimer-like phosphorylation of *tau*-protein by cyclin-dependent kinases cdk2 and cdk5. *FEBS Lett.* **336,** 417–424.

Beach, D., Durkacz, B., and Nurse, P. (1982). Functionally homologous cell cycle control genes in budding and fission yeast. *Nature* (*London*) **300,** 706–709.

Beach, D., Basilico, C., and Newport, J. (1988). Introduction. *In* "Current Communications in Molecular Biology. Cell Cycle Control in Eukaryotes" (D. Beach, C. Basilico, and J. Newport, eds.), pp. 1–8. Cold Spring Harbor Press, Cold Spring Harbor, NY.

Belenguer, P., Baldin, V., Mathieu, C., Prats, H., Bensaid, M., Bouche, G., and Amalric, F. (1989). Protein kinase NII and the regulation of rDNA transcription in mammalian cells. *Nucleic Acids Res.* **17,** 6625–6635.

Belenguer, P., Caizergues-Ferrer, M., Labbé, J.-C., Dorrée, M., and Amalric, F. (1990). Mitosis-specific phosphorylation of nucleolin by p34cdc2 protein kinase. *Mol. Cell. Biol.* **10,** 3607–3618.

Beretta, L., Dobransky, T., and Sobel, A. (1993). Multiple phosphorylation of stathmin. Identification of four sites phosphorylated in intact cells and in vitro by cyclic AMP-dependent protein kinase and $p34^{cdc2}$ *J. Biol. Chem.* **268,** 20076–20084.

Bianchi, A. B., Fischer, S. M., Robles, A. I., Rinchik, E. M., and Conti, C. J. (1993). Overexpression of cyclin D1 in mouse skin carcinogenesis. *Oncogene* **8,** 1127–1133.

Bird, R. C., Kung, T.-Y. T., Wu, G., and Young-White, R. R. (1990). Variations in c-*fos* mRNA expression during serum induction and the synchronous cell cycle. *Biochem. Cell Biol.* **68,** 858–862.

Booher, R., and Beach, D. (1989). Involvement of a type 1 protein phosphatase encoded by bws1+ in fission yeast mitotic control. *Cell* (*Cambridge, Mass.*) **57,** 1009–1016.

Bookstein, R., Shew, J. Y., Chen, B. L., Scully, B., and Lee, W.-H. (1990). Suppression of tumorigenicity of human prostate carcinoma cells by replacing a mutated RB gene. *Science* **247,** 712–715.

Bradbury, E. M., Inglis, R. J., Matthews, H. R., and Sarner, N. (1973). Phosphorylation of very-lysine-rich histone in *Physarum polycephalum.* Correlation with chromosome condensation. *Eur. J. Biochem.* **33,** 131–139.

Bradbury, E. M., Inglis, R. J., and Matthews, H. R. (1974). Control of cell division by very lysine rich histone (H1) phosphorylation. *Nature* (*London*) **247,** 257–261.

Buchkovich, K., Duffy, L. A., and Harlow, E. (1989). The retinoblastoma protein is phosphorylated during specific phases of the cell cycle. *Cell* (*Cambridge, Mass.*) **58,** 1097–1105.

Buckingham, M. (1992). Making muscle in mammals. *Trends Genet.* **8,** 144–148.

Buendia, B., Draetta, G., and Karsenti, E. (1992). Regulation of the microtubule nucleating activity of centrosomes in *Xenopus* egg extracts: Role of cyclin A-associated protein kinase. *J. Cell Biol.* **116,** 1431–1441.

Bugler, B., Bourbon, H., Lapeyre, B., Wallace, O., Chang, J. H., Amalric, F., and Olson, M. O. J. (1987). RNA binding fragments from nucleolin contain the ribonucleoprotein consensus sequence. *J. Biol. Chem.* **262,** 10922–10925.

Bullrich, F., MacLachlan, T. K., Sang, N., Druck, T., Veronese, M. L., Allen, S. L., Chiorazzi, N., Koff, A., Heubner, K., and Croce, C. M. (1995). Chromosomal mapping of members of the cdc2 family of protein kinases, cdk3, cdk6, PISSLRE, and PITALRE, and a cdk inhibitor, p27Kip1, to regions involved in human cancer. *Cancer Res.* **55,** 1199–1205.

Caizergues-Ferrer, M., Belenguer, P., Lapeyre, B., Amalric, F., Wallace, M. O., and Olson, M. O. J. (1987). Phosphorylation of nucleolin by a nucleolar type NII protein kinase. *Biochemistry* **26,** 7876–7883.

Cavanaugh, A. H., Hempel, W. M., Taylor, L. J., Rogalsky, V., Todorov, G., and Rothblum, L. I. (1995). Activity of RNA polymerase I transcription factor UBF blocked by Rb gene product. *Nature (London)* **374,** 177–180.

Chen, P.-L., Scully, P., Shew, J.-Y., Wang, J. Y.-J., and Lee, W.-H. (1989). Phosphorylation of the retinoblastoma gene product is modulated during the cell cycle and cellular differentiation. *Cell (Cambridge, Mass.)* **58,** 1193–1198.

Chen, T. A., and Allfrey, V. G. (1987). Rapid and reversible changes in nucleosome structure accompany the activation, repression, and superinduction of murine fibroblast protooncogenes c-*fos* and c-*myc*. *Proc. Natl. Acad. Sci. U.S.A.* **84,** 5252–5256.

Chou, Y. H., Bischoff, J. R., Beach, D., and Goldman, R. D. (1990). Intermediate filament reorganization during mitosis is mediated by p34cdc2 phosphorylation of vimentin. *Cell (Cambridge, Mass.)* **62,** 1063–1071.

Cobrinik, D., Dowdy, S. F., Hinds, P. W., Mittnacht, S., and Weinberg, R. A. (1992). The retinoblastoma protein and the regulation of cell cycling. *Trends Biochem. Sci.* **17,** 312–315.

Cobrinik, D., Whyte, P., Peeper, D. S., Jacks, T., and Weinberg, R. A. (1993). Cell cycle-specific association of E2F with the p130 E1A-binding protein. *Genes Dev.* **7,** 2392–2404.

Corden, J. L. (1990). Tails of RNA polymerase II. *Trends Biochem. Sci.* **15,** 383–387.

Coutavas, E., Ren, M., Oppenheim, J. D., D'Eustachio, P., and Rush, M. G. (1993). Characterization of proteins that interact with the cell-cycle regulatory protein Ran/TC4. *Nature (London)* **366,** 585–587.

Dasso, M. (1993). RCC1 in the cell cycle: The regulator of chromosome condensation takes on new roles. *Trends Biochem. Sci.* **18,** 96–101.

de Boer, C. J., Loyson, S., Kluin, P. M., Kluin-Nelemans, H. C., Schurring, E., and van Krieken, J. H. J. M. (1993). Multiple breakpoints within the *BCL-1* locus in B-cell lymphoma: Rearrangements of the cyclin D1 gene. *Cancer Res.* **53,** 4148–4152.

DeCaprio, J. A., Ludlow, J. W., Lynch, D., Furukawa, Y., Griffin, J., Piwnica-Worms, H., Huang, C.-H., and Livingstone, D. M. (1989). The product of the retinoblastoma susceptibility gene has properties of a cell cycle regulatory element. *Cell (Cambridge, Mass.)* **58,** 1085–1095.

Doonan, J. H., and Morris, N. R. (1989). The bimG gene of *Aspergillus nidulans,* required for completion of anaphase, encodes a homologue of mammalian phosphoprotein phosphatase 1. *Cell (Cambridge, Mass.)* **57,** 987–996.

Dowdy, S. F., Hinds, P. W., Louie, K., Reed, S. I., Arnold, A., and Weinberg, R. A. (1993). Physical interaction of the retinoblastoma protein with human D cyclins. *Cell (Cambridge, Mass.)* **73,** 499–511.

Doye, V., Soubrier, F., Bauw, G., Boutterin, M. C., Beretta, L., Koppel, J., Vanderkerckhove, J., and Sobel, A. (1989). A single cDNA encodes two isoforms of stathmin, a developmentally regulated neuron-enriched phosphoprotein. *J. Biol. Chem.* **264,** 12134–12137.

Doye, V., Gouvello, S., Dobransky, T., Chneiweiss, H., Beretta, L., and Sobel, A. (1992). Expression of transfected stathmin cDNA reveals novel phosphorylated forms associated with developmental and functional cell regulation. *Biochem. J.* **287,** 549–554.

Draetta, G. (1990). Cell cycle control in eukaryotes: Molecular mechanisms of cdc2 activation. *Trends Biochem. Sci.* **15,** 378–383.

Dulic, V., Lees, E., and Reed, S. T. (1992). Association of human cyclin E with a periodic G1-S phase protein kinase. *Science* **257,** 1958–1961.

Eilen, E., Hand, R., and Basilico, C. (1980). Decreased initiation of DNA synthesis in a temperature-sensitive mutant of hamster cells. *J. Cell. Physiol.* **105,** 259–266.

El-Deiry, W. S., Tokino, T., Velculescu, V. E., Levy, D. B., Parsons, R., Trent, J. M., Lin, D., Mercer, W. E., Kinzler, K. W., and Vogelstein, B. (1993). *WAF1,* a potential mediator of p53 suppression. *Cell (Cambridge, Mass.)* **75,** 817–825.

Enoch, T., Peter, M., Nurse, P., and Nigg, E. A. (1991). p34cdc2 acts as a lamin kinase in fission yeast. *J. Cell Biol.* **112,** 797–807.

Enoch, T., Carr, A. M., and Nurse, P. (1992). Fission yeast genes involved in coupling mitosis to completion of DNA replication. *Genes Dev.* **6,** 2035–2046.

Erard, M. S., Belenger, P., Caizergues-Ferrer, M., Pantaloni, A., and Amalric, F. (1988). A major nucleolar protein, nucleolin, induces chromatin decondensation by binding to histone H1. *Eur. J. Biochem.* **175,** 525–530.

Ewen, M. E., Sluss, H. K., Sherr, C. J., Matsushima, H., Kato, J.-Y., and Livingston, D. M. (1993). Functional interaction of the retinoblastoma protein with mammalian D-type cyclins. *Cell (Cambridge, Mass.)* **73,** 487–497.

Félix, M.-A., Cohen, P., and Karsenti, E. (1990). Cdc2 H1 kinase is negatively regulated by a type 2A phosphatase in the *Xenopus* early embryonic cell cycle: Evidence from the effects of okadaic acid. *EMBO J.* **9,** 675–683.

Furukawa, Y., DeCaprio, J. A., Freedman, A., Kanakura, Y., Nakamura, M., Ernst, T. J., Livingston, D. M., and Griffin, J. D. (1990). Expression and state of phosphorylation of the retinoblastoma susceptibility gene in cycling and noncycling human hematopoietic cells. *Proc. Natl. Acad. Sci. U.S.A.* **87,** 2770–2774.

Gallant, P., and Nigg, E. A. (1992). Cyclin B2 undergoes cell cycle-dependent nuclear translocation and when expressed as a non-destructible mutant, causes mitotic arrest in HeLa cells. *J. Cell Biol.* **117,** 213–224.

Gas, N., Escande, M. L., and Stevens, B. J. (1985). Immunolocalization of the 100-kDa nucleolar protein during the mitotic cycle in CHO cells. *Biol. Chem.* **53.** 209–218.

Geahlen, R. L., and Harrison, M. L. (1984). Induction of a substrate for casein kinase II during lymphocyte mitogenesis. *Biochim. Biophys. Acta* **804,** 169–175.

Goodrich, D. W., Wang, N. P., Qian, Y., Lee, E. Y.-H. P., and Lee, W.-H. (1991). The retinoblastoma gene product regulates progression through the G1 phase of the cell-cycle. *Cell (Cambridge, Mass.)* **67,** 293–302.

Gyuris, J., Golemis, E., Chertkov, H., and Brent, R. (1993). Cdi1, a human G1 and S phase protein phosphatase that associates with cdk2. *Cell (Cambridge, Mass.)* **75,** 791–803.

Hadwiger, J. A., Wittenberg, C., Richardson, H. E., de Barros Lopes, M., and Reed, S. I. (1989). A novel family of cyclin homologs that control G1 in yeast. *Proc. Natl. Acad. Sci. U.S.A.* **86,** 6255–6259.

Halevy, O., Novitch, B. G., Spicer, D. B., Skapek, S. X., Rhee, J., Hannon, G. J., Beach, D., and Lassar, A. B. (1995). Correlation of terminal cell cycle arrest of skeletal muscle with induction of p21 by MyoD. *Science* **267,** 1018–1021.

Hamel, P. A., Gallie, B. L., and Philips, R. A. (1992). The retinoblastoma protein and cell cycle regulation. *Trends Genet.* **8,** 180–185.

Harper, J. W., Adami, G. R., Wei, N., Keyomarsi, K., and Elledge, S. (1993). The p21 cdk-interacting protein Cip1 is a potent inhibitor of G1 cyclin-dependent kinases. *Cell (Cambridge, Mass.)* **75,** 805–816.

Hartwell, L. H., and Weinert, T. A. (1989). Checkpoints: Controls that ensure the order of cell cycle events. *Science* **246,** 629–635.

Hayes, J., and Nurse, P. (1989). A review of mitosis in the fission yeast *Schizosaccharomyces pombe*. *Exp. Cell Res.* **184,** 273–286.

Heald, R., and McKeon, K. (1990). Mutations of phosphorylation sites in lamin A that prevent nuclear lamina disassembly in mitosis. *Cell (Cambridge, Mass.)* **61,** 579–589.

Hensey, C. E., Hong, F., Durfee, T., Qian, Y.-W., Lee, E. Y.-H. P., and Lee, W.-H. (1994). Identification of discrete structural domains in the retinoblastoma protein. *J. Biol. Chem.* **269,** 1380–1387.

Herrera, A. H., and Olson, M. O. J. (1986). Association of protein C23 with rapidly labeled nucleolar RNA. *Biochemistry* **25,** 6258–6264.

Holloway, S. L., Glotzer, M., King, R. W., and Murray, A. W. (1993). Anaphase is initiated by proteolysis rather than by the inactivation of maturation-promoting factor. *Cell (Cambridge, Mass.)* **73,** 1393–1402.

Hosoya, N., Hosoya, H., Yamashiro, S., Mohri, H., and Matsumura, F. (1993). Localization of caldesmon and its dephosphorylation during cell division. *J. Cell Biol.* **121,** 1075–1082.

Hunt, T. (1989). Maturation promoting factor, cyclin, and the control of M-phase. *Curr. Opin. Cell Biol.* **1,** 268–274.

Hunter, T., and Karin, M. (1992). The regulation of transcription by phosphorylation. *Cell* **70,** 375–387.

Hunter, T., and Pines, J. (1994). Cyclins and cancer II: Cyclin D and cdk inhibitors come of age. *Cell (Cambridge, Mass.)* **79,** 573–582.

Jacobs, F. A., Bird, R. C., and Sells, B. H. (1985). Differentiation of myoblasts: Regulation of turnover of ribosomal proteins and their mRNAs. *Eur. J. Biochem.* **150,** 255–263.

Jiang, W., Kahn, S. M., Zhou, P., Zhang, Y.-J., Cacace, A. M., Infante, A. S., Doi, S., Santella, R. M., and Weinstein, I. B. (1993). Overexpression of cyclin D1 in rat fibroblasts causes abnormalities in growth control, cell cycle progression and gene expression. *Oncogene* **8,** 3447–3457.

Kamb, A., Gruis, N. A., Weaver-Feldhaus, J., Liu, Q., Harshman, K., Tavtigian, S. V., Stackert, E., Day, R. S., Johnson, B. E., and Skolnick, M. H. (1994). A cell cycle regulator potentially involved in genesis of many tumor types. *Science* **264,** 436–440.

Keryer, G., Luo, Z., Cavadore, J. C., Erlichman, J., and Bornens, M. (1993). Phosphorylation of the regulatory subunit of type II beta cAMP-dependent protein kinase by cyclin B/p34^{cdc2} kinase impairs its binding to microtubule-associated protein 2. *Proc. Natl. Acad. Sci. U.S.A.* **90,** 5418–5422.

Keyomarsi, K., O'Leary, N., Molnar, G., Lees, E., Fingert, H. J., and Pardee, A. B. (1993). Cyclin E, a potential prognostic marker for breast cancer. *Cancer Res.* **54,** 380–385.

Khatib, Z. A., Matsushima, H., Valentine, M., Shapiro, D. N., Sherr, C. J., and Look, A. T. (1993). Coamplification of the cdk4 gene with MDM2 and GL1 in human sarcomas. *Cancer Res.* **53,** 5535–5541.

Kim, S.-J., Lee, H.-D., Robbins, P. D., Busam, K., Sporn, M. B., and Roberts, A. B. (1991). Regulation of transforming growth factor β1 gene expression by the product of the retinoblastoma susceptibility gene. *Proc. Natl. Acad. Sci. U.S.A.* **88,** 3052–3056.

Kim, S.-J., Onwuta, U. S., Lee, Y. I., Li, R., Botchan, M. R., and Robbins, P. D. (1992). The retinoblastoma gene product regulates Sp1-mediated transcription. *Mol. Cell. Biol.* **12,** 2455–2463.

Kinoshita, N., Ohkura, H., and Yanagida, M. (1990). Distinct, essential roles of type 1 and 2A protein phosphatases in the control of the fission yeast cell division cycle. *Cell (Cambridge, Mass.)* **63,** 405–415.

Klein, G. (1987). The approaching era of the tumor suppressor genes. *Science* **238,** 1539–1545.

Koff, A., Cross, F., Fisher, A., Schumacher, J., Leguellec, K., Philippe, M., and Roberts, J. M. (1991). Human cyclin E, a new cyclin that interacts with two members of the CDC2 gene family. *Cell (Cambridge, Mass.)* **66,** 1217–1228.

Koff, A., Giordano, A., Desai, D., Yamashita, K., Harper, J. W., Elledge, S., Nishimoto, T., Morgan, D. O., Franza, B. R., and Roberts, J. M. (1992). Formation and activation of a cyclin E-cdk2 complex during the G1 phase of the human cell cycle. *Science* **257,** 1689–1694.

Koff, A., Ohtsuki, O., Polyak, K., Roberts, J. M., and Massague, J. (1993). Negative regulation of G1 in mammalian cells: Inhibition of cyclin E-dependent kinase by TGF-β. *Science* **260,** 536–539.

Kubiak, J. Z., Weber, M., de Pennart, H., Winston, N. J., and Maro, B. (1993). The metaphase II arrest in mouse oocytes is controlled through microtubule-dependent destruction of cyclin B in the presence of CSF. *EMBO J.* **12,** 3773–3778.

Langan, T. A., Gautier, J., Lohka, M., Hollingworth, R., Moreno, S., Nurse, P., Maller, J., and Sclafani, R. A. (1989). Mammalian growth-associated H1 histone kinase: A homolog of cdc2+/CDC28 protein kinases controlling mitotic entry in yeast and frog cells. *Mol. Cell. Biol.* **9,** 3860–3868.

Lapeyre, B., Bourbon, H., and Amalric, F. (1987). Nucleolin, the major nucleolar protein of growing eukaryotic cells: An unusual protein structure revealed by the nucleotide sequence. *Proc. Natl. Acad. Sci. U.S.A.* **84,** 1472–1476.

La Thangue (1994). DRTF1/E2F: An expanding family of heterodimeric transcription factors implicated in cell cycle control. *Trends Biochem. Sci.* **19,** 108–114.

Leach, F. S., Elledge, S. J., Sherr, C. J., Willson, J. K. V., Markowitz, S., Kinzler, K. W., and Vogelstein, B. (1993). Amplification of cyclin genes in colorectal carcinomas. *Cancer Res.* **53,** 1986–1989.

Lee, T. H., Solomon, M. J., Mumby, M. C., and Kirschner, M. W. (1991). INH: A negative regulator of MPF is a form of protein phosphatase 2A. *Cell (Cambridge, Mass.)* **64,** 415–423.

Lee, W. H., Shew, J. Y., Hong, F. D., Sery, T. W., Donoso, L. A., Young, L. J. Bookstein, R., and Lee, Y. P. (1987). The retinoblastoma susceptibility gene encodes a nuclear phosphoprotein associated with DNA binding activity. *Nature (London)* **329,** 642–645.

Lew, D. J., Dulic, V., and Reed, S. I. (1991). Isolation of three novel human cyclins by rescue of G1 cyclin (cln) function in yeast. *Cell (Cambridge, Mass.)* **66,** 1197–1206.

Lewin, B. (1990). Driving the cell cycle: M phase kinase, its partners, and substrates. *Cell (Cambridge, Mass.)* **61,** 743–752.

Li, C. Y., Suardet, L., and Little, J. B. (1995). Potential role of WAF1/Cip1/p21 as a mediator of TGF-beta cytoinhibitory effect. *J. Biol. Chem.* **270,** 4971–4974.

Liao, S.-M., Zhang, J., Jeffery, D. A., Koleske, A. J., Thompson, C. M., Chao D. M., Viljoen, M., van Vuuren, H. J., and Young, R. A. (1995). A kinase-cyclin pair in the RNA polymerase II holoenzyme. *Nature (London)* **374,** 193–196.

Lohka, M. J. (1989). Mitotic control by metaphase-promoting factor and cdc proteins. *J. Cell Sci.* **92**(Pt. 2), 131–135.
Lohka, M. J., Kyes, J. L., and Maller, J. L. (1987). Metaphase protein phosphorylation in *Xenopus laevis* eggs. *Mol. Cell. Biol.* **7,** 760–768.
Ludlow, J. W., Shon, J., Pipas, J. M., Livingston, J. M., and DeCaprio, J. A. (1990). The retinoblastoma susceptibility gene product undergoes cell cycle-dependent dephosphorylation and binding to and release from SV40 large T. *Cell (Cambridge, Mass.)* **60,** 387–396.
Lumpkin, C. K., McClung, J. K., Pereira-Smith, O. M., and Smith, J. M. (1986). Existence of high abundance anti-proliferative mRNA's in senescent human diploid fibroblast. *Science* **232,** 393–395.
Mahadevan, L. C., and Willis, A. C. (1990). 2-Aminopurine abolishes EGF- and TPA-stimulated pp53 phosphorylation and c-*fos* induction without affecting the activation of protein kinase C. *Oncogene* **5,** 327–335.
Mahadevan, L. C., Willis, A. C., and Barratt, M. J. (1991). Rapid histone H3 phosphorylation in response to growth factors, phorbol esters, okadaic acid, and protein synthesis inhibitors. *Cell* **65,** 775–783.
Mak, A. S., Carpenter, M., Smillie, L. B., and Wang, J. H. (1991). Phosphorylation of caldesmon by p34cdc2 kinase. Identification of phosphorylation sites. *J. Biol. Chem.* **266,** 19971–19975.
Maller, J. L., and Smith, D. W. (1985). Two-dimensional polyacrylamide gel analysis of changes in protein phosphorylation during maturation of *Xenopus* oocytes. *Dev. Biol.* **109,** 150–156.
Maridor, G., Gallant, P., Golsteyn, R., and Nigg, E. A. (1993). Nuclear localization of vertebrate cyclin A correlates with its ability to form complexes with cdk catalytic subunits. *J. Cell Sci.* **106,** 535–544.
Marin, M. C., Hsu, B., Meyn, R. E., Donehower, L. A., El-Naggar, A. K., and McDonnell, T. J. (1994). Evidence that p53 and bcl-2 are regulators of a common cell death pathway important for in vivo lymphomagenesis. *Oncogene* **9,** 3107–3112.
Matsumura, F., and Yamashiro, S. (1993). Caldesmon. *Curr. Opin. Cell Biol.* **5,** 70–76.
Matsushime, H., Ewen, M. E., Strom, D. K., Kato, J.-Y., Hanks, S. K., Roussel, M. F., and Sherr, C. J. (1992). Identification and properties of atypical catalytic subunit ($p34^{psk\text{-}j3}$/cdk4) for mammalian D type G1 cyclins. *Cell (Cambridge, Mass.)* **71,** 323–334.
Maucuer, A., Doye, V., and Sobel, A. (1990). A single amino acid difference distinguishes the human and the rat sequences of stathmin, a ubiquitous intracellular phosphoprotein associated with cell regulations. *FEBS Lett.* **264,** 275–278.
McKinney, J. D., and Heintz, N. (1991). Transcriptional regulation in the eukaryotic cell cycle. *Trends Biochem. Sci.* **16,** 430–435.
Michalides, R., van Veelen, N., Hart, A., Loftus, B., Wientjens, E., and Balm, A. (1995). Overexpression of cyclin D1 correlates with recurrence in a group of 47 operable squamous cell carcinomas of the head and neck. *Cancer Res.* **55,** 975–978.
Mihara, K., Cao, X.-R., Yen, A., Chandler, S., Driscoll, B., Murphree, A. L., T'Ang, A., and Fung, Y.-K. T. (1989). Cell cycle dependent regulation of phosphorylation of the human retinoblastoma gene product. *Science* **246,** 1300–1303.
Mitchison, J. M. (1971). "The Biology of the Cell." Cambridge University Press, Cambridge, UK.
Moreno, S., and Nurse, P. (1990). Substrates for $p34^{cdc2}$: *In vivo veritas? Cell (Cambridge, Mass.)* **61,** 549–551.

Morgan, D. O. (1995). Principles of cdk regulation. *Nature (London)* **374,** 131–133.
Morgan, D. O., Kaplan, J. M., Bishop, J. M., and Varmus, H. E. (1989). Mitosis-specific phosphorylation of p60c-src by p34cdc2-associated protein kinase. *Cell (Cambridge, Mass.)* **57,** 775–786.
Motokura, T., and Arnold, A. (1993). Cyclins and oncogenesis. *Biochim. Biophys. Acta* **1155,** 63–78.
Murray, A. W. (1992). Creative blocks: Cell-cycle checkpoints and feedback controls. *Nature (London)* **359,** 599–604.
Murray, A. W., and Kirschner, M. W. (1989). Dominoes and clocks: The union of two views of the cell cycle. *Science* **246,** 614–621.
Nagai, Y., Kaneda, S., Nomura, K., Uasuda, H., Seno, T., and Yamao, F. (1995). Ubiquitin-activating enzyme, E1, is phosphorylated in mammalian cells by protein kinase cdc2. *J. Cell Sci.* **108,** 2145–2152.
Nash, R., Tokiwa, G., Anand, S., Erickson, K., and Futcher, A. B. (1988). The *WHI1*$^{+}$ gene of *Saccharomyces cerevisiae* tethers cell division to cell size and is a cyclin homolog. *EMBO J.* **7,** 4335–4346.
Nevins, J. R. (1992). E2F: A link between the Rb tumor suppressor protein and viral oncoproteins. *Science* **258,** 424–429.
Nishimoto, T., Eilen, E., and Basilico, C. (1978). Premature chromosome condensation in a ts DNA-mutant of BHK cells. *Cell (Cambridge, Mass.)* **15,** 475–483.
Nishimoto, T., Uzawa, S., and Schlegel, R. (1992). Mitotic checkpoints. *Curr. Opin. Cell Biol.* **4,** 174–179.
Nishitani, H., Ohtsubo, M., Yamashita, K., Iida, H., Pines, J., Yasudo, H., Shibata, Y., Hunter, T., and Nishimoto, T. (1991). Loss of RCC1, a nuclear DNA-binding protein, uncouples the completion of DNA replication from the activation of cdc2 protein kinase and mitosis. *EMBO J.* **10,** 1555–1564.
Nobori, T., Miura, K., Wu, D. J., Lois, A., Takabayashi, K., and Carson, D. A. (1994). Deletions of the cyclin-dependent kinase-4 inhibitor gene in multiple human cancers. *Nature (London)* **368,** 753–756.
Nurse, P. (1975). Genetic control of cell size at cell division in yeast. *Nature (London)* **256,** 547–551.
Nurse, P. (1990). Universal control mechanism regulating onset of M-phase. *Nature (London)* **344,** 503–508.
Nurse, P. (1994). Ordering S phase and M phase in the cell cycle. *Cell (Cambridge, Mass.)* **79,** 547–550.
Nurse, P., and Bissett, Y. (1981). Gene required in G1 for commitment to cell division cycle in the fission yeast. *Nature (London)* **292,** 558–560.
Nurse, P., Thuriaux, P., and Nasmyth, K. (1976). Genetic control of the cell division cycle in the fission yeast *Schizosaccharomyces pombe. Mol. Gen. Genet.* **146,** 167–178.
Ohkura, H., Kinoshita, N., Miyatani, S., Toda, T., and Yanagida, M. (1989). The fission yeast dis2+ gene required for chromosome disjoining encodes one of two putative type 1 protein phosphatases. *Cell (Cambridge, Mass.)* **57,** 997–1007.
Ohta, K., Shiina, N., Okumura, E., Hisanaga, S., Kishimoto, T., Endo, S., Gotoh, Y., Nishida, E., and Sakai, H. (1993). Microtubule nucleating activity of centrosomes in cell-free extracts from *Xenopus* eggs: Involvement of phosphorylation and accumulation of pericentriolar material. *J. Cell Sci.* **104,** 125–137.
Okayama, H., Nagata, A., Igarashi, M., Suto, K., and Jinno, S. (1991). Mammalian G2 regulatory genes and their possible involvement in genetic instability in cancer cells. *Princess Takamatsu Symp.* **22,** 231–238.

Olson, M. O. J., and Thompson, B. A. (1983). Distribution of proteins among chromatin components of nucleoli. *Biochemistry* **22,** 3187–3193.
O'Neill, E. M., and O'Shea, E. K. (1995). Transcriptional regulation: Cyclins in initiation. *Nature (London)* **374,** 121–122.
Ookata, K., Hisanaga, S., Okano, T., Tachibana, K., and Kishimoto, T. (1992). Relocation and distinct subcellular localization of p34cdc2-cyclin B complex at meiosis reinitiation in starfish oocytes. *EMBO J.* **11,** 1763–1772.
Ookata, K., Hisanaga, S., Okumura, E., and Kishimoto, T. (1993). Association of $p34^{cdc2}$/cyclin B complex with microtubules in starfish oocytes. *J. Cell Sci.* **105,** 873–881.
Ookata, K., Hisanaga, S., Bulinski, J. C., Murofushi, H., Aizawa, H., Itoh, T. J., Hotani, H., Okumura, E., Tachibana, K., and Kishimoto, T. (1995). Cyclin B interaction with microtubule-associated protein 4 (MAP4) targets p34cdc2 kinase to microtubules and is a potential regulator of M-phase microtubule dynamics. *J. Cell Biol.* **128,** 849–862.
Pagano, M., Tam, S. W., Theodoras, A. M., Beer-Romero, P., Del Sal, G., Chau, V., Yew, P. R., Draetta, G. F., and Rolfe, M. (1995). Role of the ubiquitin-proteasome pathway in regulating abundance of the cyclin-dependent kinase inhibitor p27. *Science* **269,** 682–685.
Pai, S. R., and Bird, R. C. (1992). Growth of HeLa S3 cells cotransfected with plasmids containing a c-*fos* gene under the control of the SV40 promoter complex, pRSVcat, and G418 resistance. *Biochem. Cell Biol.* **70,** 316–323.
Pai, S. R., and Bird, R. C. (1994a). c-*fos* expression is required during all phases of the cell cycle during exponential cell proliferation. *Anticancer Res.* **14,** 985–994.
Pai, S. R., and Bird, R. C. (1994b). Overexpression of c-*fos* induces expression of the retinoblastoma tumor suppressor gene Rb in transfected cells. *Anticancer Res.* **14,** 2501–2508.
Pardee, A. B. (1989). G1 events and regulation of cell proliferation. *Science* **246,** 603–608.
Parker, S. P., Eichele, G., Zhang, P., Rawls, A., Sands, A. T., Bradley, A., Olson, E. N., Harper, J. W., and Elledge, S. J. (1995). p53-independent expression of $p21^{Cip1}$ in muscle and other terminally differentiating cells. *Science* **267,** 1024–1027.
Peeper, D. S., Parker, L. L., Ewen, M. E., Toebes, M., Hall, F. L., Xu, M., Zantema, A., van der Eb, A. J., and Piwnica-Worms, H. (1993). A- and B-type cyclins differentially modulate substrate specificity of cyclin-cdk complexes. *EMBO J.* **12,** 1947–1954.
Pelech, S. L., Sanghera, J. S., and Daya-Makin, M. (1990). Protein kinase cascades in meiotic and mitotic cell cycle control. *Biochem. Cell Biol.* **68,** 1297–1330.
Peter, M., Nakagawa, M., Dorée, M., Labbé, J. C., and Nigg, E. A. (1990). In vitro disassembly of the nuclear lamina and M phase-specified phosphorylation of lamins by cdc2 kinase. *Cell (Cambridge, Mass.)* **61,** 591–602.
Peters, G. (1994). The D-type cyclins and their role in tumorigenesis. *J. Cell Sci., Suppl* **18,** 89–96.
Pietenpol, J. A., Stein, R. W., Moran, E., Yaciuk, P., Schlegel, R., Lyons, R. M., Pittelkow, M. R., Munger, K., Howley, P. M., and Moses, H. L. (1990). TGF-β1 inhibition of c-*myc* transcription and growth in keratinocytes is abrogated by viral transforming proteins with pRB binding domains. *Cell (Cambridge, Mass.)* **61,** 777–785.
Pines, J. (1993). Cyclins and cyclin-dependent kinases: Take your partners. *Trends Biochem. Sci.* **18,** 195–197.
Pines, J. (1994). Arresting developments in cell-cycle control. *Trends Biochem. Sci.* **19,** 143–145.

Pines, J., and Hunter, T. (1991). Human cyclins A and B1 are differently located in the cell and undergo cell cycle dependent nuclear transport. *J. Cell Biol.* **115,** 1–17.
Pringle, J. R., and Hartwell, L. H. (1981). The *Saccharomyces cerevisiae* cell cycle. *In* "The Molecular Biology of the Yeast *Saccharomyces.* Life Cycle and Inheritance," (J. N. Strathern, E. W. Jones, and J. R. Broach, eds.), pp. 97–142. Cold Spring Harbor Lab., Cold Spring Harbor, NY.
Prywes, R., Fisch, T. M., and Roeder, R. G. (1988). Transcriptional regulation of c-*fos.* *Cold Spring Harb. Symp. Quant. Biol.* **53**(Pt. 2), 739–748.
Ravitz, M. J., Yan, S., Herr, K. D., and Wenner, C. E. (1995). Transforming growth factor beta-induced activation of cyclinE-cdk2 kinase and down-regulation of p27Kip1 in C3H 10T1/2 mouse fibroblasts. *Cancer Res.* **55,** 1413–1416.
Reed, S., and Nasmyth, K. A. (1981). Isolation of genes by complementation in yeast: molecular cloning of a cell cycle gene. *Proc. Natl. Acad. Sci. U.S.A.* **77,** 2119–2123.
Riabowol, K., Draetta, G., Brizuela, L., Vandre, D., and Beach, D. (1989). The cdc2 kinase is a nuclear protein that is essential for mitosis in mammalian cells. *Cell* (*Cambridge, Mass.*) **57,** 393–401.
Richardson, H. E., Wittenberg, C., Cross, F., and Reed, S. I. (1989). An essential G1 function for cyclin-like proteins in yeast. *Cell* (*Cambridge, Mass.*) **59,** 1127–1133.
Rittling, S. R., Brooks, K. M., Cristofalo, V. J., and Baserga, R. (1986). Expression of cell cycle-dependent genes in young and senescent WI-38 fibroblast. *Proc. Natl. Acad. Sci. U.S.A.* **83,** 3316–3320.
Robbins, P. D., Horowitz, J. M., and Mulligan, R. C. (1990). Negative regulation of human c-*fos* expression by the retinoblastoma gene product. *Nature* (*London*) **346,** 668–671.
Roberge, M. (1992). Checkpoint controls that couple mitosis to the completion of DNA replication. *Trends Cell Biol.* **2,** 277–281.
Schneider, H. R., Reichert, G. U., and Issinger, O. G. (1986). Enhanced casein kinase II activity during mouse embryogenesis. *Eur. J. Biochem.* **161,** 733–738.
Schubart, U. K., Das Banerjee, M., and Eng, J. (1989). Homology between the cDNAs encoding phosphoprotein p19 and SCG10 reveals a novel mammalian gene family preferentially expressed in developing brain. *DNA* **8,** 389–398.
Schwarz, J. K., Devoto, S. H., Smith, E. J., Chellappan, S. P., Jakoi, L., and Nevins, J. R. (1993). Interactions of the p107 and Rb proteins with E2F during the cell proliferation response. *EMBO J.* **12,** 1013–1020.
Serrano, M., Hannon, G. J., and Beach, D. (1993). A new regulatory motif in cell-cycle control causing specific inhibition of cyclin D/cdk4. *Nature* (*London*) **366,** 704–707.
Shan, B., Zhu, X., Chen, P. L., Durfee, T., Yang, Y., Shar, D., and Lee, W. H. (1992). Molecular cloning of cellular genes encoding retinoblastoma-associated proteins: Identification of a gene with properties of the transcription factor E2F. *Mol. Cell. Biol.* **12,** 5620–5631.
Shenoy, S., Choi, J. K., Bagrodia, S., Copeland, T. D., Maller, J. L., and Shalloway, D. (1989). Purified maturation promoting factor phosphorylates pp60c-src at the sites phosphorylated during fibroblast mitosis. *Cell* (*Cambridge, Mass.*) **57,** 763–774.
Shiekhattar, R., Mermelstein, F., Fisher, R. P., Drapkin, R., Dynlacht, B., Wessling, H. C., Morgan, D. O., and Reinberg, D. (1995). Cdk-activating kinase complex is a component of human transcription factor TFIIH. *Nature* (*London*) **374,** 283–287.
Shirodkar, S., Ewen, M., DeCaprio, J. A., Morgan, J., Livingston, D. M., and Chittenden, T. (1992). The transcription factor E2F interacts with the retinoblastoma product and

a p107-cyclin A complex in a cell cycle-regulated manner. *Cell* (*Cambridge, Mass.*) **68,** 157–166.
Skapek, S. X., Rhee, J., Spicer, D. B., and Lassar, A. B. (1995). Inhibition of myogenic differentiation in proliferating myoblasts by cyclin D1-dependent kinase. *Science* **267,** 1022–1024.
Sobel, A. (1991). Stathmin: A relay phosphoprotein for multiple signal transduction? *Trends Biochem. Sci.* **16,** 301–305.
Steel, M. (1994). Cyclins and cancer: Wheels within wheels. *Lancet* **343,** 931–932.
Stein, G. H., Beeson, M., and Gordon, L. (1990). Failure to phosphorylate the retinoblastoma gene product in senescent human fibroblast. *Science* **249,** 666–669.
Steinman, R. A., Hoffman, B., Iro, A., Guillouf, C., Liebermann, D. A., and El-Houseini, M. E. (1994). Induction of p21 (WAF-1/CIP1) during differentiation. *Oncogene* **9,** 3389–3396.
Suzuki, M. (1990). The heptad repeat in the largest subunit of RNA polymerase II binds by intercalating into DNA. *Nature* (*London*) **334,** 562–565.
Suzuki, N., Matsui, H., and Hosoya, T. (1985). Effects of androgen and polyamines on the phosphorylation of nucleolar proteins from rat ventral prostates with particular reference to 110-kDa phosphoprotein. *J. Biol. Chem.* **260,** 8050–8055.
Suzuki, N., Saito, T., and Hosoya, T. (1987). In vivo effects of dexamethazone and cycloheximide on the phosphorylation of 110-kDa proteins and protein kinase activities of rat liver nucleoli. *J. Biol. Chem.* **262,** 4696–4700.
Tombes, R. M., Peloquin, J. G., and Borisy, G. G. (1991). Specific association of an M-phase kinase with isolated mitotic spindles and identification of two of its substrates as MAP4 and MAP1B. *Cell Regul.* **2,** 861–874.
Tsujimura, K., Ogawara, M., Takeuchi, Y., Imajoh-Ohmi, T., Ha, M. H., and Inagaki, M. (1994). Visualization and function of vimentin phosphorylation by cdc2 kinase during mitosis. *J. Biol. Chem.* **269,** 31097–31106.
Tsuruta, H., Sakamoto, H., Onda, M., and Terada, M. (1993). Amplification and overexpression of *EXP1* and *EXP2*/cyclin D1 genes in human esophageal carcinomas. *Biochem. Biophys. Res. Commun.* **196,** 1529–1536.
Udvadia, A. J., Rogers, K. T., and Horowitz, J. M. (1992). A common set of nuclear factors bind to promoter elements regulated by the retinoblastoma protein. *Cell Growth Differ.* **3,** 597–608.
Uzvolgyi, E., Classon, M., Henriksson, M., Huang, H. J., Szekely, L., Lee, W.-H., Klein, G., and Sunmegi, J. (1991). Reintroduction of a normal retinoblastoma gene into retinoblastoma and osteosarcoma cells inhibits the replication associated function of SV40 large T antigen. *Cell Growth Differ.* **2,** 297–303.
van den Heuvel, S., and Harlow, E. (1993). Distinct roles for cyclin-dependent kinases in cell cycle control. *Science* **262,** 2050–2054.
Verde, F., Labbé, J.-C., Dorée, M., and Karsenti, E. (1990). Regulation of microtubule dynamics by cdc2 protein kinase in cell-free extracts of *Xenopus* eggs. *Nature* (*London*) **343,** 233–238.
Vogelstein, B., and Kinzler, K. W. (1992). p53 function and dysfunction. *Cell* (*Cambridge, Mass.*) **70,** 523–526.
Wang, N. P., Chen, P.-L., Huang, S., Donoso, L. A., Lee, W.-H., and Lee, E. Y.-H. P. (1990). DNA-binding activity of retinoblastoma protein is intrinsic to its carboxyl-terminal region. *Cell Growth Differ.* **1,** 233–239.

Ward, G. E., and Kirschner, M. W. (1990). Identification of cell cycle-regulated phosphorylation sites on nuclear lamin C. *Cell (Cambridge, Mass.)* **61,** 561–577.

Warren, G. (1993). Membrane partitioning during cell division. *Annu. Rev. Biochem.* 323–348.

Weinberg, R. A. (1989). The molecular basis of retinoblastoma in genetic analysis of tumor suppression. *Ciba Found. Symp.* **142,** 99–111.

Weinberg, R. A. (1991). Tumor suppressor genes. *Science* **254,** 1138–1146.

Wells, D., and McBride, C. (1986). A comprehensive compilation and alignment of histones and histone genes. *Nucleic Acids Res.* **17,** Suppl., 311–346.

Westendorf, J. M., Swenson, K. I., and Ruderman, J. V. (1989). The role of cyclin B in meiosis I. *J. Cell Biol.* **108,** 1431–1444.

Xiong, Y., Zhang, H., and Beach, D. (1993). Subunit rearrangement of the cyclin-dependent kinases is associated with cellular transformation. *Genes Dev.* **7,** 1572–1583.

Xu, H.-J., Hu, S.-X., Hashimoto, T., Takahashi, R., and Benedict, W. F. (1989). The retinoblastoma susceptibility gene product: A characteristic pattern in normal cells and abnormal expression in malignant cells. *Oncogene* **4,** 807–812.

Yamakita, Y. Yamashiro, S., and Matsumura, F. (1992). Characterization of mitotically phosphorylated caldesmon. *J. Biol. Chem.* **267,** 12022–12029.

Yamashiro, S., and Matsumura, F. (1991). Mitosis-specific phosphorylation of caldesmon: Possible molecular mechanism of cell rounding during mitosis. *BioEssays* **13,** 563–568.

Yamashiro, S., Yamakita, Y., Ishikawa, R., and Matsumura, F. (1990). Mitosis-specific phosphorylation causes 83K non-muscle caldesmon to dissociate from microfilaments. *Nature (London)* **344,** 675–678.

Yamashiro, S., Yamakita, Y., Hosoya, H., and Matsumura, F. (1991). Phosphorylation of non-muscle caldesmon by p34cdc2 kinase during mitosis. *Nature (London)* **349,** 169–172.

Yatsunami, J., Komori, A., Ohta, T., Suganuma, M., and Fujiki, H. (1993). Hyperphosphorylation of retinoblastoma protein and p53 by okadaic acid, a tumor promoter. *Cancer Res.* **53,** 239–241.

You, J., and Bird, R. C. (1995). Selective induction of cell cycle regulatory genes cdk1 ($p34^{cdc2}$), cyclins A/B, and the tumor suppressor gene Rb in transformed cells by okadaic acid. *J. Cell. Physiol.* **164,** 424–433.

Zhang, Y.-J., Chen, C.-J., Lee, C. S., Kahn, S. M., Santella, R. M., and Weinstein, I. B. (1993). Amplification and overexpression of cyclin D1 in human hepatocellular carcinoma. *Biochem. Biophys. Res. Commun.* **196,** 1010–1016.

III

Chromatin Structure and Transcription Complexes

6

Nuclear Architecture in Developmental Transcriptional Control of Cell Growth and Tissue-Specific Genes

GARY S. STEIN, ANDRÉ J. VAN WIJNEN, JANET L. STEIN, JANE B. LIAN, AND MARTIN MONTECINO[1]

Department of Cell Biology and Cancer Center
University of Massachusetts Medical Center
Worcester, Massachusetts

[1] Present address: Department of Molecular Biology, University of Concepción, Casilla 2407, Concepción, Chile.

I. INTRODUCTION: CONTRIBUTIONS OF NUCLEAR ARCHITECTURE TO TRANSCRIPTIONAL CONTROL

During the past several years, we have gained great insight into the complexities of transcriptional control in eukaryotic cells. Our concept of a promoter has evolved from the initial expectation of a single regulatory sequence which determines transcriptional competency and level of expression. We now appreciate that transcriptional control is mediated by an interdependent series of regulatory sequences which reside 5′, 3′ and within transcribed regions of genes. Rather than focusing on the minimal sequences required for transcriptional control to support biological activity, investigators are working to define functional limits. Consequently, contributions of distal flanking sequences to regulation of transcription are being addressed experimentally. This is necessary for understanding the mechanisms by which activities of multiple promoter elements respond to a broad spectrum of regulatory signals and the activities of these regulatory sequences are functionally integrated. Cross-talk between a series of regulatory domains must be understood in diverse biological circumstances in which expression of genes supports cell and tissue functions. The overlapping binding sites for transcription factors within promoter regulatory elements and protein–protein interactions which influence transcription factor activity provide further components of the requisite diversity to accommodate regulatory options for physiologically responsive gene expression.

As the intricacies of gene organization and regulation are elucidated, the implications of a fundamental biological paradox become strikingly evident. How, with a limited representation of gene-specific regulatory elements and low abundance of cognate transactivation factors, can sequence-specific interactions occur to support a threshold for initiation of transcription within nuclei of intact cells? Viewed from a quantitative perspective, the *in vivo* regulatory challenge is to account for the formation of functional transcription initiation complexes with a nuclear concentration of regulatory sequences that is approximately 20 nucleotides per 2.5 yards of DNA and a similarly restricted level of DNA binding proteins.

It is becoming increasingly apparent that nuclear architecture provides a basis for the support of stringently regulated modulation of cell growth and tissue-specific transcription which is necessary for the onset and progression of differentiation. Here multiple lines of evidence point to contributions by three levels of nuclear organizations to *in vivo* transcriptional control in which structural parameters are functionally coupled to regulatory events (Table I). The primary level of gene organization establishes a linear ordering of promoter regulatory elements. This representation of regulatory sequences reflects competency for responsiveness to physiological regulatory signals. However, interspersion of sequences between promoter elements which exhibit coordinate and synergistic activities indicates the requirement of a structural basis for integration of activities at independent regulatory domains. Parameters of chromatin structure and nucleosome organization are a second level of genome architecture that reduces the distance between promoter elements, thereby supporting interactions between the modular components of transcriptional control. Each nucleosome (approximately 140 nucleotide base pairs wound around a core complex of two each of H3, H4, H2, and H2B histone proteins) contracts linear spacing by sevenfold. Higher-order chromatin structure further reduces nucleotide distances between regulatory sequences. Folding of nucleosome arrays into solenoid-type structures provides a potential for interactions which support synergism between promoter elements and responsiveness to multiple signaling pathways. A third level of nuclear architecture which contributes to transcriptional control is provided by the nuclear matrix (summarized in Table II). The anastomosing network of fibers and filaments which

TABLE I

Levels of Nuclear Organization That Contribute to Transcriptional Control

Structural Parameters	Regulatory Events
1. Organization of promoter regulatory elements	Representation of physiologically responsive transcriptional regulatory sequences
2. Chromatin structure and nucleosome organization	Integration of activities at independent promoter domains
	Responsiveness to multiple signaling pathways
3. Composition and organization of nuclear architecture	Gene localization
	Concentration and targeting of transcription factors
	Imprinting and modulation of chromatin structure

TABLE II

Contributions of Nuclear Matrix to Regulation of Gene Expression

Parameter	References
Involvement of nuclear matrix in DNA replication	Pardoll *et al.*, 1980; Belgrader *et al.*, 1991; Berezney and Coffey, 1974, 1975, 1977; Vaughn *et al.*, 1990; Jackson and Cook, 1986; Nakayasu and Berezney, 1989
Involvement of nuclear matrix in transcriptional control	
Biologically relevant modifications in representation of nuclear matrix proteins	
Cell type and tissue-specific nuclear matrix proteins.	Fey and Penman, 1988; Fey *et al.*, 1986, 1991; Pienta *et al.*, 1991; He *et al.*, 1990; Capco *et al.*, 1982; Bidwell *et al.*, 1993; Nickerson *et al.*, 1990a
Developmental stage-specific nuclear matrix proteins.	Dworetzky *et al.*, 1992
Tumor-specific nuclear matrix proteins.	Bidwell *et al.*, 1994a; Getzenberg and Coffey, 1991; Pienta and Coffey, 1991
Steroid hormone-responsive nuclear matrix proteins.	Barrack and Coffey, 1983; Kumara-Siri *et al.*, 1986; Getzenberg and Coffey, 1990
Polypeptide hormone-responsive nuclear matrix proteins.	Bidwell *et al.*, 1994b
Association of actively transcribed genes with nuclear matrix.	
Matrix-associated regions of genes (MARs)	Dworetzky *et al.*, 1992; Nelkin *et al.*, 1980; Robinson *et al.*, 1982; Schaack *et al.*, 1990; Stief *et al.*, 1989; Bode and Maass, 1988; Ciejek *et al.*, 1983; Cockerill and Garrard, 1986; Gasser and Laemmli, 1986; Jackson and Cook, 1985; Jarman and Higgs, 1988; Käs and Chasin, 1987; Keppel, 1986; Mirkovitch *et al.*, 1984, 1988; Phi-Van and Strätting, 1988; Phi-Van *et al.*, 1990; Robinson *et al.*, 1982; Thorburn *et al.*, 1988; Farache *et al.*, 1990; de Jong *et al.*, 1990; Jackson, 1991; Nelson *et al.*, 1986; Targa *et al.*, 1994; Kay and Bode, 1994; Dietz *et al.*, 1994; Klehr *et al.*, 1991

MAR-binding proteins.	Von Kries *et al.*, 1991, 1994; Buhrmester *et al.*, 1995; Nakagomi *et al.*, 1994; Fishel *et al.*, 1993; Tsutsui *et al.*, 1993; Dickinson *et al.*, 1992; Dickinson and Kohwi-Shigematsu, 1995; Nakagomi *et al.*, 1994; Hakes and Berezney, 1991; Cunningham *et al.*, 1994
Nuclear matrix localization of transcription factors	
Steroid hormone receptors	Barrack and Coffey, 1983; Kumara-Siri *et al.*, 1986; Landers and Spelsberg, 1992; van Steensel *et al.*, 1995
Ubiquitous transcription factors	Dworetzky *et al.*, 1992; Guo *et al.*, 1995; Zenk *et al.*, 1990
Tissue-specific transcription factor	Bidwell *et al.*, 1993; Merriman *et al.*, 1995
Viral regulatory proteins	Schaack *et al.*, 1990; Abulafia *et al.*, 1984
Selective partitioning of transcription factors between nuclear matrix and nonmatrix nuclear fractions	Van Wijnen *et al.*, 1993
Involvement of nuclear matrix in posttranscriptional control	
RNA processing	Van Eeklen and van Venrooij, 1981; Zeitlin *et al.*, 1987; Nickerson *et al.*, 1990b; Nickerson and Penman, 1992; Xing *et al.*, 1993; Ben-Ze'ev and Aloni, 1983; Marriman *et al.*, 1982; Ross *et al.*, 1982; Schroder *et al.*, 1987a,b; Jackson *et al.*, 1981; Herman *et al.*, 1978; Durfee *et al.*, 1994; Blencowe *et al.*, 1994; Huang and Spector, 1992; O'Keefe et al., 1994
Posttranslational modifications of chromosomal proteins	Hendzel *et al.*, 1994
Phosphorylation	Tawfic and Ahmed, 1994

constitutes the nuclear matrix supports the structural properties of the nucleus as a cellular organelle and accommodates structural modifications associated with proliferation, differentiation, and changes necessary to sustain the phenotypic requirements of specialized cells. Regulatory functions of the nuclear matrix include but are by no means restricted to gene localization, imposition of physical constraints on chromatin structure which support formation of loop domains, concentration and targeting of transcription factors, RNA processing and transport of gene transcripts, concentration and targeting of transcription factors, and imprinting and modifications of chromatin structure. Taken together these components of nuclear architecture facilitate biological requirements for physiologically responsive modifications in gene expression within the contexts of (1) homeostatic control involving rapid, short-term, transient responsiveness; (2) developmental control which is progressive and stage-specific; and (3) differentiation-related control which is associated with long-term phenotypic commitments to gene expression for support of the structural and functional properties of cells and tissues.

We are just beginning to appreciate the significance of nuclear domains in the control of gene expression. However, it is already apparent that local nuclear environments which are generated by the multiple aspects of nuclear structure are intimately tied to developmental expression of cell growth and tissue-specific genes. From a broader perspective, it is becoming increasingly evident that, reflecting the diversity of regulatory requirements as well as the phenotype-specific and physiologically responsive representation of nuclear structural proteins, there is a reciprocally functional relationship between nuclear structure and gene expression. Nuclear structure is a primary determinant of transcriptional control, and the expressed genes modulate the regulatory components of nuclear architecture. Thus, the power of addressing gene expression within the three-dimensional context of nuclear structure would be difficult to overestimate. Membrane-mediated initiation of signaling pathways that ultimately influence transcription has been recognized for some time. Here the mechanisms which sense, amplify, dampen, and/or integrate regulatory signals involve structural as well as functional components of cellular membranes. Extending the structure–regulation paradigm to nuclear architecture expands the cellular context in which cell structure–gene expression interrelationships are operative.

II. DEVELOPMENTAL TRANSCRIPTIONAL CONTROL DURING PROLIFERATION AND DIFFERENTIATION: REGULATION OF CELL CYCLE-DEPENDENT HISTONE GENES AND THE BONE-SPECIFIC OSTEOCALCIN GENE DURING PROGRESSIVE EXPRESSION OF THE OSTEOBLAST PHENOTYPE

It has been well documented that differentiation is a multistep process, orchestrated by a complex and interdependent series of stringently controlled regulatory events. Expression of genes supporting proliferation, and those which mediate both development and maintenance of cell and tissue phenotypic properties, are responsive to intracellular as well as extracellular regulatory cues. Osteoblast differentiation is a striking example of a process in which progression of temporally compartmentalized gene expression, together with cross-talk between regulatory events requisite for each developmental period, is necessary for establishment of bone tissue organization.

The sequential expression of cell growth and tissue-specific genes during osteoblast differentiation is supported by developmental transcriptional control (Stein *et al.*, 1990; Stein and Lian, 1993). Both *in vivo* and *in vitro* cultures of normal diploid osteoblasts, proliferating cells express genes which mediate competency for cell growth as well as extracellular matrix biosynthesis. Postproliferatively, genes functionally related to the organization and mineralization of the bone extracellular matrix are expressed (Stein *et al.*, 1990; Stein and Lian, 1993). From the perspective of growth control, there is a requirement to support expression of genes for proliferation in early-stage cells. At the same time, expression of genes associated with the postproliferative acquisition of bone cell and tissue phenotypic properties must be repressed. Then, at a key transition point which marks completion of the initial developmental period, proliferation is downregulated and gene expression which supports the establishment and maintenance of bone tissue organization is upregulated.

In this chapter we will focus on regulatory mechanisms controlling transcription of the cell cycle-regulated genes in proliferating osteoblasts and those controlling transcription of the bone-specific osteocalcin (OC) gene in mature osteoblasts during extracellular matrix mineralization. Transcription of these cell growth and tissue-specific genes will be considered within the context of regulatory contributions from principal components of nuclear architecture.

A. Histone Gene Promoter as Model for Integration of Regulatory Signals Mediating Cell Cycle Control and Proliferation/Differentiation Interrelationships

The histone gene promoter is a paradigm for cell cycle-mediated transcriptional control (viewed in Stein *et al.,* 1992, 1994). Transcription is constitutive throughout the cell cycle, upregulated at the onset of S phase and completely suppressed in quiescent cells or following the onset of differentiation (Baumbach *et al.,* 1984; Detke *et al.,* 1979; Plumb *et al.,* 1983a,b). Consequently, activity of the promoter is responsive to regulatory signals which contribute to transcriptional competency for cell cycle progression at the G_1/S phase transition point and to transcriptional downregulation postproliferatively (Fig. 1). The modularly organized promoter regulatory elements of the histone H4 gene promoter and the cognate transcription factors have been characterized within the context of cell cycle-dependent regulatory parameters (van Wijnen *et al.,* 1989, 1992; Ramsey-Ewing *et al.,* 1994; Pauli *et al.,* 1987). There is a direct indication that the cell cycle regulatory element exhibits phosphorylation-dependent modifications in transcription factor interactions which parallel and are functionally related to cell cycle as well as growth control of histone gene expression. The S phase transcription factor complexes assembling at the H4 promoter include cdc2, cyclin A, retinoblastoma (Rb)-related protein, and interferon regulatory factor-2 (IRF-2) (van Wijnen *et al.,* 1994; Vaughan *et al.,* 1995), reflecting an integration of phosphorylation-mediated control of histone gene expression, enzymes involved in DNA replication, and growth stimulation and suppression at the G_1/S phase transition point (Fig. 2).

This regulatory mechanism is not restricted to cell cycle-dependent transcriptional control of the histone H4 gene. There is an analogous representation and organization of regulatory motifs in the histone H3 and histone H1 gene promoters supporting coordinate control of histone genes which are coexpressed (van den Ent *et al.,* 1994). In a broader biological context, there are similarities between histone gene cell cycle regulatory element sequences and those in the proximal promoter of the thymidine kinase gene which exhibits enhanced transcription during

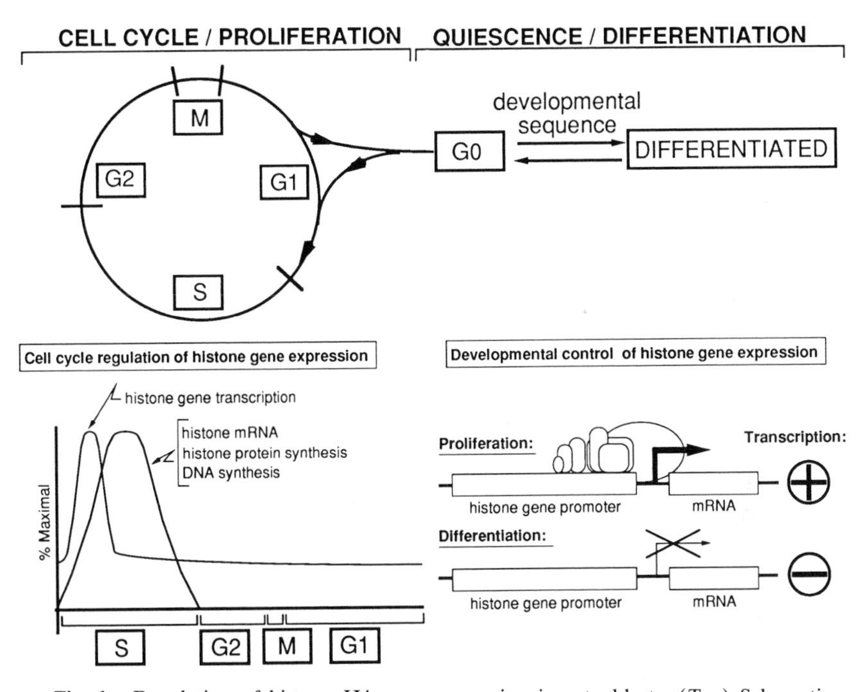

Fig. 1. Regulation of histone H4 gene expression in osteoblasts. (*Top*) Schematic representation of the cell cycle (G_1, S, G_2, mitosis), indicating the pathway associated with the postproliferative onset of differentiation initiated following completion of mitosis. (*Lower left*) Representation of data defining the principal biochemical parameters of histone gene expression, indicating restriction of histone protein synthesis and the presence of histone mRNA to S phase cells (DNA synthesis). Constitutive transcription of histone genes occurs throughout the cell cycle, with an enhanced transcriptional level during the initial 2 hr of S phase. These results establish the combined contribution of transcription and mRNA stability to the S phase-specific regulation of histone biosynthesis in proliferating cells, with histone mRNA levels as the rate-limiting step. (*Lower right*) In contrast, the completion of proliferative activity at the onset of differentiation is mediated by transcriptional downregulation of histone gene expression, supported by a parallel decline in rate of transcription and cellular mRNA levels.

S phase (Dou *et al.,* 1992). It may therefore be possible that genes functionally linked to DNA replication may, at least in part, be coordinately controlled. Support for such a mechanism is provided by analogous promoter domains of the histone and thymidine kinase promoters which both interact with transcription factor complexes that include cyclins, cyclin-dependent kinases, and Rb-related proteins.

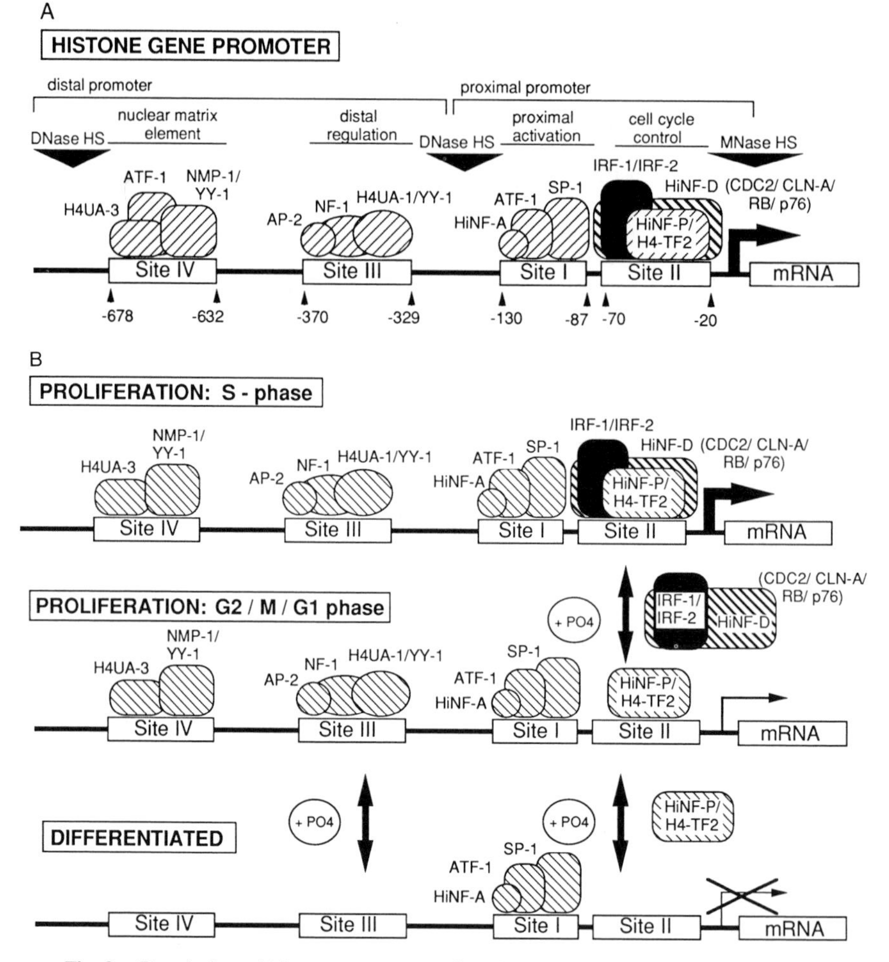

Fig. 2. Regulation of histone gene expression during the cell cycle. (A) Organization of the human histone H4 gene promoter regulatory elements (sites I–IV) is illustrated. The transcription factors which exhibit sequence-specific interactions with these domains are indicated during the S phase of the cell cycle when the gene is maximally transcribed. Site II contributes to cell cycle regulation of transcription. Site IV binds a nuclear matrix protein complex (NMP-1/YY-1), while the protein–DNA interactions at sites III and I support general transcriptional enhancement. The site II complex includes cyclin A, cyclin-dependent kinase cdc2, an RB-related protein, and IRF growth regulatory factors, reflecting integration of phosphorylation-mediated control of histone gene expression. (B) (*Top*) Occupancy of the four principal regulatory elements of the histone H4 gene promoter during the S phase of the cell cycle when transcription is maximal is schematically

B. Transcriptional Control of Bone-Specific Osteocalcin Gene at Onset of Extracellular Matrix Mineralization in Postproliferative Osteoblasts

Influences of promoter regulatory elements that are responsive to basal and tissue-restricted transactivation factors, steroid hormones, growth factors, and other physiological mediators have provided the basis for understanding regulatory mechanisms contributing to developmental expression of osteocalcin, tissue specificity, and biological activity (Stein *et al.,* 1990; Stein and Lian, 1993; Lian and Stein, 1996). These regulatory elements and cognate transcription factors support postproliferative transcriptional activation and steroid hormone (e.g., vitamin D) enhancement at the onset of extracellular matrix mineralization during osteoblast differentiation (Fig. 3). Thus, the bone-specific OC gene is organized in a manner which supports responsiveness to homeostatic physiological mediators and developmental expression in relation to bone cell differentiation.

The regulatory sequences illustrated in Fig. 3 have been established in the OC gene promoter and coding region by one or more criteria that include (1) demonstration of an influence on transcriptional activity by deletion, substitution, or site-specific mutagenesis *in vitro* and *in vivo;* (2) identification and characterization of sequence-specific regulatory element occupancy by cognate transcription factors *in vitro* and *in vivo;* (3) modifications in protein–DNA interactions as a function of biological activity; and (4) consequential modifications in functional activity following overexpression or suppression of factors which exhibit sequence-specific recognition for regulatory domains. A series of elements contributing to basal expression (Lian *et al.,* 1989) include a TATA sequence (located at −42 to −39) and the osteocalcin box (OC box), a 24-nucleotide element with a CCAAT motif as a central core, both

---

illustrated. (*Middle*) The site II transcription factor complex is modified by phosphorylation during the G_1/G_2/mitotic periods of the cell cycle, resulting in modifications in levels of transcription. Phosphorylation-dependent dissociation of the IRF and HiNF-D (cdc2, cyclin A, and RB-related protein) factors occurs in non-S phase cells. (*Bottom*) The complete loss of transcription factor complexes at sites II, III, and IV occurs following exit from the cell cycle with the onset of differentiation. At this time, transcription is completely downregulated.

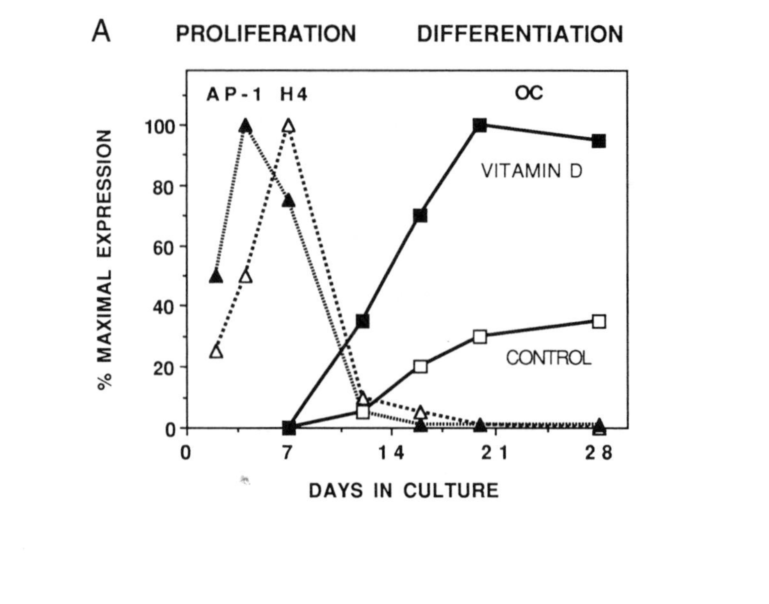
A
PROLIFERATION
DIFFERENTIATION
AP-1
H4
OC
VITAMIN D
CONTROL
% MAXIMAL EXPRESSION
100
80
60
40
20
0
0
7
14
21
28
DAYS IN CULTURE

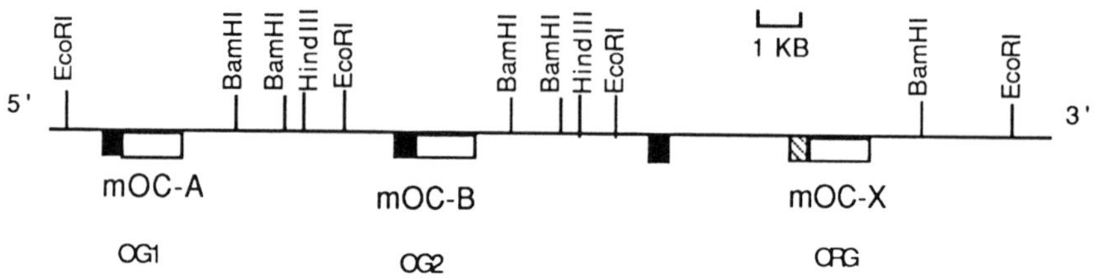
B
EcoRI
BamHI
BamHI
HindIII
EcoRI
BamHI
BamHI
HindIII
EcoRI
1 KB
BamHI
EcoRI
5'
3'
mOC-A
mOC-B
mOC-X
OG1
OG2
ORG

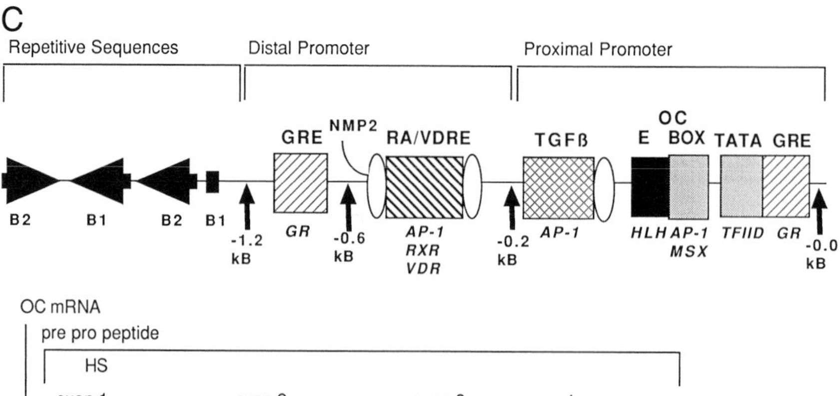
C
Repetitive Sequences
Distal Promoter
Proximal Promoter
GRE
NMP2
RA/VDRE
TGFß
E
OC BOX
TATA
GRE
B2
B1
B2
B1
-1.2 kB
GR
-0.6 kB
AP-1
RXR
VDR
-0.2 kB
AP-1
HLH
AP-1
MSX
TFIID
GR
-0.0 kB
OC mRNA
pre pro peptide
HS

exon 1
SIL
exon 2
exon 3
exon 4

required to render the gene transcribable (Towler *et al.*, 1994; Kawaguchi *et al.*, 1992). The OC box is a highly conserved regulatory sequence required for basal expression of the rat, mouse, and human OC genes. The OC box additionally serves a regulatory function in defining the threshold for initiation of transcription and contributes to bone tissue-specific expression of the OC gene (Heinrichs *et al.*, 1993a,b, 1995). However, caution must be exercised in attributing tissue-specific transcriptional control to a single element. Contributions of multiple sequences appear to be operative in tissue-specific regulation, thereby providing opportunities for expression of the osteocalcin gene in bone under diverse biological circumstances. Multiple glucocorticoid-responsive elements (GREs) with sequences that exhibit both strong and weak affinities for glucocorticoid receptor binding have been identified in the proximal promoter (Stromstedt *et al.*, 1991; Heinrichs *et al.*, 1993a; Aslam *et al.*, 1995). Interactions of other transcription factors with the proximal GREs including NF-IL6 have been reported (Towler and Rodan, 1994), further expanding the potential of the OC gene to be transcriptionally regulated by glucocorticoids. It is reasonable to consider

---

Fig. 3. Organization and expression of the osteocalcin gene during osteoblast differentiation. (A) Expression of the osteocalcin gene (open squares) and vitamin D enhancement (solid squares) during the postproliferative period of the osteoblast developmental sequence is shown. H4 histone mRNA (open triangles) and AP-1 activity (solid triangles) are indications of proliferative activity. (B) A 25-kb segment of the genome is shown which includes three osteocalcin genes designated MOCA, MOCB, and MOCX. MOCA and MOCB are expressed in bone, while MOCX is expressed during the late prenatal and early postnatal periods in nonskeletal tissues. (C) The organization of the bone-specific osteocalcin gene is schematically illustrated, indicating the regulatory domains within the initial 700 nucleotides 5′ to the transcription start site. In the proximal promoter, several classes of transcription factors are represented which bind to key regulatory elements. The OC box is the primary tissue-specific transcriptional element that binds homeodomain proteins (MSX). Fos/jun-related proteins form heterodimers at AP-1 sites or a cryptic tissue-specific complex. HLH proteins bind to the contiguous E box motif. The nuclear matrix protein-binding sites interact with an AML-related transcription factor. Several glucocorticoid response elements (GREs) are indicated, as well as the vitamin D response element (VDRE), which is a primary enhancer sequence. Additionally, AP-1 sites are indicated that overlap the TGFβ (TGRE) and the VDRE. The combined and integrated activities of overlapping regulatory elements and associated transcription factors provide a mechanism for developmental control of expression during osteoblast growth and differentiation.

that the OC gene GREs may be selectively utilized in a developmentally and/or physiologically responsive manner. The possibility of functional interactions with transcription factors other than glucocorticoid receptors under certain conditions should not be dismissed.

The vitamin D-responsive element (VDRE) functions as an enhancer (Morrison *et al.,* 1989; Demay *et al.,* 1990; Markose *et al.,* 1990; Kerner *et al.,* 1989; Terpening *et al.,* 1991). The VDRE transcription factor complex appears to be a target for modifications in vitamin D-mediated transcription by other physiological factors including tumor necrosis factor α (TNF-α) (Kuno *et al.,* 1994) and retinoic acid (Scule *et al.,* 1990; Bortell *et al.,* 1993; Schrader *et al.,* 1993; Macdonald *et al.,* 1993; Kliewer *et al.,* 1992). Additional regulatory sequences include an NF-κB site also reported to be involved in regulation mediated by TNF-α (Li and Stashenko, 1993); a series of AP-1 sites (Ozono *et al.,* 1991; Lian *et al.,* 1991; Demay *et al.,* 1992), one of which mediates transforming growth factor β (TGFβ) responsiveness (Lian and Stein, 1993; Banerjee *et al.,* 1995); an E box (Tamura and Noda, 1994) that binds HLH containing transcription factor complexes; and a sequence in the proximal promoter that binds a multisubunit complex containing a CP1/NR-Y-CBF-like CAAT factor complex. Two OC gene promoter regulatory domains which exhibit recognition for transcription factors that mediate developmental pattern formation are an MSX binding site within the OC box (Towler *et al.,* 1994a; Heinrichs *et al.,* 1993b, 1995; Hoffmann *et al.,* 1994) and an AML-1 site (*runt* homology) sequence (Merriman *et al.,* 1995; van Wijnen *et al.,* 1994b). These sequences may represent components of regulatory mechanisms that contribute to pattern formation associated with bone tissue organization during initial developmental stages and subsequently during tissue remodeling. Involvement of other homeodomain-related genes that are expressed during skeletal development in the control of osteoblast proliferation and differentiation is worthy of consideration. These include but are not restricted to the families of *Dlx* and *PAX* genes (Lufkin *et al.,* 1991, 1992; Krumlauf, 1993; Gruss and Waltner, 1992; Tabin, 1991; Niehrs and DeRobertis, 1992; McGinnis and Krumlauf, 1992; Simeone *et al.,* 1994; Cohen *et al.,* 1989). Although a majority of the responses identified reside in the region of the promoter which spans the VDRE domain to the first exon, upstream sequences that must be further defined may contribute to both basal and enhancer-mediated control of transcription (Lian and Stein, 1993; Terpening *et al.,* 1991; Yoon *et al.,* 1988; Bortell *et al.,* 1992; Morrison and Eisman, 1993; Aslam *et al.,* 1994; Owen *et al.,* 1993). A GRE reading at −683 to −697 is an example of such an upstream regulatory element (Aslam *et al.,* 1994). Additionally, intragenic and downstream regulatory sequences are a viable candidate for regulatory involvement in OC gene expression.

The overlapping and contiguous organization of regulatory elements, as illustrated by the TATA/GRE, E box/AP-1/CCAAT/homeodomain and TNF-α/VDRE (Kuno *et al.,* 1994) AP-1/VDRE, provides a basis for combined activities that support responsiveness to physiological mediators (Owen *et al.,* 1990; Nanes *et al.,* 1990, 1991; Taichman and Hauschka, 1992; Li and Stashenko, 1992; Evans *et al.,* 1990; Guidon *et al.,* 1993; Jenis *et al.,* 1993; Vaishnav *et al.,* 1988; Fanti *et al.,* 1992; Schedlich *et al.,* 1994). Additionally, hormones modulate binding of transcription factors other than the cognate receptor to nonsteroid regulatory sequences. For example, vitamin D-induced interactions occur at the basal TATA domain (Owen *et al.,* 1993) and 1,25-$(OH)_2D_3$ upregulates MSX-2 binding to the OC box homeodomain motif, as well as supporting increased MSX-2 expression (Towler *et al.,* 1994; Kawaguchi *et al.,* 1992). It is this complexity of OC gene promoter element upregulation that allows for hormone responsiveness in relation to either basal or enhanced levels of expression. The protein–DNA interactions at the principal promoter regulatory elements which mediate levels of OC gene transcription are schematically summarized in Fig. 3.

III. NUCLEAR STRUCTURE SUPPORTING CELL CYCLE STAGE-SPECIFIC HISTONE GENE TRANSCRIPTION IN PROLIFERATING OSTEOBLASTS

A. Chromatin Structure and Nucleosome Organization

A synergistic contribution of activities by sites I, II, III, and IV H4 histone gene promoter elements to the timing and extent of H4-FO108 gene transcription has been established experimentally (Kroeger *et al.,* 1987; Wright *et al.,* 1992; Birnbaum *et al.,* 1995; Ramsey-Ewing, *et al.,* 1994). The integration of intracellular signals that act independently on these multiple elements may occur partly in the three-dimensional organization of the promoter within the spatial context of nuclear architecture (Fig. 4).

The presence of nucleosomes in the H4 promoter when the gene is transcriptionally active (Moreno *et al.,* 1986, 1987; Chrysogelos *et al.,*

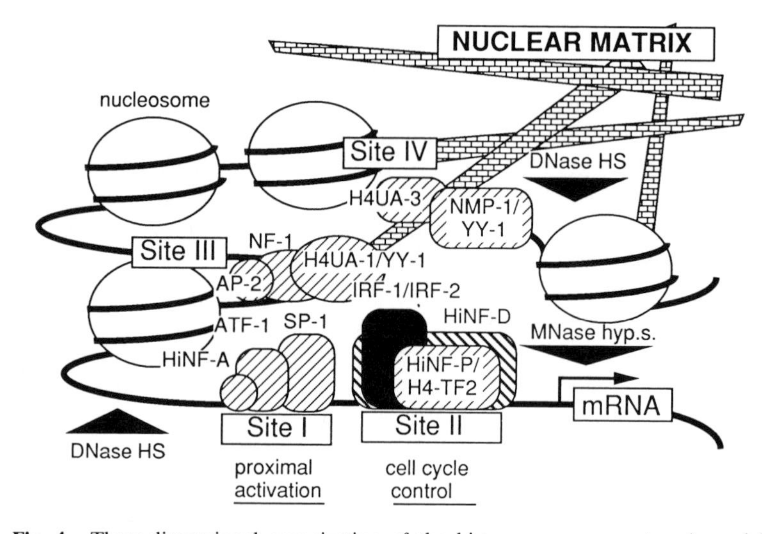

Fig. 4. Three-dimensional organization of the histone gene promoter. A model is schematically presented for the spatial organization of the rat osteocalcin gene promoter based on evidence of nucleosome placement and the interaction of DNA-binding sequences with the nuclear matrix. These components of chromatin structure and nuclear architecture restrict mobility of the promoter and impose physical constraints that reduce distances between proximal and distal promoter elements. Such a postulated organization of the osteocalcin gene promoter can facilitate cooperative interactions for cross-talk between elements that mediate transcription factor binding and consequently determine the extent to which the gene is transcribed.

1985, 1989) (see Fig. 6) may increase the proximity of independent regulatory elements, in addition to supporting synergistic and/or antagonistic cooperative interactions between histone gene DNA-binding activities. In addition, chromatin structure and nucleosomal organization vary as a function of the cell cycle (Moreno *et al.*, 1986, 1987; Chrysogelos *et al.*, 1985, 1989), which may enhance and restrict accessibility of transcription factors and modulate the extent to which DNA-bound factors are phosphorylated.

Parameters of chromatin structure and nucleosome organization were experimentally established by accessibility of DNA sequences within nuclei of intact cells to a series of nucleases that include micrococcal nuclease for establishing nucleosome placement, deoxyribonuclease (DNase) I for mapping nuclease hypersensitive sites, S1 nuclease for determining single-stranded DNA sequences, and restriction endonucleases for determining protein–DNA interactions within specific sites at single-nucleotide resolution (Fig. 5). Findings from these experiments illustrate that nucleosome spacing exhibits cell cycle-dependent variations in response to levels of histone gene expression and/or nuclear

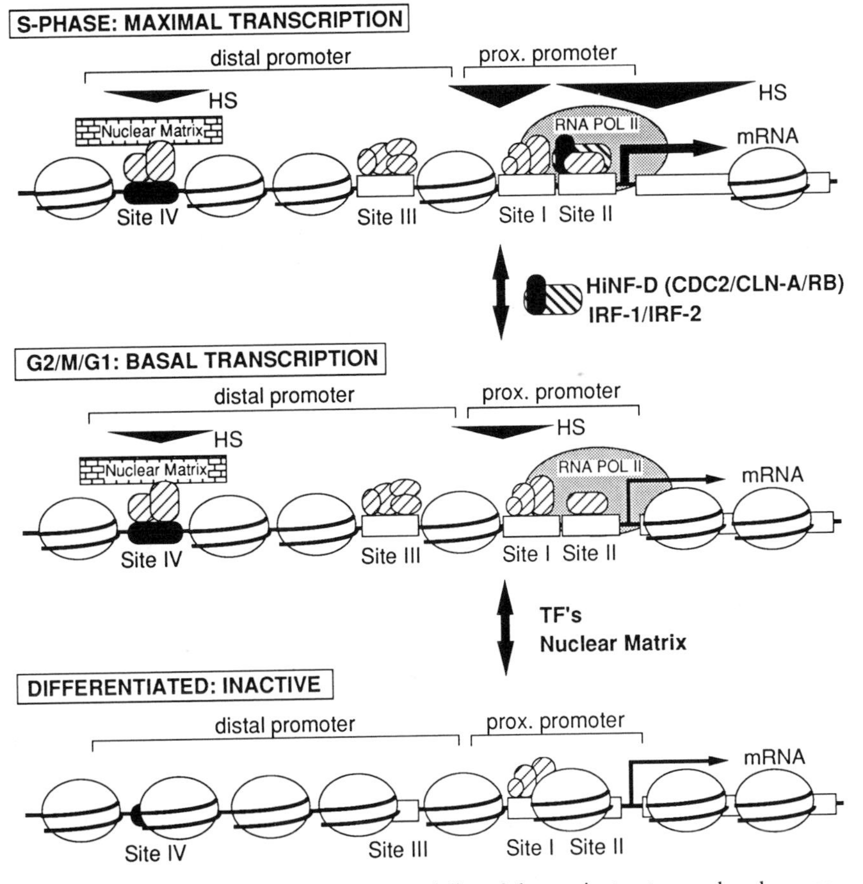

Fig. 5. Schematic illustration of the remodeling of chromatin structure and nucleosome organization which accommodates cell cycle stage-specific and developmental parameters of histone gene promoter architecture to support modifications in the level of expression. Placement of nucleosomes and representation as well as magnitude of nuclease hypersensitive sites (solid triangles) are designated. The principal regulatory elements and transcription factors are shown.

organization related to mitotic division. A combined modification is observed in accessibility of histone gene sequences to micrococcal nuclease, DNase I, S1 nuclease, and restriction endonucleases, reflecting a remodeling of chromatin architecture related to what extent the histone gene is transcribed to support proliferative activity and cell cycle progression.

B. The Nuclear Matrix

A nuclear matrix attachment site has been identified in the upstream region (−0.8 kb) of the H4-FO108 promoter (Dworetzky *et al.,* 1992), which may serve two functions: (1) imposing constraints on chromatin structure and (2) concentrating and localizing transcription factors. Such roles for the nuclear matrix in the regulation of histone gene expression are supported by distinct modifications in the composition of nuclear matrix proteins observed when proliferation-specific genes are downregulated during differentiation (Dworetzky *et al.,* 1990) and, more directly, by the isolation of ATF-related and YY1 transcription factors from the nuclear matrix (Dworetzky *et al.,* 1992; Guo *et al.,* 1995) which interact with site IV or the H4-FO108 gene promoter.

The specific mechanisms by which the 5′ histone gene promoter elements and sequence-specific transactivating factors participate in regulating transcription of the histone H4-FO108 gene remain to be determined. However, regulation is unquestionably operative within the context of the complex series of spatial interactions which are responsive to a broad spectrum of biological signals (Figs. 4, 5).

IV. NUCLEAR STRUCTURE SUPPORTING DEVELOPMENTAL AND STEROID HORMONE-RESPONSIVE OSTEOCALCIN GENE TRANSCRIPTION DURING OSTEOBLAST DIFFERENTIATION

A. Chromatin Structure and Nucleosome Organization

Modifications in parameters of chromatin structure and nucleosome organization parallel both competency for transcription and the extent to which the OC gene is transcribed. Changes are observed in response to physiological mediators of basal expression and steroid hormone responsiveness. This remodeling of chromatin provides a conceptual and experimental basis for the involvement of nuclear architecture in developmental, homeostatic, and physiological control of OC gene expression during establishment and maintenance of bone tissue structure and activity (Figs. 6–11).

In both normal diploid osteoblasts and osteosarcoma cells, basal expression and enhancement of OC gene transcription are accompanied by two alterations in the structural properties of chromatin. DNase I hypersensitivity of sequences flanking the tissue-specific OC box and the VDRE enhancer domain are observed (Montecino *et al.*, 1994a,b; Breen *et al.*, 1994). Together with modifications in nucleosome placement (Montecino *et al.*, 1994b), a basis for accessibility of transactivation factors to basal and steroid hormone-dependent regulatory sequences can be explained. In early-stage proliferating normal diploid osteoblasts, when the OC gene is repressed, nucleosomes are placed in the OC box and in VDRE promoter sequences; nuclease-hypersensitive sites are not present in the vicinity of these regulatory elements. In contrast, when OC gene expression is transcriptionally upregulated postproliferatively and vitamin D-mediated enhancement of transcription occurs, the OC box and VDRE become nucleosome free and these regulatory domains are flanked by DNase I-hypersensitive sites (Figs. 6, 7).

Functional relationships between structural modifications in chromatin and OC gene transcription are observed in response to 1,25-$(OH)_2D_3$ in ROS 17/2.8 osteosarcoma cells which exhibit vitamin D-responsive transcriptional upregulation. There are marked changes in nucleosome placement at the VDRE and OC box, as well as DNase I hypersensitivity of sequences flanking these basal and enhancer OC gene promoter sequences (Montecino *et al.*, 1994a,b; Breen *et al.*, 1994). The complete absence of hypersensitivity, and the presence of nucleosomes in the VDRE and OC box domains of the OC gene promoter in ROS 24/1 cells which lack the vitamin D receptor, additionally corroborate these findings (Montecino *et al.*, 1994a; Breen *et al.*, 1994) (Figs. 8, 9). These steroid hormone-sensitive alterations in chromatin structure have been confirmed by restriction enzyme accessibility of promoter sequences within intact nuclei (Montecino *et al.*, 1994a) and by LMPCR (ligation-mediated polymerase chain reaction) (Montecino *et al.*, 1996b) at single-nucleotide resolution.

We have found that agents which induce histone hyperacetylation (sodium butyrate) promote reorganization of the nucleosomal structure

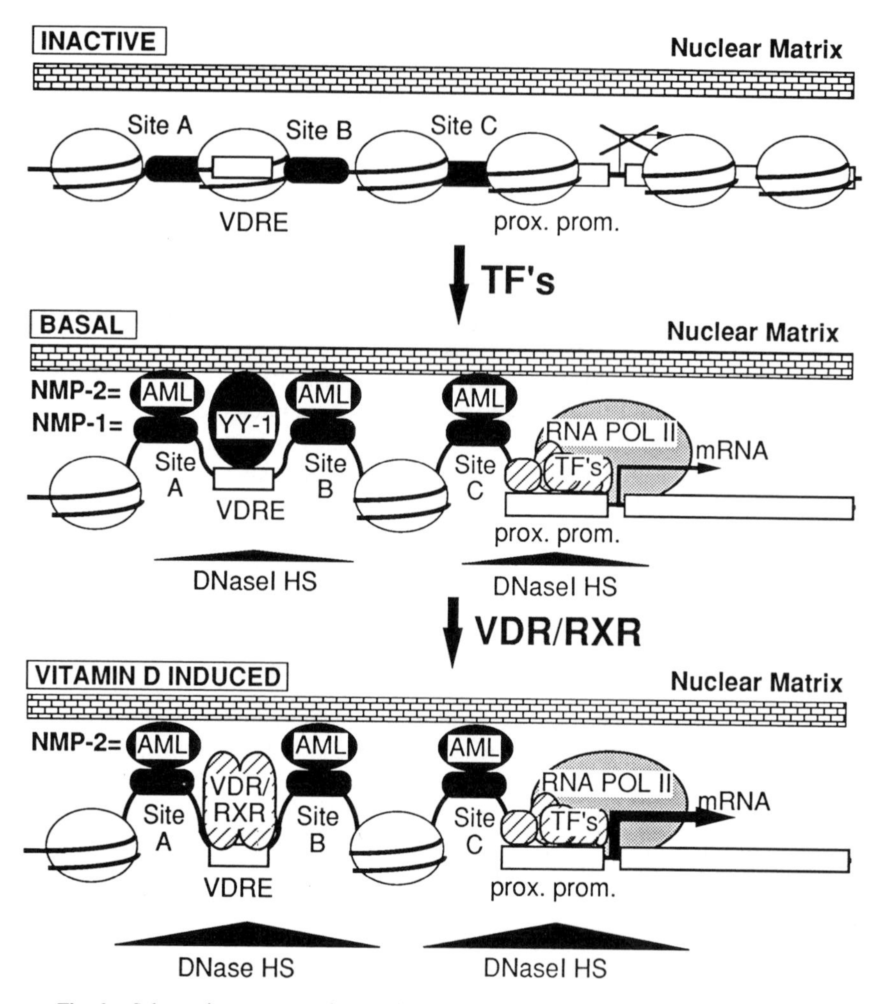

Fig. 6. Schematic representation of the osteocalcin gene promoter's organization and occupancy of regulatory elements by cognate transcription factors paralleling and supporting functional relationships to either (*top*) suppression of transcription in proliferating osteoblasts, (*middle*) activation of expression in differentiated cells, or (*bottom*) enhancement of transcription by vitamin D. The placement of nucleosomes is indicated. Remodeling of chromatin structure in nucleosome organization to support suppression, basal, and vitamin D-induced transcription of the osteocalcin gene is indicated. The representation and magnitude of DNase I hypersensitive sites are designated by solid triangles, and gene–nuclear matrix interactions are shown.

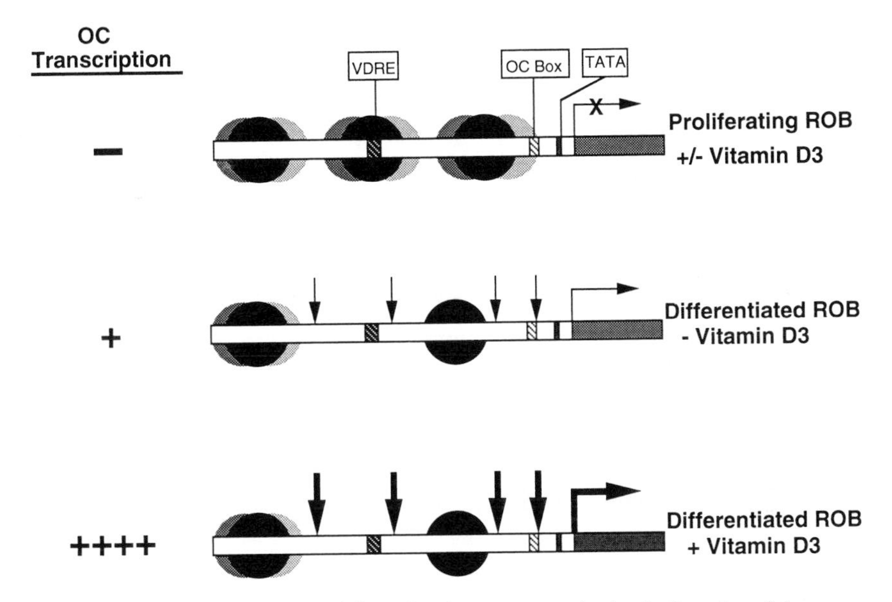

Fig. 7. Developmental remodeling of nucleosome organization in the osteocalcin gene promoter correlates with transcriptional activity during differentiation of normal diploid osteoblasts. The positioning of nucleosomes in the osteocalcin gene promoter was determined by combining DNaseI, micrococcal nuclease, and restriction endonuclease digestions with indirect endlabeling. The filled circles represent the placement of nuclesomes, with the shadows indicating movement within nuclease-protected segments. The vertical arrows correspond to the limits of the distal (−600 to −400) and proximal (−170 to −70) DNase I hypersensitive sites, which increase in differentiated normal diploid osteoblasts in response to vitamin D. The VDRE (−465 to −437) and the osteocalcin box (−99 to −77) are designated.

in the distal region of the OC gene promoter (including VDRE). This transition results in inhibition of the vitamin D-induced upregulation of basal transcription in ROS 17/2.8 cells (Fig. 10). Additionally, we have established an absolute requirement for sequences residing in the proximal region of the OC gene promoter for both formation of the proximal DNase I hypersensitive site and basal transcriptional activity. Our approach was to assay nuclease accessibility (DNase I and restriction endonucleases) in ROS 17/2.8 cell lines stably transfected with promoter deletion constructs driving expression of a CAT reporter gene (Fig. 11).

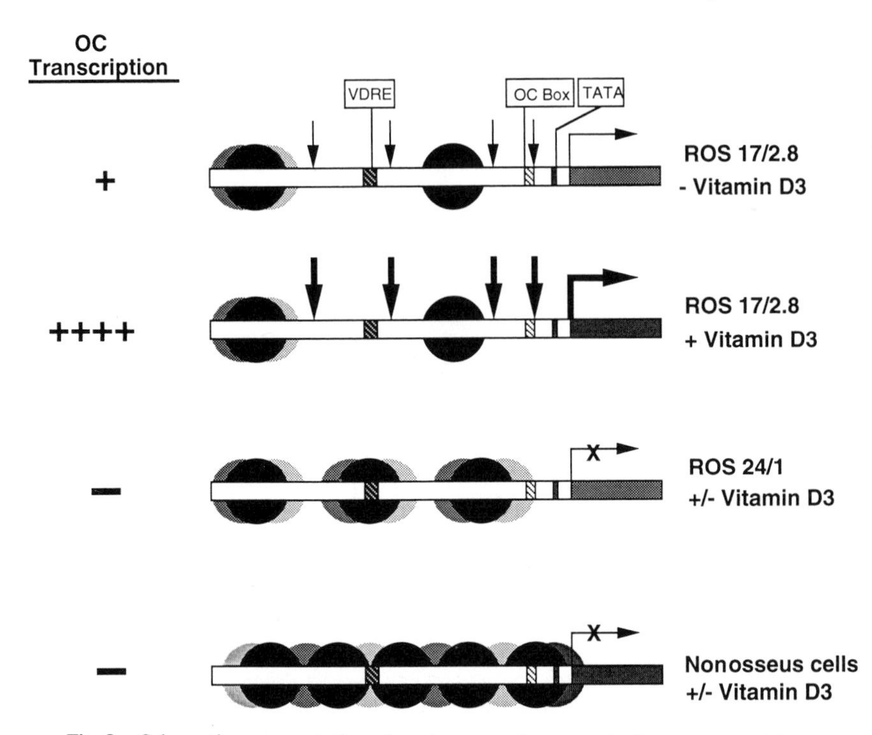

Fig. 8. Schematic representation of nucleosome placement in the rat osteocalcin gene promoter in osteosarcoma cells in relation to osteocalcin gene transcription. The presence and positioning of nucleosomes in the osteocalcin gene promoter of ROS 17/2.8 cells (constitutively expressing osteocalcin) and ROS 24/1 cells (not expressing osteocalcin) were determined by combining nuclease sensitivity (DNase I, micrococcal nuclease, and restriction endonuclease) with indirect endlabeling. The filled circles represent positioned nucleosomes, and the shadows indicate movement of nucleosomes within protected segments. The vertical arrows correspond to the limits of the distal (−600 to −400) and proximal (−170 to −70) DNase I hypersensitive sites, which are increased following vitamin D treatment. The VDRE (−465 to −437) and the OC box (−99 to −77) are designated within the distal and proximal nuclease hypersensitive domains, respectively.

B. The Nuclear Matrix

Involvement of the nuclear matrix in the control of OC gene transcription is provided by several lines of evidence. One of the most compelling is association of a bone-specific nuclear matrix protein designated NMP2

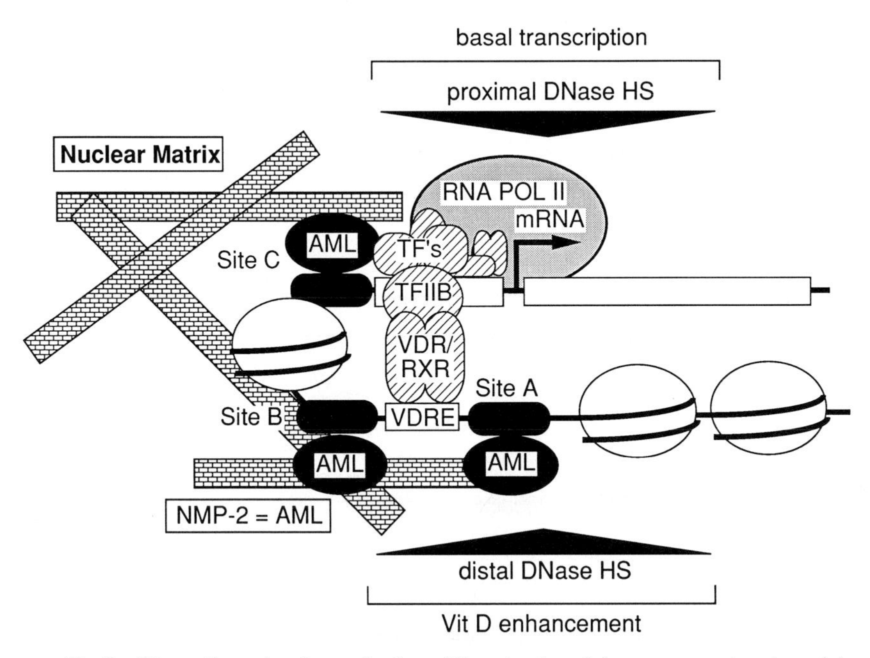

Fig. 9. Three-dimensional organization of the rat osteocalcin gene promoter. A model is schematically presented for the spatial organization of the rat osteocalcin gene promoter based on evidence of nucleosome placement and the interaction of DNA binding sequences with the nuclear matrix. These components of chromatin structure and nuclear architecture restrict mobility of the promoter and impose physical constraints that reduce distances between proximal and distal promoter elements. Such postulated organization of the osteocalcin gene promoter can facilitate cooperative interactions for cross-talk between elements that mediate transcription factor binding and consequently determine the extent to which the gene is transcribed.

with sequences flanking the VDRE of the OC gene promoter (Bidwell *et al.,* 1993). Initial characterization of the NMP2 factor has revealed that a component is an AML-1-related transactivation protein which is a runt homology factor associated with developmental pattern formation in *Drosophila* (van Wijnen *et al.,* 1994b). These results implicate the nuclear matrix in regulating events that mediate structural properties of the VDRE domain and may also contribute to linkage of OC gene expression, with modulation of the pattern formation requisite for skeletal tissue organization during bone formation and remodeling. Here the possibility of multiple regulatory sequences in the OC gene promoter that

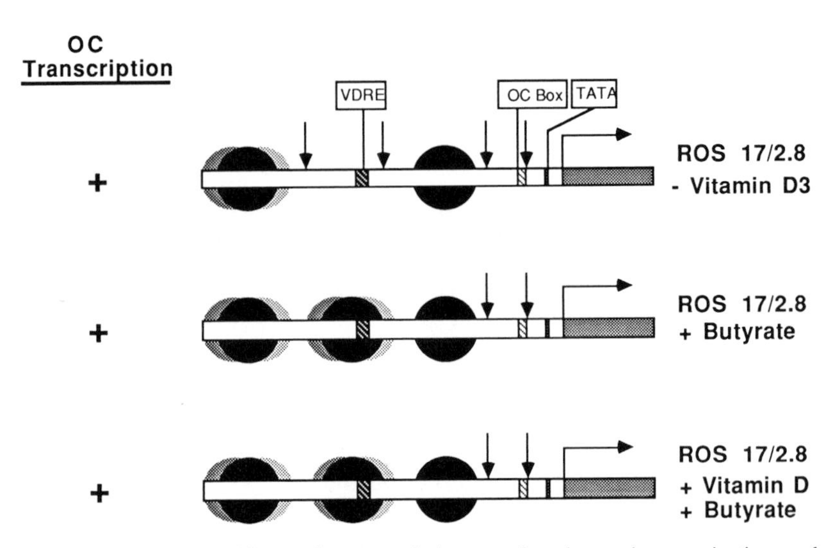

Fig. 10. Effect of histone hyperacetylation on the chromatin organization and transcriptional activity of the osteocalcin gene in osteoblastic cells. Changes in chromatin structure of the osteocalcin gene promoter of ROS 17/2.8 cells following treatment with a histone deacetylase inhibitor (sodium butyrate) were determined by combining incubation of isolated nuclei with nucleases (DNase I, micrococcal nuclease, and restriction endonucleases) and detection by the indirect end-labeling method. The filled circles represent the putative nucleosomes, and the vertical arrows correspond to the limits of the distal (−600 to −400) and proximal (−170 to −70) DNase I hypersensitive sites. The result of this treatment is that the VDRE is no longer within a DNase I hypersensitive site, which also correlates with a block in the vitamin D-induced upregulation of osteocalcin gene transcription.

are functionally associated with pattern formation is indicated because of binding by MSX homeodomain proteins at the OC box (Hoffmann *et al.*, 1994; Towler and Rodan, 1994).

It is apparent from available findings that the linear organization of gene regulatory sequences is necessary but insufficient to accommodate the requirements for physiological responsiveness to homeostatic, developmental, and tissue-related regulatory signals. It would be presumptuous to propose a formal model for the three-dimensional organization of the OC gene promoter. However, the working model presented in Figs. 6 and 9 represents postulated interactions between OC gene promoter elements that reflect the potential for integration of activities by nuclear architecture to support requirements at the cell and tissue levels.

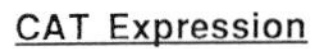

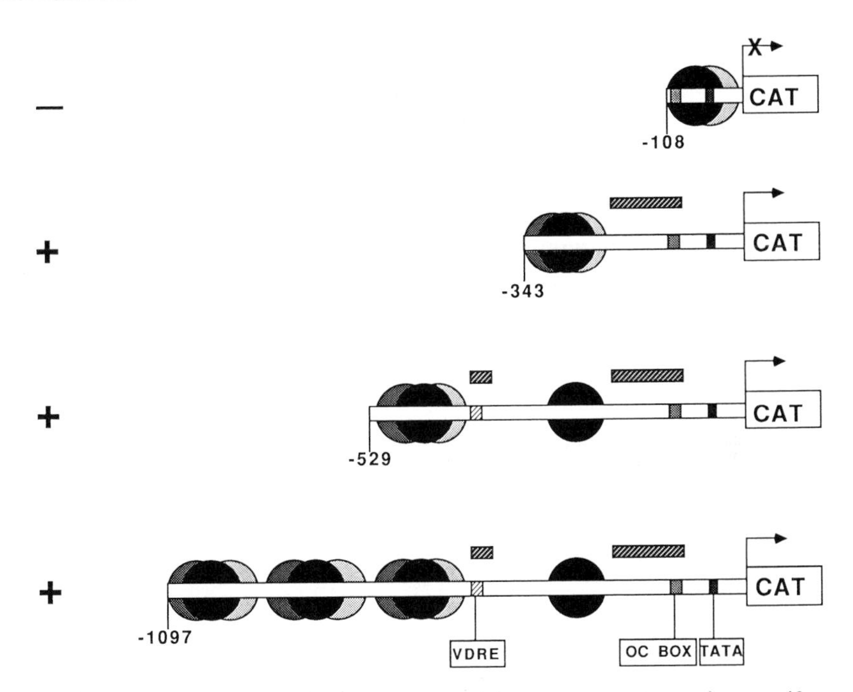

Fig. 11. Chromatin organization of the osteocalcin gene promoter requires specific promoter sequences. Constructs including sequential deletions from the rat osteocalcin gene promoter were stably transfected in ROS 17/2.8 cells. Expression was determined by CAT activity and chromatin organization, analyzed as described in Figs. 8, 9, and 10. Horizontal bars represent the distal and proximal DNase I hypersensitive sites and the filled circles the putative nucleosomes.

A functional role of the nuclear matrix in steroid hormone-mediated transcriptional control of the OC gene is further supported by overlapping binding domains within the VDRE for the VDR and the NMP-1 nuclear matrix protein, which we have shown to be a YY1 transcription factor. One can speculate that reciprocal interactions of NMP-1 and VDR complexes may contribute to the competency of the VDRE to support transcriptional enhancement. Binding of NMP-2 at the VDRE flanking sequence may establish permissiveness for VDR interactions by gene–nuclear matrix associations which facilitate conformational modifications in the transcription factor recognition sequences.

Taken together, these findings provide a basis for involvement of both the nuclear matrix and the chromatin structure in modulating accessibility of promoter sequences to cognate transcription factors and facilitating the integration of activities at multiple regulatory domains. *In vivo* studies support functional contributions of nuclear matrix proteins to steroid hormone-mediated transcription. Overexpression of acute myelogenous leukemia factor (AML) transcription factors which flank the OC gene VDRE upregulates expression. In contrast, overexpression of YY1 which binds to a site overlapping the OC gene vitamin D receptor binding sequences abrogates the vitamin D enhancement of transcription. Additionally, functional data supporting cross-talk between the gene VDRE and the GATA domain are provided by the demonstration that the transcription factor TF2B and the VDR cooperatively coactivate ligand-dependent transcription (Blanco *et al.,* 1995). Functional interrelationships between the VDRE and THEA domains are further supported by the demonstration that TF2B and the vitamin D receptor are partner proteins by the two-hybrid system (Macdonald *et al.,* 1995).

V. CONCLUSIONS AND PROSPECTS

It is becoming increasingly evident that developmental transcriptional control and modification in transcription to accommodate homeostatic regulation of cell and tissue function are modulated by the integration of a complex series of physiological regulatory signals. Fidelity of responsiveness necessitates the convergence of activities mediated by multiple regulatory elements of gene promoters. Our knowledge of promoter organization and of the repertoire of transcription factors which mediate activities provides a single-dimension map of options for biological control. We are beginning to appreciate the additional structural and functional dimensions provided by chromatin structure, nucleosome organization, and subnuclear localization and targeting of both genes and transcription factors. Particularly exciting is increasing evidence of dynamic modifications in nuclear structure which parallel developmental expression of genes. The extent to which nuclear structure regulates and/or is regulated by modifications in gene expression remains to be experimentally established.

It is necessary to define mechanistically how nuclear matrix-mediated subnuclear distribution of actively transcribed genes is responsive to nuclear matrix association of transcriptional and post-transcriptional regulatory factors. In addition, we cannot dismiss the possibility that association of regulatory factors with the nuclear matrix is consequential to sequence-specific interactions of transcriptionally active genes with the nuclear matrix. As these issues are resolved, we will gain additional insight into determinants of the cause and/or effect relationships which interrelate specific components of nuclear architecture with gene expression at the transcriptional and posttranscriptional levels. However, it is justifiable to anticipate that while nuclear structure–gene expression interrelationships are operative under all biological conditions, situation-specific variations are the rule rather than the exception. As subtleties in the functional component of nuclear architecture are further defined, the significance of nuclear domains to DNA splicing, transcription, and processing of RNA transcripts will be further understood.

A challenge that we face is to establish experimentally the rate-limiting nucleotide sequences and factors which integrate nuclear structure and gene expression. What are the sequence determinants and regulatory proteins which facilitate remodeling of chromatin structure and nucleosome organization to facilitate developmental and homeostatic requirements for transcription? How are structurally and functionally dynamic modifications in chromatin organization related to interactions of genes with the nuclear matrix? There is confidence that these issues will be resolved in part by studies being carried out with lower eukaryotes in which gene content is limited and our knowledge of the genetics is encyclopedic, as well as easily applied to mapping structural and regulatory parameters of gene expression. However, caution must be exercised in extrapolating results from such studies to generalizations which apply to all eukaryotic cells and organisms. While the size of the genome and the content of genetic information are lower in yeast and *Drosophila* than in mammalian cells, the nuclear organization and proteins which package DNA as chromatin are different. These variations may reflect the increased regulatory components in mammalian cells which support both structure and function.

ACKNOWLEDGMENTS

Studies reported in this chapter were supported by grants from the National Institutes of Health (GM32010, AR39588, AR42262) and the March of Dimes Birth Defects Foundation. The authors appreciate the editorial assistance of Elizabeth Bronstein in the preparation of the chapter.

REFERENCES

Abulafia, R., Ben-Ze'Ev, A., Hay, N., and Aloni, Y. (1984). Control of late SV40 transcription by the attenuation mechanism and transcriptionally active ternary complexes are associated with the nuclear matrix. *J. Mol. Biol.* **172,** 467–487.

Aslam, F., Lian, J. B., Stein, G. S., Stein, J. L., Litwack, G., van Wijnen, A. J., and Shalhoub, V. (1994). Glucocorticoid responsiveness of the osteocalcin gene by multiple distal and proximal elements. *J. Bone Miner. Res.* **9,** S125.

Aslam, F., Shalhoub, V., van Wijnen, A. J., Banerjee, C., Bortell, R., Shakoori, A. R., Litwack, G., Stein, J. L., Stein, G. S., and Lian, J. B. (1995). Contributions of distal and proximal promoter elements to glucocorticoid regulation of osteocalcin gene transcription. *Mol. Endocrinol.* **9,** 679–690.

Banerjee, C., Stein, J. L., van Wijnen, A. J., Frenkel, B., Lian J. B., and Stein, G. S. (1996). TGF-β1 responsiveness of the rat osteocalcin gene is mediated by an AP-1 binding site. *Endocrinology* **137,** 1991–2000.

Barrack, E. R., and Coffey, D. S. (1983). Hormone receptors and the nuclear matrix. *In* "Gene Regulation by Steroid Hormones II" (A. K. Roy and J. H. Clark, eds.), pp. 239–266. Springer-Verlag, New York.

Baumbach, L., Marashi, F., Plumb, M., Stein, G. S., and Stein, J. L. (1984). Inhibition of DNA replication coordinately reduces cellular levels of core and H1 histone mRNAs: Requirement for protein synthesis. *Biochemistry* **23,** 1618–1625.

Belgrader, P., Siegel, A. J., and Berezney, R. (1991). A comprehensive study on the isolation and characterization of the HeLa S3 nuclear matrix. *J. Cell Sci.* **98,** 281–291.

Ben-Ze'ev, A., and Aloni, Y. (1983). Processing of SV40 RNA is associated with the nuclear matrix and is not followed by the accumulation of low-molecular weight RNA products. *Virology* **125,** 475–479.

Berezney, R., and Coffey, D. S. (1974). Identification of a nuclear protein matrix. *Biochem. Biophys. Res. Commun.* **60,** 1410–1417.

Berezney, R., and Coffey, D. S. (1975). Nuclear protein matrix: Association with newly synthesized DNA. *Science* **189,** 291–292.

Berezney, R., and Coffey, D. S. (1977). Nuclear matrix: Isolation and characterization of a framework structure from rat liver nuclei. *J. Cell Biol.* **73,** 616–637.

Bidwell, J. P., van Wijnen, A. J., Fey, E. G., Dworetzky, S., Penman, S., Stein, J. L., Lian, J. B., and Stein, G. S. (1993). Osteocalcin gene promoter-binding factors are tissue-specific nuclear matrix components. *Proc. Natl. Acad. Sci. U.S.A.* **90,** 3162–3166.

Bidwell, J. P., Fey, E. G., van Wijnen, A. J., Penman, S., Stein, J. L., Lian, J. B., and Stein, G. S. (1994a). Nuclear matrix proteins distinguish normal diploid osteoblasts from osteosarcoma cells. *Cancer Res.* **54,** 28–32.

Bidwell, J. P., van Wijnen, A. J., Banerjee, C., Fey, E. G., Merriman, H., Penman, S., Stein, J. L., Lian, J. B., and Stein, G. S. (1994b). PTH-responsive modifications in the nuclear matrix of ROS 17/2.8 rat osteosarcoma cells. *Endocrinology* (*Baltimore*) **134,** 1738–1744.

Birnbaum, M. J., Wright, K. L., van Wijnen, A. J., Ramsey-Ewing, A. L., Bourke, M. T., Last, T. J., Aziz, F., Frenkel, B., Rao, B. R., Aronin, N., Stein, G. S., and Stein, J. L. (1995). Functional role for Sp1 in the transcriptional amplification of a cell cycle regulated human histone H4 gene. *Biochemistry* **34,** 7648–7658.

Blanco, J. C. G., Wang, I.-M., Tsai, S. Y., Tsai, M.-J., O'Malley, B. W., Jurutka, P. W., Haussler, M. R., and Ozato, K. (1995). Transcription factor TFIIB and the vitamin D receptor cooperatively activate ligand-dependent transcription. *Proc. Natl. Acad. Sci. U.S.A.* **92,** 1535–1539.

Blencowe, B. J., Nickerson, J. A., Issner, R., Penman, S., and Sharp, P. A. (1994). Association of nuclear matrix antigens with exon-containing splicing complexes. *J. Cell Biol.* **127,** 593–607.

Bode, J., and Maass, K. (1988). Chromatin domain surrounding the human interferon-β gene as defined by scaffold-attached regions. *Biochemistry* **27,** 4706–4711.

Bortell, R., Owen, T. A., Bidwell, J. P., Gavazzo, P., Breen, E., van Wijnen, A. J., DeLuca, H. F., Stein, J. L., Lian, J. B., and Stein, G. S. (1992). Vitamin D-responsive protein-DNA interactions at multiple promoter regulatory elements that contribute to the level of rat osteocalcin gene expression. *Proc. Natl. Acad. Sci. U.S.A.* **89,** 6119–6123.

Bortell, R., Owen, T. A., Shalhoub, V., Heinrichs, A., Aronow, M. A. B., and Stein, G. S. (1993). Constitutive transcription of the osteocalcin gene in osteosarcoma cells is reflected by altered protein-DNA interactions at promoter regulatory elements. *Proc. Natl. Acad. Sci. U.S.A.* **90,** 2300–2304.

Breen, E. C., van Wijnen, A. J., Lian, J. B., Stein, G. S., and Stein, J. L. (1994). In vivo occupancy of the vitamin D responsive element in the osteocalcin gene supports vitamin D dependent transcriptional upregulation in intact cells. *Proc. Natl. Acad. Sci. U.S.A.* **91,** 12902–12906.

Buhrmester, H., von Kries, J. P., and Strätling, W. H. (1995). Nuclear matrix protein ARBP recognizes a novel DNA sequence motif with high affinity. *Biochemistry* **34,** 4108–4117.

Capco, D. G., Wan, K. M., and Penman, S. (1982). The nuclear matrix: Three-dimensional architecture and protein composition. *Cell* (*Cambridge, Mass.*) **29,** 847–858.

Chrysogelos, S., Riley, D. E., Stein, G. S., and Stein, J. L. (1985). A human histone H4 gene exhibits cell cycle-dependent changes in chromatin structure that correlate with its expression. *Proc. Natl. Acad. Sci. U.S.A.* **82,** 7535–7539.

Chrysogelos, S., Pauli, U., Stein, G. S., and Stein, J. L. (1989). Fine mapping of the chromatin structure of a cell cycle-regulated human H4 histone gene. *J. Biol. Chem.* **264,** 1232–1237.

Ciejek, E. M., Tsai, M.-J., and O'Malley, B. W. (1983). Actively transcribed genes are associated with the nuclear matrix. *Nature* (*London*) **306,** 607–609.

Cockerill, P. N., and Garrard, W. T. (1986). Chromosomal loop anchorage of the kappa immuno-globulin gene occurs next to the enhancer in a region containing topoisomerase II sites. *Cell* (*Cambridge, Mass.*) **44,** 273–282.

Cohen, S. M., Broner, G., Kuttner, F., Jurgens, G., and Jackle, H. (1989). *Distal-less* encodes a homeodomain protein required for limb development in *Drosophila*. *Nature* (*London*) **338,** 432–434.

Cunningham, J. M., Purucker, M. E., Jane, S. M., Safer, B., Vanin, E. F., Ney, P. A., Lowrey, C. H., and Nienhuis, A. W. (1994). The regulatory element 3′ to the A gamma-globin gene binds to the nuclear matrix and interacts with special A-T-rich binding protein 1 (SATB1), an SAR/MAR-associating region DNA binding protein. *Blood* **84,** 1298–1308.

de Jong, L., van Driel, R., Stuurman, N., Meijne, A. M. L., and van Renswoude, J. (1990). Principles of nuclear organization. *Cell Biol. Int. Rep.* **14,** 1051–1074.

Demay, M. B., Gerardi, J. M., DeLuca, H. F., and Kronenberg, H. M. (1990). DNA sequences in the rat osteocalcin gene that bind the 1,25-dihydroxyvitamin D_3 receptor and confer responsiveness to 1,25-dihydroxyvitamin D_3. *Proc. Natl. Acad. Sci. U.S.A.* **87,** 369–373.

Demay, M. B., Kiernan, M. S., DeLuca, H. F., and Kronenberg, H. M. (1992). Characterization of 1,25-dihydroxyvitamin D_3 receptor interactions with target sequences in the rat osteocalcin gene. *Mol. Endocrinol.* **6,** 557–562.

Detke, S., Lichtler, A., Phillips, I., Stein, J. L., and Stein, G. S. (1979). Reassessment of histone gene expression during the cell cycle in human cells by using homologous H4 histone cDNA. *Proc. Natl. Acad. Sci. U.S.A.* **76,** 4995–4999.

Dickinson, L. A., and Kohwi-Shigematsu, T. (1995). Nucleolin is a matrix attachment region DNA-binding protein that specifically recognizes a region with high base-unpairing potential. *Mol. Cell. Biol.* **15,** 456–465.

Dickinson, L. A., Joh, T., Kohwi, Y., and Kohwi-Shigematsu, T. (1992). A tissue-specific MAR/SAR DNA-binding protein with unusual binding site recognition. *Cell (Cambridge, Mass.)* **70,** 631–645.

Dietz, A., Kay, V., Schlake, T., Landsmann, J., and Bode, J. (1994). A plant scaffold attached region detected close to a T-DNA integration site is active in mammalian cells. *Nucleic Acids Res.* **22,** 2744–2751.

Dou, Q., Markell, P. J., and Pardee, A. B. (1992). Retinoblastoma-like protein and cdc2 kinase are in complexes that regulate a G1/S event. *Proc. Natl. Acad. Sci. U.S.A.* **89,** 3256–3260.

Durfee, T., Mancini, M. A., Jones, D., Elledge, S. J., and Lee, W. H. (1994). The amino-terminal region of the retinoblastoma gene product binds a novel nuclear matrix protein that co-localizes to centers for RNA processing. *J. Cell Biol.* **127,** 609–622.

Dworetzky, S. I., Fey, E. G., Penman, S., Lian, J. B., Stein, J. L., and Stein, G. S. (1990). Progressive changes in the protein composition of the nuclear matrix during rat osteoblast differentiation. *Proc. Natl. Acad. Sci. U.S.A.* **87,** 4605–4609.

Dworetzky, S. I., Wright, K. L., Fey, E. G., Penman, S., Lian, J. B., Stein, J. L., and Stein, G. S. (1992). Sequence-specific DNA binding proteins are components of a nuclear matrix attachment site. *Proc. Natl. Acad. Sci. U.S.A.* **89,** 4178–4182.

Evans, D. B., Thavarajah, M., and Kanis, J. A. (1990). Involvement of prostaglandin E_2 in the inhibition of osteocalcin synthesis by human osteoblast-like cells in response to cytokines and systemic hormones. *Biochem. Biophys. Res. Commun.* **167,** 194–202.

Fanti, P., Kindy, M. S., Mohapatra, S., Klein, J., Colombo, G., and Malluche, H. H. (1992). Dose-dependent effects of aluminum on osteocalcin synthesis in osteoblast-like ROS 17/2 cells in culture. *Am. J. Physiol.* **263,** E1113–E1118.

Farache, G., Razin, S. V., Rzeszowska-Wolny, J., Moreau, J., Targa, F. R., and Scherrer, K. (1990). Mapping of structural and transcription-related matrix attachment sites in the a-globin gene domain of avian erythroblasts and erythrocytes. *Mol. Cell. Biol.* **10,** 5349–5358.

Fey, E. G., and Penman, S. (1988). Nuclear matrix proteins reflect cell type of origin in cultured human cells. *Proc. Natl. Acad. Sci. U.S.A.* **85,** 121–125.

Fey, E. G., Krochmalnic, G., and Penman, S. (1986). The nonchromatin substructures of the nucleus: The ribonucleoprotein (RNP)-containing and RNP-depleted matrices analyzed by sequential fractionation and resinless section electron microscopy. *J. Cell Biol.* **102,** 1654–1665.

Fey, E. G., Bangs, P., Sparks, C., and Odgren, P. (1991). The nuclear matrix: Defining structural and functional roles. *Crit. Rev. Eukaryotic Gene Expression* **1,** 127–143.

Fishel, B. R., Sperry, A. O., and Garrard, W. T. (1993). Yeast calmodulin and a conserved nuclear protein participate in the in vivo binding of a matrix association region. *Proc. Natl. Acad. Sci. U.S.A.* **90,** 5623–5627.

Gasser, S. M., and Laemmli, U. K. (1986). Cohabitation of scaffold binding regions with upstream/enhancer elements of three developmentally regulated genes of *D. melanogaster. Cell (Cambridge, Mass.)* **46,** 521–530.

Getzenberg, R. H., and Coffey, D. S. (1990). Tissue specificity of the hormonal response in sex accessory tissues is associated with nuclear matrix protein patterns. *Mol. Endocrinol.* **4,** 1336–1342.

Getzenberg, R. H., and Coffey, D. S. (1991). Identification of nuclear matrix proteins in the cancerous and normal rat prostate. *Cancer Res.* **51,** 6514–6520.

Gruss, P., and Walther, C. (1992). Pax in development. *Cell (Cambridge, Mass.)* **69,** 719–722.

Guidon, P. T., Salvatori, R., and Bockman, R. S. (1993). Gallium nitrate regulates rat osteoblast expression of osteocalcin protein and mRNA levels. *J. Bone Miner. Res.* **8,** 103–110.

Guo, B., Odgren, P. R., van Wijnen, A. J., Last, T. J., Fey, E. G., Penman, S., Stein, J. L., Lian, J. B., and Stein, G. S. (1995). The nuclear matrix protein NMP-1 is the transcription factor YY-1: Subnuclear localization with numatrin/B23. *Proc. Natl. Acad. Sci. U.S.A.* **92,** 10526–10530.

Hakes, D. J., and Berezney, R. (1991). Molecular cloning of matrin F/G: A DNA binding protein of the nuclear matrix that contains putative zinc finger motifs. *Proc. Natl. Acad. Sci. U.S.A.* **88,** 6186–6190.

He, D., Nickerson, J. A., and Penman, S. (1990). Core filaments of the nuclear matrix. *J. Cell Biol.* **110,** 569.

Heinrichs, A. A. J., Bortell, R., Rahman, S., Stein, J. L., Alnemri, E. S., Litwack, G., Lian, J. B., and Stein, G. S. (1993a). Identification of multiple glucocorticoid receptor binding sites in the rat osteocalcin gene promoter. *Biochemistry* **32,** 11436–11444.

Heinrichs, A. A. J., Banerjee, C., Bortell, R., Owen, T. A., Stein, J. L., Stein, G. S., and Lian, J. B. (1993b). Identification and characterization of two proximal elements in the rat osteocalcin gene promoter that may confer species-specific regulation. *J. Cell. Biochem.* **53,** 240–250.

Heinrichs, A. A. J., Bortell, R., Bourke, M., Lian, J. B., Stein, G. S., and Stein, J. L. (1995). Proximal promoter binding protein contributes to developmental, tissue-restricted expression of the rat osteocalcin gene. *J. Cell. Biochem.* **57,** 90–100.

Hendzel, M. J., Sun, J.-M., Chen, H. Y., Rattner, J. B., and Davie, J. R. (1994). Histone acetyltransferase is associated with the nuclear matrix. *J. Biol. Chem.* **269,** 22894–22901.

Herman, R., Weymouth, L., and Penman, S. (1978). Heterogeneous nuclear RNA-protein fibers in chromatin-depleted nuclei. *J. Cell Biol.* **78,** 663–674.

Hoffmann, H. M., Catron, K. M., van Wijnen, A. J., McCabe, L. R., Lian, J. B., Stein, G. S., and Stein, J. L. (1994). Transcriptional control of the tissue-specific developmentally regulated osteocalcin gene requires a binding motif for the MSX-family of homeodomain proteins. *Proc. Natl. Acad. Sci. U.S.A.* **91,** 12887–12891.

Huang, S., and Spector, D. L. (1992). U1 and U2 small nuclear RNAs are present in nuclear speckles. *Proc. Natl. Acad. Sci. U.S.A.* **89,** 305–308; published erratum appears in *ibid.,* pp. 4218–4219.

Jackson, D. A. (1991). Structure-function relationships in eukaryotic nuclei. *BioEssays* **13,** 1–10.

Jackson, D. A., and Cook, P. R. (1985). Transcription occurs at a nucleoskeleton. *EMBO J.* **4,** 919–925.

Jackson, D. A., and Cook, P. R. (1986). Replication occurs at a nucleoskeleton. *EMBO J.* **5,** 1403–1410.

Jackson, D. A., McCready, S. J., and Cook, P. R. (1981). RNA is synthesized at the nuclear cage. *Nature* (*London*) **292,** 552–555.

Jarman, A. P., and Higgs, D. R. (1988). Nuclear scaffold attachment sites in the human globin gene complexes. *EMBO J.* **7,** 3337–3344.

Jenis, L. G., Waud, C. E., Stein, G. S., Lian, J. B., and Baran, D. T. (1993). Effect of gallium nitrate in vitro and in normal rats. *J. Cell. Biochem.* **52,** 330–336.

Käs, E., and Chasin, L. A. (1987). Anchorage of the Chinese hamster dihydrofolate reductase gene to the nuclear scaffold occurs in an intragenic region. *J. Mol. Biol.* **198,** 677–692.

Kawaguchi, N., DeLuca, H. F., and Noda, M. (1992). Id gene expression and its suppression by 1,25-dihydroxyvitamin D_3 in rat osteoblastic osteosarcoma cells. *Proc. Natl. Acad. Sci. U.S.A.* **89,** 4569–4572.

Kay, V., and Bode, J. (1994). Binding specificity of a nuclear scaffold: Supercoiled, single-stranded, and scaffold-attached-region DNA. *Biochemistry* **33,** 367–374.

Keppel, F. (1986). Transcribed human ribosomal RNA genes are attached to the nuclear matrix. *J. Mol. Biol.* **187,** 15–21.

Kerner, S. A., Scott, R. A., and Pike, J. W. (1989). Sequence elements in the human osteocalcin gene confer basal activation and inducible response to hormonal vitamin D_3. *Proc. Natl. Acad. Sci. U.S.A.* **86,** 4455–4459.

Klehr, D., Maass, K., and Bode, J. (1991). Scaffold-attached regions from the human interferon beta domain can be used to enhance the stable expression of genes under the control of various promoters. *Biochemistry* **30,** 1264–1270.

Kliewer, S. A., Umesono, K., Mangelsdorf, D. J., and Evans, R. M. (1992). Retinoic X receptor interacts with nuclear receptors in retinoic acid, thyroid hormone an vitamin D signalling. *Nature (London)* **355,** 441–446.

Kroeger, P., Stewart, C., Schaap, T., van Wijnen, A., Hirshman, J., Helms, S., Stein, G. S., and Stein, J. L. (1987). Proximal and distal regulatory elements that influence in vivo expression of a cell cycle dependent H4 histone gene. *Proc. Natl. Acad. Sci. U.S.A.* **84,** 3982–3986.

Krumlauf, R. (1993). *Hox* genes and pattern formation in the branchial region of the vertebrate head. *Trends Genet.* **9,** 106–112.

Kumara-Siri, M. H., Shapiro, L. E., and Surks, M. I. (1986). Association of the 3,5,3′-triiodo L-thyronine nuclear receptor with the nuclear matrix of cultured growth hormone-producing rate pituitary tumor cells (GC cells). *J. Biol. Chem.* **261,** 2844–2852.

Kuno, H., Kurian, S. M., Hendy, G. N., White, J., DeLuca, H. F., Evans, C.-O., and Nanes, M. S. (1994). Inhibition of 1,25-dihydroxyvitamin D_3 stimulated osteocalcin gene transcription by tumor necrosis factor-α: Structural determinants within the vitamin D response element. *Endocrinology* (*Baltimore*) **134,** 2524–2531.

Landers, J. P., and Spelsberg, T. C. (1992). New concepts in steroid hormone action: Transcription factors, proto-oncogenes, and the cascade model for steroid regulation of gene expression. *Crit. Rev. Eukaryotic Gene Expression* **2,** 19–63.

Li, Y., and Stashenko, P. (1992). Proinflammatory cytokines tumor necrosis factor and IL-6, but not IL-1, down-regulate the osteocalcin gene promoter. *J. Immunol.* **148,** 788–794.

Li, Y., and Stashenko, P. (1993). Characterization of a tumor necrosis factor-responsive element which down-regulates the human osteocalcin gene. *Mol. Cell. Biol.* **13,** 3714–3721.

Lian, J. B., and Stein, G. S. (1993). Proto-oncogene mediated control of gene expression during osteoblast differentiation. *Ital. J. Miner. Electron Metab.* **7,** 175–183.

Lian, J. B., and Stein, G. S. (1996). Osteocalcin gene expression: A molecular blueprint for developmental and steroid hormone mediated regulation of osteoblast growth and differentiation. *Endocr. Rev.* (in press).

Lian, J. B., Stewart, C., Puchacz, E., Mackowiak, S., Shalhoub, V., Collart, D., Zambetti, G., and Stein, G. S. (1989). Structure of the rat osteocalcin gene and regulation of vitamin D-dependent expression. *Proc. Natl. Acad. Sci. U.S.A.* **86,** 1143–1147.

Lian, J. B., Stein, G. S., Bortell, R., and Owen, T. A. (1991). Phenotype suppression: A postulated molecular mechanism for mediating the relationship of proliferation and differentiation by fos/jun interactions at AP-1 sites in steroid responsive promoter elements of tissue-specific genes. *J. Cell. Biochem.* **45,** 9–14.

Lufkin, T., Dierich, A., LeMeur, M., Mark, M., and Chambon, P. (1991). Disruption of the *Hox*-1.6 homeobox gene results in defects in a region corresponding to its rostral domain of expression. *Cell (Cambridge, Mass.)* **66,** 1105–1119.

Lufkin, T., Mark, M., Hart, C. P., LeMeur, M., and Chambon, P. (1992). Homeotic transformation of the occipital bones of the skull by ectopic expression of a homeobox gene. *Nature (London)* **359,** 835–841.

Macdonald, P. N., Dowd, D. R., Nakajima, S., Galligan, M. A., Reeder, M. C., Haussler, C. A., Ozato, K., and Haussler, M. R. (1993). Retinoid X receptors stimulate and 9-*cis* retinoic acid inhibits 1,25-dihydroxyvitamin D_3-activated expression of the rat osteocalcin gene. *Mol. Cell. Biol.* **13,** 5907–5917.

Macdonald, P. N., Sherman, D. R., Dowd, D. R., Jefcoat, S. C., Jr., and DeLisle, R. K. (1995). The vitamin D receptor interacts with general transcription factor IIB. *J. Biol. Chem.* **270,** 4748–4752.

Mariman, E. C. M., van Eekelen, C. A. G., Reinders, J., Berns, A. J. M., and van Venrooij, W. J. (1982). Adenoviral heterogenous nuclear RNA is associated with the host nuclear matrix during splicing. *J. Mol. Biol.* **154,** 103–119.

Markose, E. R., Stein, J. L., Stein, G. S., and Lian, J. B. (1990). Vitamin D-mediated modifications in protein-DNA interactions at two promoter elements of the osteocalcin gene. *Proc. Natl. Acad. Sci. U.S.A.* **87,** 1701–1705.

McGinnis, W., and Krumlauf, R. (1992). Homeobox genes and axial patterning. *Cell (Cambridge, Mass.)* **68,** 283–302.

Merriman, H. L., van Wijnen, A. J., Hiebert, S., Bidwell, J. P., Fey, E., Lian, J. B., Stein, J. L., and Stein, G. S. (1995). The tissue-specific nuclear matrix protein, NMP-2, is a member of the AML/PEBP2/*runt domain* transcription factor family: Interactions with the osteocalcin gene promoter. *Biochemistry* **34,** 13125–13132.

Mirkovitch, J., Mirault, M-.E., and Laemmli, U. K. (1984). Organization of the higher-order chromatin loop: Specific DNA attachment sites on nuclear scaffold. *Cell (Cambridge, Mass.)* **39,** 223–232.

Mirkovitch, J., Gasser, S. M., and Laemmli, U. K. (1988). Scaffold attachment of DNA loops in metaphase chromosomes. *J. Mol. Biol.* **200,** 101–109.

Montecino, M., Pockwinse, S., Lian, J. B., Stein, G. S., and Stein, J. L. (1994a). DNase I hypersensitive sites in promoter elements associated with basal and vitamin D dependent transcription of the bone-specific osteocalcin gene. *Biochemistry* **33,** 348–353.

Montecino, M., Lian, J. B., Stein, G. S., and Stein, J. L. (1994b). Specific nucleosomal organization supports developmentally regulated expression of the osteocalcin gene. *J. Bone Miner. Res.* **9,** S352.

Montecino, M., Lian, J. B., Stein, G. S., and Stein, J. L. (1996a). Changes in chromatin structure support constitutive and developmentally regulated transcription of the bone specific osteocalcin gene in osteoblastic cells. Submitted for publication.

Montecino, M. *et al.* (1996b). In preparation.

Moreno, M. L., Chrysogelos, S. A., Stein, G. S., and Stein, J. L. (1986). Reversible changes in the nucleosomal organization of a human H4 histone gene during the cell cycle. *Biochemistry* **25,** 5364–5370.

Moreno, M. L., Stein, G. S., and Stein, J. L. (1987). Nucleosomal organization of a BPV minichromosome containing a human H4 histone gene. *Mol. Cell. Biochem.* **74,** 173–177.

Morrison, N. A., and Eisman, J. A. (1993). Role of the negative glucocorticoid regulatory element in glucocorticoid repression of the human osteocalcin promoter. *J. Bone Miner. Res.* **8,** 969–975.

Morrison, N. A., Shine, J., Fragonas, J. C., Verkest, V., McMenemy, L., and Eisman, J. A. (1989). 1,25-Dihydroxyvitamin D-responsive element and glucocorticoid repression in the osteocalcin gene. *Science* **246,** 1158–1161.

Nakagomi, K., Kohwi, Y., Dickinson, L. A., and Kohwi-Shigematsu, T. (1994). A novel DNA-binding motif in the nuclear matrix attachment DNA-binding protein SATB1. *Mol. Cell. Biol.* **14,** 1852–1860.

Nakayasu, H., and Berezney, R. (1989). Mapping replicational sites in the eukaryotic nucleus. *J. Cell Biol.* **108,** 1–11.

Nanes, M. S., Rubin, J., Titus, L., Hendy, G. N., and Catherwood, B. D. (1990). Interferon-γ inhibits 1,25-dihydroxyvitamin D_3-stimulated synthesis of bone G1a protein in rat osteosar-coma cells by a pretranslational mechanism. *Endocrinology* (*Baltimore*) **127,** 588–594.

Nanes, M. S., Rubin, J., Titus, L., Hendy, G. N., and Catherwood, B. D. (1991). Tumor necrosis factor alpha inhibits 1,25-dihydroxyvitamin D_3-stimulated bone gla protein synthesis in rat osteosarcoma cells (ROS 17/2.8) by a pretranslational mechanism. *Endocrinology* (*Baltimore*) **128,** 2577–2582.

Nelkin, B. D., Pardoll, D. M., and Vogelstein, B. (1980). Localization of SV40 genes with supercoiled loop domains. *Nucleic Acids Res.* **8,** 5623–5633.

Nelson, W. G., Pienta, K. J., Barrack, E. R., and Coffey, D. S. (1986). The role of the nuclear matrix in the organization and function of DNA. *Nucleic Acids Res.* **14,** 6433–6451.

Nickerson, J. A., and Penman, S. (1992). The nuclear matrix: Structure and involvement in gene expression. *In* "Molecular and Cellular Approaches to the Control of Proliferation and Differentiation" (G. Stein and J. Lian, eds.), pp. 434–380. Academic Press, San Diego, CA.

Nickerson, J. A., Krockmalnic, G., He, D., and Penman, S. (1990a). Immunolocalization in three dimensions: Immunogold staining of cytoskeletal and nuclear matrix proteins in resinless electron microscopy sections. *Proc. Natl. Acad. Sci. U.S.A.* **87,** 2259–2263.

Nickerson, J. A., He, D., Fey, E. G., and Coffey, D. S. (1990b). The nuclear matrix. *In* "The Eukaryotic Nucleus: Molecular Biochemistry and Macromolecular Assemblies" (P. R. Strauss and S. H. Wilson, eds.), Vol. 2, p. 763. Telford Press, Caldwell, NJ.

Niehrs, C., and DeRobertis, E. M. (1992). Vertebrate axis formation. *Curr. Opin. Gene. Dev.* **2,** 550–555.

O'Keefe, R. T., Mayeda, A., Sadowski, C. L., Krainer, A. R., and Spector, D. L. (1994). Disruption of pre-mRNA splicing in vivo results in reorganization of splicing factors. *J. Cell Biol.* **124,** 249–260.

Owen, T. A., Bortell, R., Yocum, S. A., Smock, S. L., Zhang, M., Abate, C., Shalhoub, V., Aronin, N., Wright, K. L., van Wijnen, A. J., Stein, J. L., Curran, T., Lian, J. B., and Stein, G. S. (1990). Coordinate occupancy of AP-1 sites in the vitamin D responsive and CCAAT box elements by fos-jun in the osteocalcin gene: A model for phenotype suppression of transcription. *Proc. Natl. Acad. Sci. U.S.A.* **87,** 9990–9994.

Owen, T. A., Bortell, R., Shalhoub, V., Heinrichs, A., Stein, J. L., Stein, G. S., and Lian, J. B. (1993). Postproliferative transcription of the rat osteocalcin gene is reflected by vitamin D-responsive developmental modifications in protein-DNA interactions at basal and enhancer promoter elements. *Proc. Natl. Acad. Sci. U.S.A.* **90,** 1503–1507.

Ozono, K., Sone, T., and Pike, J. W. (1991). The genomic mechanism of action of 1,25-dihydroxyvitamin D_3. *J. Bone Miner. Res.* **6,** 1021–1027.

Pardoll, D. M., Vogelstein, B., and Coffey, D. S. (1980). A fixed site of DNA replication in eukaryotic cells. *Cell (Cambridge, Mass.)* **19,** 527–536.

Pauli, U., Chrysogelos, S., Stein, J. L., Stein, G. S., and Nick, H. (1987). Protein-DNA interactions in vivo upstream of a cell cycle regulated human H4 histone gene. *Science* **236,** 1308–1311.

Phi-Van, L., and Strätling, W. H. (1988). The matrix attachment regions of the chicken lysozyme gene co-map with the boundaries of the chromatin domain. *EMBO J.* **7,** 655–664.

Phi-Van, L., von Kries, J. P., Ostertag, W., and Strätling, W. H. (1990). The chicken lysozyme 5′ matrix attachment region increases transcription from a heterologous promoter in heterologous cells and dampens positional effects on the expression of transfected genes. *Mol. Cell. Biol.* **10,** 2302–2307.

Pienta, K. J., and Coffey, D. S. (1991). Correlation of nuclear morphometry with progression of breast cancer. *Cancer (Philadelphia)* **68,** 2012–2016.

Pienta, K. J., Getzenberg, R. H., and Coffey, D. S. (1991). Cell structure and DNA organization. *Crit. Rev. Eukaryotic Gene Expression* **1,** 355–385.

Plumb, M. A., Stein, J. L., and Stein, G. S. (1983a). Coordinate regulation of multiple histone mRNAs during the cell cycle in HeLa cells. *Nucleic Acids Res.* **11,** 2391–2410.

Plumb, M. A., Stein, J. L., and Stein, G. S. (1983b). Influence of DNA synthesis inhibition on the coordinate expression of core human histone genes. *Nucleic Acids Res.* **11,** 7927–7945.

Ramsey-Ewing, A., van Wijnen, A., Stein, G. S., and Stein, J. L. (1994). Delineation of a human histone H4 cell cycle element in vivo: The master switch for H4 gene transcription. *Proc. Natl. Acad. Sci. U.S.A.* **91,** 4475–4479.

Robinson, S. I., Nelkin, B. D., and Vogelstein, B. (1982). The ovalbumin gene is associated with the nuclear matrix of chicken oviduct cells. *Cell (Cambridge, Mass.)* **28,** 99–106.

Ross, D. A., Yen, R. W., and Chae, C. B. (1982). Association of globin ribonucleic acid and its precursors with the chicken erythroblast nuclear matrix. *Biochemistry* **21,** 764–771.

Schaack, J., Ho, W. Y.-W., Friemuth, P., and Shenk, T. (1990). Adenovirus terminal protein mediates both nuclear-matrix association and efficient transcription of adenovirus DNA. *Genes Dev.* **4,** 1197–1208.

Schedlich, L. J., Flanagan, J. L., Crofts, L. A., Gillies, S. A., Goldberg, D., Morrison, N. A., and Eisman, J. A. (1994). Transcriptional activation of the human osteocalcin gene by basic fibroblast growth factor. *J. Bone Miner. Res.* **9,** 143–152.

Schrader, M., Bendik, I., Becker-Andre, M., and Carlberg, C. (1993). Interaction between retinoic acid and vitamin C signaling pathways. *J. Biol. Chem.* **268,** 17830–17836.

Schroder, H. C., Trolltsch, D., Friese, U., Bachmann, M., and Muller, W. E. G. (1987a). Mature mRNA is selectively released from the nuclear matrix by an ATP/dATP-dependent mechanism sensitive to topoisomerase inhibitors. *J. Biol. Chem.* **262,** 8917–8925.

Schroder, H. C., Trolltsch, D., Wenger, R., Bachmann, M., Diehl-Seifert, B., and Muller, W. E. G. (1987b). Cytochalasin B selectively releases ovalbumin mRNA precursors but not the mature ovalbumin mRNA from hen oviduct nuclear matrix. *Eur. J. Biochem.* **167,** 239–245.

Scule, R., Umesono, K., Mangelsdorf, D. J., Bolado, J., Pike, J. W., and Evans, R. M. (1990). Jun-Fos and receptors for vitamins A and D recognize a common response element in the human osteocalcin gene. *Cell (Cambridge, Mass.)* **61,** 497–504.

Simeone, A., Acampora, D., Pannese, M., D'Esposito, M., Stornaiuolo, A., Gulisano, M., Mallamaci, A., Kastury, K., Druck, T., Huebner, K., and Boncinelli, E. (1994). Cloning and characterization of two new members of the vertebrate Dlx family. *Proc. Natl. Acad. Sci. U.S.A.* **91,** 2250–2254.

Stein, G. S., and Lian, J. B. (1993). Molecular mechanisms mediating proliferation/differentiation interrelationships during progressive development of the osteoblast phenotype. *Endocr. Rev.* **14,** 424–442.

Stein, G. S., Lian, J. B., and Owen, T. A. (1990). Relationship of cell growth to the regulation of tissue-specific gene expression during osteoblast differentiation. *FASEB J.* **4,** 3111–3123.

Stein, G. S., Stein, J. L., van Wijnen, A. J., and Lian, J. B. (1992). Regulation of histone gene expression. *Curr. Opin. Cell Biol.* **4,** 166–173.

Stein, G. S., Stein, J. L., van Wijnen, A. J., and Lian, J. B. (1994). Histone gene transcription: A model for responsiveness to an integrated series of regulatory signals mediating cell cycle control and proliferation/differentiation interrelationships. *J. Cell. Biochem.* **54,** 393–404.

Stief, A., Winter, D. M., Strätling, W. H., and Sippel, A. E. (1989). A nuclear attachment element mediates elevated and position-independent gene activity. *Nature (London)* **341,** 343–345.

Stromstedt, P. E., Poellinger, L., Gustafsson, J. A., and Cralstedt-Duke, J. (1991). The glucocorticoid receptor binds to a sequence overlapping the TATA box of the human osteocalcin promoter: A potential mechanism for negative regulation. *Mol. Cell. Biol.* **11,** 3379–3383.

Tabin, C. J. (1991). Retinoids, homeoboxes, and growth factors: Towards molecular models for limb development. *Cell (Cambridge, Mass.)* **66,** 199–217.

Taichman, R. S., and Hauschka, P. V. (1992). Effects of interleukin-1 beta and tumor necrosis factor-alpha on osteoblastic expression of osteocalcin and mineralized extracellular matrix in vitro. *J. Inflamm.* **16**(6), 587–601.

Tamura, M., and Noda, M. (1994). Identification of a DNA sequence involved in osteoblast-specific gene expression via interaction with helix-loop-helix (HLH)-type transcription factors. *J. Cell Biol.* **126,** 773–782.

Targa, F. R., Razin, S. V., de Moura Gallo, C. V., and Scherrer, K. (1994). Excision close to matrix attachment regions of the entire chicken alpha-globin gene domain by nuclease S1 and characterization of the framing structures. *Proc. Natl. Acad. Sci. U.S.A.* **91,** 4422–4426.

Tawfic, S., and Ahmed, K. (1994). Growth stimulus-mediated differential translocation of casein kinase 2 to the nuclear matrix. *J. Biol. Chem.* **269,** 24615–24620.

Terpening, C. M., Haussler, C. A., Jurutka, P. W., Galligan, M. A., Komm, B. S., and Haussler, M. R. (1991). The vitamin D-responsive element in the rat bone G1a protein gene is an imperfect direct repeat that cooperates with other cis-elements in 1,25-dihydroxyvitamin D_3-mediated transcriptional activation. *Mol. Endocrinol.* **5,** 373–385.

Thorburn, A., Moore, R., and Knowland, J. (1988). Attachment of transcriptionally active sequences to the nucleoskeleton under isotonic conditions. *Nucleic Acids Res.* **16,** 7183.

Towler, D. A., and Rodan, G. A. (1994). Cross-talk between glucocorticoid and PTH signaling in the regulation of the rat osteocalcin promoter. *J. Bone Miner. Res.* **9,** S282.

Towler, D. A., Bennett, C. D., and Rodan, G. A. (1994). Activity of the rat osteocalcin basal promoter in osteoblastic cells is dependent upon homeodomain and CP1 binding motifs. *Mol. Endocrinol.* **8,** 614–624.

Tsutsui, K., Tsutsui, K., Okada, S., Watarai, S., Seki, S., Yasuda, T., and Shohmori, T. (1993). Identification and characterization of a nuclear scaffold protein that binds the matrix attachment region DNA. *J. Biol. Chem.* **268,** 12886–12894.

Vaishnav, R., Beresford, J. N., Gallagher, J. A., and Russell, R. G. G. (1988). Effects of the anabolic steroid stanozolol on cells derived from human bone. *Clin. Sci.* **74,** 455–460.

van den Ent, F. M. I., van Wijnen, A. J., Lian, J. B., Stein, J. L., and Stein, G. S. (1994). Cell cycle controlled histone H1, H3 and H4 genes share unusual arrangements of recognition motifs for HiNF-D supporting a coordinate promoter binding mechanism. *J. Cell. Physiol.* **159,** 515–530.

van Eeklen, C. A. G., and van Venrooij, W. J. (1981). hnRNA and its attachment to a nuclear-protein matrix. *J. Cell Biol.* **88,** 554–563.

van Steensel, B., Jenster, G., Damm, K., Brinkmann, A. O., and van Driel, R. (1995). Domains of the human androgen receptor and glucocorticoid receptor involved in binding to the nuclear matrix. *J. Cell. Biochem.* **57,** 465–478.

van Wijnen, A. J., Wright, K. L., Lian, J. B., Stein, J. L., and Stein, G. S. (1989). Human H4 histone gene transcription requires the proliferation-specific nuclear factor HiNF-D. *J. Biol. Chem.* **264,** 15034–15042.

van Wijnen, A. J., van den Ent, F. M. I., Lian, J. B., Stein, J. L., and Stein, G. S. (1992). Overlapping and CpG methylation-sensitive protein/DNA interactions at the histone H4 transcriptional cell cycle domain: Distinctions between two human H4 gene promoters. *Mol. Cell. Biol.* **12,** 3273–3287.

van Wijnen, A. J., Bidwell, J. P., Fey, Edward, G., Penman, S., Lian, J. B., Stein, J. L., and Stein, G. S. (1993). Nuclear matrix association of multiple sequence specific DNA binding activities related to SP-1, ATF, CCAAT, C/EBP, OCT-1 and AP-1. *Biochemistry* **32,** 8397–8402.

van Wijnen, A. J., Aziz, F., Grana, X., De Luca, A., Desai, R. K., Jaarsveld, K., Last, T. J., Soprano, K., Giordano, A., Lian, J. B., Stein, J. L., and Stein, G. S. (1994a). Transcription of histone H4, H3 and H1 cell cycle genes: Promoter factor HiNF-D contains CDC2, cyclin A and an RB-related protein. *Proc. Natl. Acad. Sci. U.S.A.* **91,** 12882–12886.

van Wijnen, A. J., Merriman, H., Guo, B., Bidwell, J. P., Stein, J. L., Lian, J. B., and Stein, G. S. (1994b). Nuclear matrix interactions with the osteocalcin gene promoter: Multiple binding sites for the runt-homology related protein NMP-2 and for NMP-1, a heteromeric CREB-2 containing transcription factor. *J. Bone Miner. Res.* **9,** S148.

Vaughan, P. S., Aziz, F., van Wijnen, A. J., Wu, S., Harada, H., Taniguchi, T., Soprano, K., Stein, G. S., and Stein, J. L. (1995). Activation of a cell cycle regulated histone gene by the oncogenic transcription factor IRF2. *Nature* **377,** 362–365.

Vaughn, J. P., Dijkwel, P. A., Mullenders, L. H. F., and Hamlin, J. L. (1990). Replication forks are associated with the nuclear matrix. *Nucleic Acids Res.* **18,** 1965–1969.

von Kries, J. P., Buhrmester, H., and Strätling, W. H. (1991). A matrix/scaffold attachment region binding protein: Identification, purification and mode of binding. *Cell (Cambridge, Mass.)* **64,** 123–135.

von Kries, J. P., Buck, F., and Strätling, W. H. (1994). Chicken MAR binding protein p120 is identical to human heterogeneous nuclear ribonucleoprotein (hnRNP)U. *Nucleic Acids Res.* **22,** 1215–1220.

Wright, K. L., Dell'Orco, R. T., van Wijnen, A. J., Stein, J. L., and Stein, G. S. (1992). Multiple mechanisms regulate the proliferation specific histone gene transcription factor, HiNF-D, in normal human diploid fibroblasts. *Biochemistry* **31,** 2812–2818.

Xing, Y., Johnson, C. V., Dobner, P. R., and Lawrence, J. B. (1993). Higher level organization of individual gene transcription and RNA splicing. *Science* **259,** 1326–1330.

Yoon, K., Rutledge, S. J. C., Buenaga, R. F., and Rodan, G. A. (1988). Characterization of the rat osteocalcin gene: Stimulation of promoter activity by 1,25-dihydroxyvitamin D_3. *Biochemistry* **17,** 8521–8526.

Zeitlin, S., Parent, A., Silverstein, S., and Efstratiadis, A. (1987). Pre-mRNA splicing and the nuclear matrix. *Mol. Cell. Biol.* **7,** 111–120.

Zenk, D. W., Ginder, G. D., and Brotherton, T. W. (1990). A nuclear-matrix protein binds very tightly to DNA in the avian β-globin gene enhancer. *Biochemistry* **29,** 5221–5226.

7

Transcription by RNA Polymerase II and Nuclear Architecture

DERICK G. WANSINK, LUITZEN DE JONG, AND ROEL VAN DRIEL

E. C. Slater Institute
University of Amsterdam
Amsterdam, The Netherlands

I. INTRODUCTION

Nuclear transcription is carried out by three different DNA-dependent RNA polymerases, termed RNA polymerase I, II, and III (RPI,

RPII, RPIII; Sentenac, 1985; Archambault and Friesen, 1993). These three enzyme complexes differ in their template specificity, sensitivity to specific inhibitors, and nuclear localization. RPI synthesizes the 45S rRNA precursor in the nucleolus (Sollner-Webb and Mougey, 1991). RPII is localized in the nucleoplasm and is responsible for the synthesis of heterogeneous nuclear RNA (hnRNA) (i.e., predominantly pre-mRNA), most small nuclear RNAs (snRNAs), and at least one repetitive RNA sequence (Shafit-Zagardo *et al.*, 1983; Young, 1991). RPIII is also localized in the nucleoplasm and synthesizes several small RNAs, like 5S rRNA, tRNA, U6 snRNA, and 7SL RNA (Willis, 1993).

Most studies concerning RNA synthesis are done in solution in test tube experiments. In such systems, the nuclear context of RNA synthesis is usually disregarded (for a comprehensive review on nuclear organization, see Spector, 1993). It is becoming increasingly clear that the nuclear context (e.g., chromatin structure, nuclear matrix) plays an important role in the regulation of gene expression. In soluble, cell-free systems, specific higher-order effects in the nucleus may be overlooked. For a complete understanding of gene expression, it is therefore essential to study transcription in its nuclear environment. This chapter focuses on transcription by RPII, with special emphasis on its relationship to nuclear architecture.

II. LOCALIZATION OF RNA POLYMERASE II TRANSCRIPTION SITES IN THE NUCLEUS

A. Introduction

Transcription by RPII covers the majority of extranucleolar RNA synthesis in most eukaryotic cells. Like the nucleolus, where RNA synthesis by RPI is concentrated, the nucleoplasm contains distinct ultrastructural domains. These are clusters of interchromatin granules, clusters of perichromatin fibrils, coiled bodies, and nuclear bodies. Their role in RNA synthesis (and/or RNA processing) has remained elusive, however. Through the years, only a few techniques were available to visualize sites of RNA synthesis. The spatial distribution of RPII transcription sites in the nucleus has therefore remained unclear for a long time. The different methods that are used to study the localization of (potential) transcription sites in the nucleus are discussed in Sections II,B–G. A general picture of how actively transcribed chromatin is distributed throughout the nucleus is now emerging.

B. Electron Microscopic Autoradiography

The first technique that was developed to visualize transcription sites was electron microscopy combined with high-resolution autoradiography (EMARG) on [^{3}H]uridine-labeled cells (reviewed by Fakan and Puvion, 1980; Fakan, 1986). It is often combined with the regressive ethylenediaminetetraacetic acid (EDTA) technique (Bernhard, 1969), which stains specifically ribonucleoprotein (RNP) structures in the nucleus. After a short [^{3}H]uridine pulse, strong labeling is observed in the nucleolus, corresponding to rRNA synthesis (Fakan and Bernhard, 1971; Fakan *et al.*, 1976; Fakan and Nobis, 1978). Extranucleolar labeling, corresponding to RNA synthesis by RPII and RPIII, is weaker and is found throughout the nucleoplasm at the border of condensed chromatin domains in association with so-called perichromatin fibrils. Perichromatin fibrils have a diameter varying between 3 and 5 nm, sometimes up to 20 nm (Fakan and Puvion, 1980). Their exact nature is not known. They may correspond to (nascent) transcripts that are associated with hnRNP proteins or snRNPs (Bachellerie *et al.*, 1975; Fakan, 1994). Interchromatin granules are RNP-containing structures, 20–25 nm in diameter, that appear in clusters distributed randomly throughout the nucleoplasm (Fakan and Puvion, 1980). Such clusters of interchromatin granules are only weakly labeled by [^{3}H]uridine or remain unlabeled, even after long incubation periods. This indicates that they probably do not represent sites of transcription.

A significant drawback of EMARG is its low spatial resolution, which is less than that of the electron microscope (Fakan, 1986). In addition, EMARG is a complex method and often requires long exposure times. A great advantage, however, is the preservation of nuclear ultrastructure due to fixation of [^{3}H]uridine-labeled cells without permeabilization. EMARG has contributed greatly to our knowledge of RNA synthesis in relation to nuclear ultrastructure, especially at a time when the study of nuclear organization was still in its infancy.

C. Antibodies against RNA Polymerase II Molecules

A method of studying the distribution of potential sites of transcription consists of labeling RPII molecules with specific antibodies. Two independent immunofluorescence studies, using different antibodies and different cell types, report diffuse nucleoplasmic staining (Bona *et al.*, 1981; Jiménez-García and Spector, 1993). Using the same antibody used by Jiménez-García and Spector (1993), called 8WG16 (Thompson *et al.*, 1989), we found a finely punctuated nucleoplasmic staining (M. Grande and R. van Driel, submitted). This slight discrepancy in distribution may be due to differences in the immunolabeling procedure or the cell types used.

A different type of distribution of RPII molecules was reported by Clark *et al.* (1991). They found RPII molecules predominantly in the periphery of the nucleus. When nascent RNA is labeled with biotin-conjugated uridine triphosphate (UTP) during run-on transcription, RPII molecules engaged in transcription are also found in the nuclear periphery. The significance of this localization study is questionable, however, because nuclear morphology was not well preserved under the labeling conditions used, i.e., the nuclear lamina was often not intact, and peripheral chromatin was dispersed into a halo around the nucleus (Clark *et al.*, 1991).

It should be noted that anti-RPII antibodies may not discriminate between RPII molecules that are engaged in transcription and those that are inactive (e.g., stalled RPII molecules or RPII molecules not bound to DNA). Therefore, the RPII–antigen distribution may be an overestimation of the number of RPII transcription sites. On the other hand, it is conceivable that an antibody does discriminate between active and inactive RPII molecules because the epitope in either of the two populations is masked or absent. A double-labeling study combining anti-RPII labeling with either [^{3}H]uridine or BrUTP labeling of nascent pre-mRNA (see Section II,G) to localize active sites of transcription will solve this problem. For comparison, on polytene chromosomes RPII molecules are found predominantly in so-called interbands and puffs, the same sites where [^{3}H]uridine is incorporated (Sass, 1982). This shows that, at least in polytene nuclei, chromatin domains containing RPII are also sites of transcription.

D. Antibodies against RNA–DNA Hybrids

An entirely new approach to map transcription sites in the nucleus has been introduced by Testillano *et al.* (1994). This method is based on the notion that DNA–RNA hybrids are intermediates in the transcription process (see references cited in Testillano *et al.*, 1994). Antibodies directed against DNA–RNA hybrids label specific structures in the nucleoplasm and the nucleolus. These structures correspond largely to structures that become labeled by [^{3}H]uridine. It cannot be excluded, however, that DNA–RNA hybrids unrelated to transcription are also detected by these antibodies.

E. *In Situ* Nick Translation

Attempts have been made to localize transcription sites by *in situ* nick translation. These experiments were based on the notion that mainly actively transcribed genes are nuclease sensitive (Weintraub and Groudine, 1976; Levitt *et al.*, 1979). When *in situ* nick translation is used, DNase-sensitive sites are found predominantly, although not exclusively, in the nuclear periphery (Hutchison and Weintraub, 1985; de Graaf *et al.*, 1990; Krystosek and Puck, 1990). The suggestion that transcription takes place preferentially in the nuclear periphery is not consistent with [^{3}H]uridine- and BrUTP-labeling experiments (see Section II,G), which show that RPII transcription sites are distributed throughout the nucleoplasm. The reason for this discrepancy is not known. *In situ* nick translation may not be a suitable technique to examine the distribution of active chromatin in the nucleus. The relationship between active chromatin and deoxyribonuclease (DNase) sensitivity may be more complex than was initially thought. Double-labeling experiments combining *in situ* nick translation and BrUTP labeling (see Section II,G), preferably at the electron microscopic level, will give new clues on how nuclease-sensitive chromatin is related to transcriptionally active loci.

F. *In Situ* Hybridization

In situ hybridization can be used to visualize sites of transcription of individual genes (Lawrence and Singer, 1991). In principle, the position of any gene can be localized if a specific probe is available. The Epstein-Barr virus (EBV) genome in Namalwa cells is integrated preferentially in the inner 50% of the nuclear volume (Lawrence *et al.,* 1988). Also, the *neu* oncogene is found in the nuclear interior. In contrast, the dystrophin gene is usually localized near the nuclear envelope (Lawrence and Singer, 1991). It has been suggested that each gene occupies a specific position in the interphase nucleus (Carter and Lawrence, 1991). Thus far, however, a clear correlation between the three-dimensional location of a gene and its transcriptional activity has not been recognized.

The c-*fos* gene (Huang and Spector, 1991) and the fibronectin gene (Xing *et al.,* 1993) are usually found close to domains that are highly enriched in the splicing factor SC-35. These domains correspond to interchromatin granule clusters (see Spector *et al.,* 1991). *In situ* hybridization with a poly(dT) probe has shown that these interchromatin granule clusters contain high concentrations of poly(A) RNA (Carter *et al.,* 1991, 1993; Visa *et al.,* 1993). This does not mean, however, that these domains are sites of transcription, as suggested by Carter *et al.* (1991), because (1) probably the entire poly(A) RNA population is detected with a poly(dT) probe, i.e., newly synthesized hnRNA, hnRNA en route to the cytoplasm, and hnRNA stored in the nucleus, and (2) it is uncertain whether polyadenylation occurs directly at the site of transcription. Moreover, earlier EMARG studies on [^{3}H]thymidine-labeled cells (Fakan and Hancock, 1974) and [^{3}H]uridine-labeled cells (Fakan and Bernhard, 1971; Fakan *et al.,* 1976) showed that interchromatin granule clusters do not contain DNA and are not transcriptionally active.

Summarizing, *in situ* hybridization is a very useful technique to localize individual genes and specific transcripts in the interphase nucleus. The next step is to combine *in situ* hybridization with electron microscopy (e.g., Huang and Spector, 1991; Sibon *et al.,* 1994). The combination of high specificity and high resolution offers interesting opportunities to investigate the ultrastructural localization of specific active or inactive genes. Furthermore, intron- and exon-specific probes can be used to study the localization of successive steps in RNA processing and transport of pre-mRNA and mature RNA from the site of synthesis to the nuclear envelope (Sibon *et al.,* 1995).

G. Labeling Nascent RNA with BrUTP

Cook and co-workers and we have independently developed a new method to visualize sites of transcription in the nucleus (Jackson *et al.*, 1993; Wansink *et al.*, 1993, 1994a). This method is based on the incorporation of the UTP analog 5-bromouridine 5′-triphosphate (BrUTP) *in situ* into nascent RNA. Sites of BrUTP incorporation are visualized by a monoclonal antibody (MAb) that specifically recognizes bromouridine, followed by immunofluorescence microscopy. We observe a punctated nuclear staining composed of hundreds of domains scattered throughout the nucleus, except for the nucleolus (Fig. 1; optical section). No staining is observed when the transcription reaction is done in the presence of 1 μg/ml α-amanitin. This shows that the staining represents

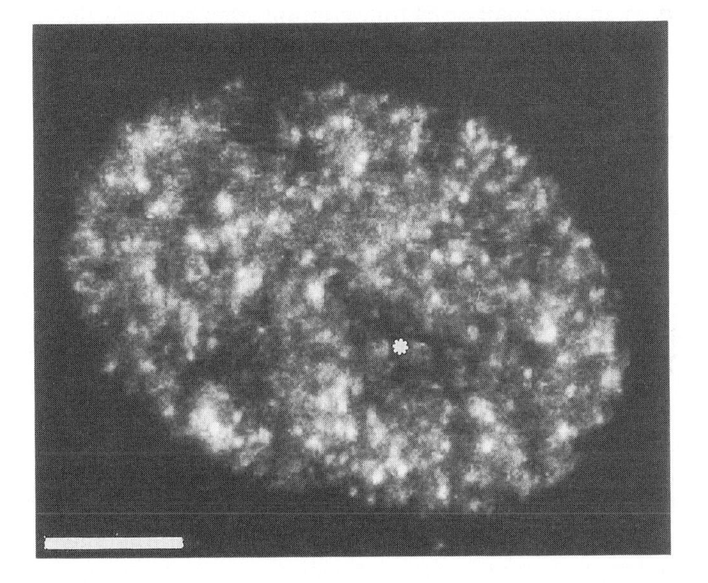

Fig. 1. Localization of transcription sites in the nucleus. Human skin fibroblasts were permeabilized, and run-on transcription was done in the presence of BrUTP for 15 min to label nascent RNA. Cells were fixed in 2% formaldehyde and processed for immunofluorescence microscopy using anti-BrUTP antibodies. Shown is an optical section through the middle of a BrUTP-labeled nucleus. Sites of RPII transcription are not confined to the nuclear periphery but are distributed throughout the nucleoplasm. In this particular image, nucleolar transcription (caused by RPI) is also apparent (asterisk). Bar: 5 μm.

RNA synthesized by RPII. The punctated pattern is observed under conditions in which most BrUTP-labeled RNA is nascent, i.e., (1) after labeling *in vitro* during run-on transcription in permeabilized cells and (2) after short-term labeling *in vivo* after microinjection of BrUTP directly into the cell. Therefore, we conclude that nuclear domains that contain BrUTP-labeled RNA represent genuine sites of transcription.

We initially concluded that the BrUTP-labeled nuclear staining observed by immunofluorescence microscopy, consisting of hundreds of transcription domains per nucleus, may indicate that each domain contains several actively transcribed genes grouped closely together during transcription (Wansink *et al.,* 1993). However, electron microscopic studies on BrUTP-labeled nuclei show that this is probably not the case (Wansink *et al.,* 1996). In our electron microscopic preparations, we observe as many as several thousand gold-labeled transcription sites distributed more or less uniformly throughout the nucleus. The apparent discrepancy in the number of transcription sites between the immunofluorescence labeling and the immunogold labeling is probably the result of the difference in resolution between light microscopy and electron microscopy. Most likely, each light microscopically defined transcription domain corresponds to several gold-labeled transcription sites and therefore to several actively transcribed genes.

In contrast to the nucleospasm, nucleoli are often not stained after BrUTP labeling (Wansink *et al.,* 1993; Jackson *et al.,* 1993). Intrinsic properties of the nucleolus and nucleolar RNA probably can explain this phenomenon: (1) BrUTP is less efficiently incorporated by RPI than by RPII (Wansink *et al.,* 1993); (2) rRNA contains little uridine, and therefore will contain less BrU than RPII transcripts; (3) the nucleolus has a dense ultrastructure, which may hamper access of antibodies; and (4) packaging of rRNA into ribonucleoprotein particles may occur during synthesis, making the BrU epitope inaccessible for antibodies. The latter two explanations are supported by the observation that nucleoli are clearly labeled in nuclear matrices isolated from BrUTP-labeled nuclei. This indicates that under these conditions—most of the nuclear DNA and proteins are absent from nuclear matrices—antibodies are able to reach BrUTP-labeled rRNA in the nucleolus. In nuclear matrices the extranucleolar labeling is very similar to that observed in whole nuclei. No additional transcription domains seem to be uncovered in the nucleoplasm. This implies that all nucleoplasmic transcription sites are already visible in whole nuclei.

In conclusion, labeling nascent RNA with BrUTP is a simple, sensitive, nonradioactive method of localizing active sites of transcription in the nucleus *in vivo* and *in vitro.* It is the first technique that visualizes transcription at high resolution in the light microscope and provides unique possibilities in double-labeling experiments to study the localization of transcription sites in relation to other nuclear activities and components (see Sections IV, V; Jackson *et al.,* 1993; Wansink *et al.,* 1993, 1994b; Hassan *et al.,* 1994). Because the presence of BrU in pre-mRNA inhibits splicing (Wansink *et al.,* 1994c; see also Sierakowska *et al.,* 1989), the use of BrUTP in long-term *in vivo* experiments is limited. Extension of the BrUTP technique to the electron microscopic level has been successful and offers new possibilities of studying RNA synthesis and nuclear ultrastructure (Dundr and Raška, 1993; Hozák *et al.,* 1994; Wansink *et al.,* 1996).

H. Concluding Remarks

After our discussion of six different techniques used to visualize (potential) sites of RPII transcription, it appears that the ones that directly label nascent RNA, i.e., either by [^{3}H]uridine or by BrUTP, are the most reliable. These techniques show that RPII transcription sites are distributed throughout the nucleus, without any preference for either the nuclear interior or the periphery. The observation that actively transcribed genes are located predominantly at the periphery of domains of condensed chromatin (Fakan and Bernhard, 1971; Fakan *et al.,* 1976) is corroborated by our electron microscopic studies (Wansink *et al.,* 1996). The significance of this intriguing finding is not yet clear and awaits further investigation.

III. TRANSCRIPTION AND THE NUCLEAR MATRIX

The nuclear matrix is operationally defined as the structure that remains after nuclease treatment and subsequent extraction of most of the lipids, DNA, proteins, and sometimes RNA from the interphase nucleus. Depending on the isolation procedure used, this nuclear substructure is called nuclear matrix, nucleoskeleton, nuclear scaffold, nuclear matrix–intermediate filament complex, or karyoskeleton. Here we will consistently use the term nuclear matrix. The nuclear matrix is thought to play an important role in maintaining the functional and structural integrity of the nucleus, including the organization of RNA synthesis (reviewed by Verheijen *et al.*, 1988; Cook, 1988, 1989; de Jong *et al.*, 1990; van Driel *et al.*, 1991; Stuurman *et al.*, 1992). However, the exact role of the nuclear matrix in the regulation and organization of transcription by RPII is still not clear.

Many experiments, mainly biochemical, have shown that the RPII transcription machinery is associated with the nuclear matrix. (1) Actively transcribed class II genes are highly enriched in nuclear matrix preparations, whereas inactive genes are not (Robinson *et al.*, 1982; Ciejek *et al.*, 1983; Jost and Seldran, 1984; Jackson and Cook, 1985; Buttyan and Olsson, 1986; Andreeva *et al.*, 1992). (2) RNA is identified as a component of the nuclear matrix (van Eekelen and van Venrooij, 1981; Ciejek *et al.*, 1982; Mariman *et al.*, 1982; Nickerson *et al.*, 1989; He *et al.*, 1990). Also, nascent RNA is associated with the nuclear matrix (Jackson and Cook, 1985). Using BrUTP to label nascent RNA, we have shown that the distribution of sites containing nascent RNA (i.e., transcription sites) is largely maintained in nuclear matrices (Fig. 2; Wansink *et al.*, 1993; see also Jackson *et al.*, 1993). This finding agrees with the preservation of the "track"-like distribution of EBV primary transcripts in nuclear matrix preparations of Namalwa cells (Xing and Lawrence, 1991). (3) Nuclear matrix preparations retain much of the RPII transcriptional activity found in isolated nuclei (Abulafia *et al.*, 1984; Jackson and Cook, 1985; Razin and Yarovaya, 1985; Razin *et al.*, 1985; D. G. Wansink and J. Groenink, unpublished observations). (4) Many transcription factors are found enriched in the nuclear matrix (Barrack, 1987; Klempnauer, 1988; Waitz and Loidl, 1991; Stein *et al.*, 1991; Bidwell *et al.*, 1993; van Wijnen *et al.*, 1993). The molecular interactions that form the basis for the association of the different components of the RPII transcription complex with the nuclear matrix are not known and need further investigation. In the human androgen receptor and the human glucocorticoid receptor, specific receptor domains are responsible for nuclear matrix association (van Steensel *et al.*, 1995).

In contrast to the papers just cited, some reports suggest that transcription is not associated with the nuclear matrix (Kirov *et al.*, 1984; Roberge *et al.*, 1988; Fisher *et al.*, 1989). The reason for this discrepancy is not known but may be due to differences in the isolation procedure of nuclear matrices (e.g., loss of the internal nuclear matrix).

Besides the nuclear distribution of transcription sites and EBV transcripts, the distribution of replication domains (Nakayasu and Berezney,

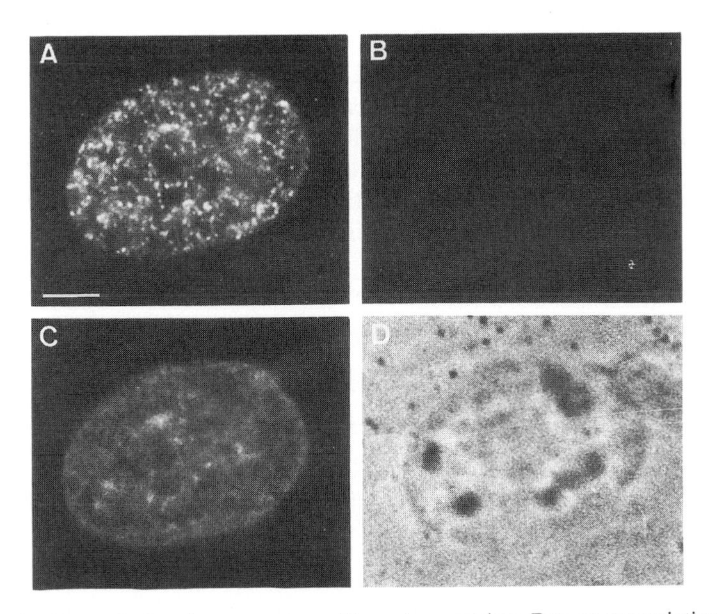

Fig. 2. Transcription sites are preserved in nuclear matrices. Run-on transcription was carried out in permeabilized human skin fibroblasts in the presence of BrUTP for 30 min. Labeled cells were treated with sodium tetrathionate to stabilize the internal nuclear matrix structure, digested with RNase-free DNase, and extracted with 0.25 *M* ammonium sulfate. Nuclear matrices were fixed in 2% formaldehyde and immunolabeled using anti-BrUTP antibodies. (A) Nascent RNA; (B) DNA stained with Hoechst 33258, 0.25 s exposure; under these conditions, whole nuclei give a bright image; (C) as in (B) with 10 s exposure; (D) phase contrast. Note that most DNA has been extracted. In the nuclear matrices nucleolar RNA staining is clearly visible (compare D with A), whereas the nucleoplasmic staining remains unaltered compared to the staining in unextracted nuclei. This shows that the distribution of transcription sites is preserved in nuclear matrices. Bar: 5 μm. Reproduced from the *Journal of Cell Biology,* **122:**283–293 (1993), by copyright permission of The Rockefeller University Press.

1989; Hozák *et al.,* 1993), epidermal growth factor (EGF)-receptor transcripts (Sibon *et al.,* 1993, 1994), and the splicing factor SC-35 (D. G. Wansink and O. C. M. Sibon, unpublished observations) are preserved in nuclear matrices. This suggests that the nuclear matrix rather than the chromatin plays a structural role in maintaining the organization of the interphase nucleus. The nuclear matrix may act as the active structure at which most nuclear processes take place (see Cook, 1989). For instance, the nuclear matrix may represent the *in vivo* structure to which

RNA is bound when it is synthesized, during subsequent RNA processing, until it is transported through the nuclear pore to the cytoplasm (Xing and Lawrence, 1991). Furthermore, the nuclear matrix may form the physical connection between different nuclear domains involved in pre-mRNA processing (see Section IV). Clearly, much needs to be learned about the structure and function of the nuclear matrix and about the molecular interactions that obviously exist between matrix components and pre-mRNA.

IV. TRANSCRIPTION AND PRE-mRNA PROCESSING

Pre-mRNAs form the major product of transcription by RPII. Most pre-mRNAs are processed to mature mRNAs in the nucleus and are then exported to the cytoplasm. Knowledge of the spatial and functional relationship between sites of pre-mRNA synthesis and sites of pre-mRNA processing is crucial in understanding the organization of the nucleus. Pre-mRNA processing must occur at the site of synthesis or en route to the nuclear pore complex. At present, there is an interesting controversy in the literature on how pre-mRNA synthesis, processing, and intranuclear transport are spatially related and functionally interdependent (see Rosbash and Singer, 1993; Xing and Lawrence, 1993; Kramer *et al.,* 1994).

Most pre-mRNAs are modified by (1) the addition of a 5′-cap soon after initiation of transcription, (2) the addition of a 3′-poly(A) tail after 3′-cleavage, and (3) the removal of introns during pre-mRNA splicing. The capping enzyme and probably also poly(A) polymerase are located at the site of transcription. In contrast, the temporal and spatial organization of splicing is still not clear. Splicing factors are concentrated essentially in three different types of domains in the nucleus: (1) interchromatin granule clusters, (2) clusters of perichromatin fibrils, and (3) coiled bodies (Fakan and Puvion, 1980; Fakan, 1986; Lamond and Carmo-Fonseca, 1993a). It is obvious that these domains must somehow be involved in splicing. The molecular composition of these domains has been studied extensively to understand their individual roles in pre-mRNA synthesis and splicing.

1. *Interchromatin granule clusters.* Clusters of interchromatin granules most likely do not contain DNA because they are not labeled by

[^{3}H]thymidine (Fakan and Hancock, 1974) and only slowly by [^{3}H]uridine (Fakan and Bernhard, 1971; Fakan *et al.,* 1976; Fakan, 1986). Interchromatin granule clusters are highly enriched in snRNAs, snRNP proteins, and non-snRNP splicing factors, e.g, SC-35 (Fakan *et al.,* 1984; Spector *et al.,* 1991; Carmo-Fonseca *et al.,* 1991). In contrast, they contain only few hnRNP proteins (Fakan *et al.,* 1984). Obviously, poly(A) RNAs, which are present in high concentration in interchromatin granule clusters (Carter *et al.,* 1991, 1993; Visa *et al.,* 1993), bind little or no hnRNP proteins, unlike probably most hnRNAs in perichromatin fibrils. This suggests that poly(A) RNA in interchromatin granule clusters either contains RNA species other than poly(A) RNA in perichromatin fibrils or is in a different stage of maturation than that in perichromatin fibrils. Together these results indicate that interchromatin granule clusters do not represent sites of transcription. Evidence favors the hypothesis that interchromatin granule clusters are also not major sites of RNA processing (see the following discussion).

2. *Perichromatin fibrils.* Perichromatin fibrils represent sites of transcription because they become labeled immediately after a short pulse of [^{3}H]uridine (Fakan and Bernhard, 1971; Fakan *et al.,* 1976). SnRNAs, snRNP proteins, and non-snRNP proteins are associated with perichromatin fibrils (Fakan *et al.,* 1984; Spector *et al.,* 1991; Carmo-Fonseca *et al.,* 1991). They also contain poly(A) RNA (Carter *et al.,* 1991, 1993; Visa *et al.,* 1993). Perichromatin fibrils are the major location of hnRNP proteins in the nucleus (Fakan *et al.,* 1984). Therefore, it is likely that perichromatin fibrils represent (or contain) nascent transcripts (Fakan, 1994).

3. *Coiled bodies.* Coiled bodies are spherical structures with a diameter of 0.5–1 μm (reviewed by Lamond and Carmo-Fonseca, 1993b). Their number varies from cell to cell and ranges between zero and six per nucleus. Coiled bodies are highly enriched in snRNAs and snRNP proteins (Fakan *et al.,* 1984; Raška *et al.,* 1991; Carmo-Fonseca *et al.,* 1992; Spector *et al.,* 1992). The non-snRNP protein U2AF is also present in coiled bodies (Zhang *et al.,* 1992), whereas SC-35, also a non-snRNP splicing factor, is not (Răska *et al.,* 1991). Poly(A) RNA is not detected in coiled bodies (Visa *et al.,* 1993). It is not known whether transcription occurs in coiled bodies. Generally, coiled bodies are observed only in transformed or immortal cell lines and not in cell lines of defined passage (Spector *et al.,* 1992). This suggests that coiled bodies do not play a general role in RNA processing. They may be involved in processing of specific transcripts. It has also been suggested that coiled bodies are

involved in post-splicing events, e.g., recycling of snRNPs or degradation of introns (Lamond and Carmo-Fonseca, 1993b).

In which of these three nuclear structures does splicing take place? A crucial observation is that many, if not all, splicing factors are present at the site of pre-mRNA synthesis. We found, for example, SC-35 (Fu and Maniatis, 1990) in low concentrations at many transcription sites (Fig. 3; Wansink *et al.*, 1993). So, in principle, splicing can occur at the same site where transcription takes place. Several observations indicate that splicing is initiated cotranscriptionally (LeMaire and Thummel, 1990; Beyer and Osheim, 1991; Kopczynski and Muskavitch, 1992; Baurén and Wieslander, 1994). This conclusion agrees with the finding that hnRNP proteins and snRNPs become associated with pre-mRNAs during transcription (Fakan *et al.*, 1986; Amero *et al.*, 1992; Matunis *et al.*, 1993). In conclusion, splicing probably starts at the site of pre-mRNA synthesis, i.e., in close association with perichromatin fibrils. Whether splicing is also completed there is not clear.

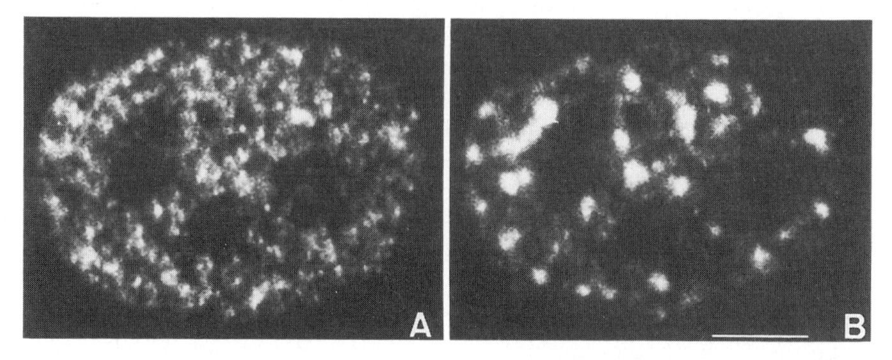

Fig. 3. Double labeling of nascent RNA and SC-35 domains enriched in splicing components. Permeabilized human skin fibroblasts were labeled with BrUTP for 15 min. Subsequently, cells were fixed in 2% formaldehyde and immunolabeled. In the same cells, domains enriched in snRNPs were visualized with a MAb against the essential splicing component SC-35 (Fu and Maniatis, 1990). Optical sections through the middle of the nucleus are shown. (A) Nascent RNA labeled with BrUTP. (B) Domains enriched in splicing components stained by anti-SC-35 (same optical section as in A). No relationship was observed between sites containing nascent RNA and intensely labeled SC-35 domains (i.e., interchromatin granule clusters). Much colocalization was observed between sites containing nascent RNA and areas showing weak SC-35 staining (i.e., perichromatin fibril clusters). Bar: 5μm. Reproduced from the *Journal of Cell Biology,* **122:**283–293 (1993), by copyright permission of The Rockefeller University Press.

The dynamic distribution of splicing factors in the nucleus has been discussed in a few papers. First, virus-infected cells provide an interesting model because virus infection causes a dramatic redistribution of ribonucleoproteins (Martin *et al.,* 1987). Jiménez-García and Spector (1993) used adenovirus-infected HeLa cells and report that soon after virus infection, snRNPs and SC-35 are recruited from interchromatin granule clusters, the presumed storage places of these factors, to the sites of viral RNA synthesis, where splicing is thought to take place. Bridge *et al.* (1993), however, claim that this finding reflects only the situation in a minority of the population of infected cells. Second, it was shown in HeLa cells that on inhibition of the splicing reaction interchromatin granule clusters round up, decrease in number, and increase in the size and content of splicing factors (O'Keefe *et al.,* 1994). In addition, the physical connections that are normally observed between interchromatin granule clusters (i.e., probably perichromatin fibrils) disappear. Third, the distribution of snRNPs was investigated in differentiating murine erythroleukemia cells (Antoniou *et al.,* 1993). It was shown that snRNPs aggregate in large clusters of interchromatin granules when transcription ceases at later stages of cell differentiation. Together these results suggest that interchromatin granule clusters function primarily as storage or assembly sites for splicing factors rather than as sites where splicing is concentrated. In addition, they may play a role in degradation of introns, degradation of "old" or flawed poly(A) RNA, or regeneration of spliceosomes. Still, it cannot be fully excluded that splicing, which was initiated at the site of pre-mRNA synthesis, may be completed in or near interchromatin granule clusters.

Suggestive evidence for the direct involvement of interchromatin granule clusters in splicing came from *in situ* hybridization studies. Nascent c-*fos* and fibronectin transcripts are found in tracks closely associated, but mostly not coinciding, with interchromatin granule clusters (Huang and Spector, 1991; Xing *et al.,* 1993). In the case of fibronectin transcripts, spatial separation of intron-containing and spliced transcripts is observed in the track (Xing *et al.,* 1993). This was interpreted to indicate that RNA is transported along the track, while splicing occurs close to or in the interchromatin granule cluster (Xing and Lawrence, 1993). However, because *in situ* hybridization only shows the steady-state localization of (accessible) transcripts in the nucleus, it cannot be excluded that the introns are already excised at the beginning of the track, i.e., at the site of synthesis, and that they are transported with the spliced transcript along the track. In that case, the interchromatin granule cluster would not be directly involved in the splicing process. As for the c-*fos* and fibronectin transcripts, we too observe transcription sites near interchromatin granule clusters after labeling nascent RNA with BrUTP (Fig. 3; Wansink *et al.,* 1993). However, in our preparations, many RPII transcription sites are also seen that are not localized near interchromatin granule clusters (Fig. 3). Although the genes that correspond to these transcription sites are not known, this observation shows that the localization of c-*fos* and fibronectin transcription is not representative of all class II transcription sites.

Summarizing, the current model is that splicing initiates cotranscriptionally, so that the first introns are removed at the site of pre-mRNA synthesis. Splicing may be completed there but may also be finished during transport of the pre-mRNA to the nuclear envelope. If so, such splicing may occur close to or in an interchromatin granule cluster.

V. TRANSCRIPTION AND DNA REPLICATION

In S phase the entire genome is replicated while large parts of it are still transcribed. DNA replication, like transcription by RPII, occurs in specific domains in the nucleus. Replication domains are defined by labeling nascent DNA with specific nucleotide analogs (e.g., Nakamura *et al.,* 1986; Nakayasu and Berezney, 1989; Manders *et al.,* 1992; O'Keefe *et al.,* 1992). In early S phase, replication domains are relatively small and scattered throughout the nucleus. In late S phase they increase in size, decrease in number, and are located preferentially in the nuclear periphery and around the nucleolus. Generally, actively transcribed genes are replicated early, whereas untranscribed DNA is replicated late in S phase (Goldman, 1988). Replication domains and transcription domains may play a role in this tight coordination of gene duplication and gene expression.

Because DNA polymerases and RNA polymerases use the same DNA as a template, the question arises regarding the spatial relationship between replication and transcription in S-phase nuclei. How are transcription and replication related at the resolution of the light microscope? To address this question, we have visualized replication and transcription simultaneously by labeling permeabilized cells with two different nucleotide analogs (Wansink *et al.,* 1994b). Our analysis focused specifically on replication and transcription domains, defined as nuclear compartments that display relatively high replication and transcription activity, respectively. We found that in late S phase transcription domains clearly do not colocalize with replication domains (Fig. 4C,D). This agrees with

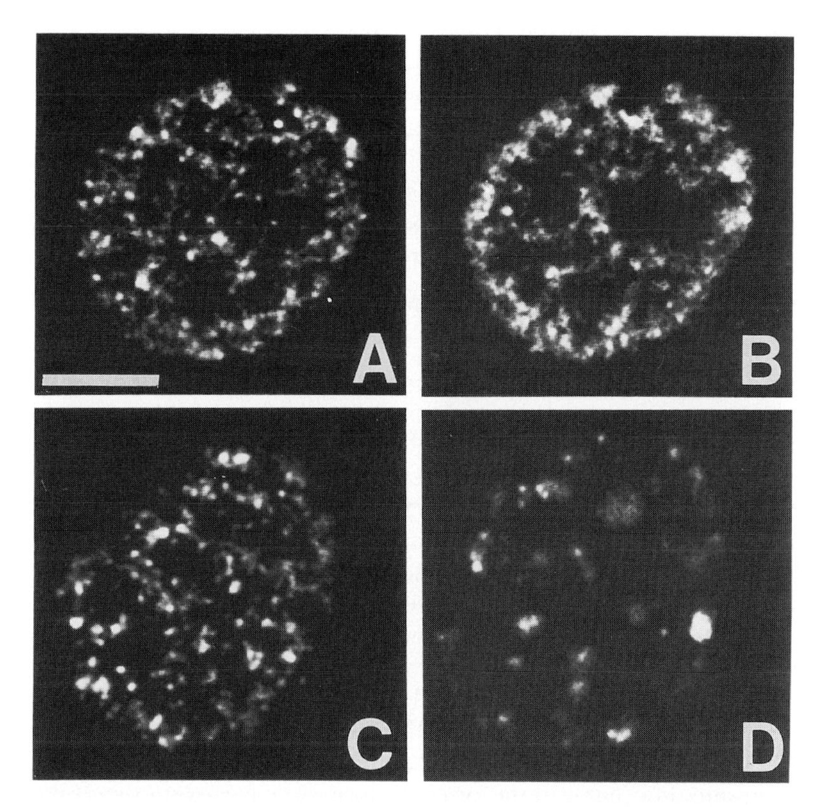

Fig. 4. RPII transcription is concentrated outside replication domains in S phase. Human bladder carcinoma cells were permeabilized and incubated in the presence of BrUTP as well as digoxigenin-dUTP for 30 min. Cells were fixed and processed for immunofluorescence microscopy using antibodies against BrU and digoxigenin, respectively. Local background was subtracted as described (Wansink *et al.,* 1994b). The transcription component (A and C) and the replication component (B and D) of optical sections through the middle of two doubly labeled nuclei are shown separately. (A) corresponds to the same optical section as (B); (C) corresponds to (D). (A/B) is an early S-phase nucleus, and (C/D) is a late S-phase nucleus. Virtually no colocalization (i.e., complete overlap) between transcription domains and replication domains is observed (see Wansink *et al.,* 1994b, for color images). Bar: 5 μm.

the notion that late-replicating DNA is largely transcriptionally silent. In addition, electron microscopic studies correlating replication with nuclear ultrastructure have shown that in late S phase predominantly heterochromatin, corresponding to transcriptionally inactive chromatin, is replicated (O'Keefe *et al.,* 1992). Replication patterns in early S phase are more complex, and their spatial relationship with transcription domains is difficult. However, we find that in early S phase transcription domains also generally do not colocalize with replication domains (Fig. 4A,B). We conclude, therefore, that no nuclear domains exist that are highly active in both replication and transcription at the same time.

What is the biological significance when transcription domains and replication domains do not colocalize? Absence of colocalization suggests strong regulation and coordination between replication and transcription in S phase. It implies that when the DNA in a nuclear domain is being replicated, transcription in that domain stops or is strongly reduced until replication in the entire domain is completed (Wansink *et al.,* 1994b). The decrease in transcription activity in a domain at the onset of replication may result from the concentration of replication-specific factors [e.g., DNA polymerases, proliferating cell nuclear antigen (PCNA), methyltransferase] at the expense of transcription factors. Concentrating either transcription or replication in a domain and excluding potentially disturbing elements will increase the efficiency and reliability of the process.

Our conclusions differ from those of Cook and co-workers, who also investigated the relationship between replication and transcription (Hassan *et al.,* 1994). The reason for the differences between our findings and theirs is difficult to understand because similar labeling techniques were used. The occurrence of a RNase-insensitive transcription signal (Hassan *et al.,* 1994), cell synchronization, different image analysis procedures, and different criteria for colocalization may explain the discrepancy. It is clear that relationships between complex labeling patterns in microscopy are very difficult. Objective criteria for colocalization need to be defined (e.g., Taneja *et al.,* 1992; Manders *et al.,* 1993). Such standards will help to determine unequivocally the relationship between complex distributions in localization studies in the nucleus. In addition, the high resolving power of electron microscopy will be very useful to investigate further how transcription and replication are spatially related in S-phase nuclei.

VI. OUTLOOK AND QUESTIONS

Our knowledge of the functional distribution of active chromatin in the nucleus, and of the relationship between RNA processing and nuclear organization, has increased considerably in the past decade. Many questions, however, are still unanswered. First, the higher-order structure of chromatin, which is responsible for the specific localization of active genes on the surface of chromatin domains, needs to be studied in more detail. This chromatin organization is likely to play an important role in the regulation of gene activation and gene silencing. Second, the dynamics of factors involved in RNA metabolism, e.g., between putative "storage sites" (interchromatin granule clusters) and "active sites" (perichromatin fibril clusters), is still elusive. Besides, what is the molecular composition of the entire RNA synthesis and processing machinery at an actively transcribed gene? Third, information about intranuclear RNA transport is virtually lacking. What is the route of transcripts from their site of synthesis to a nuclear pore? RNA transport and RNA processing are probably closely linked processes. Is the nuclear matrix involved in either of them?

Both light microscopy and electron microscopy are important to answer these and further questions. Electron microscopy is essential for high resolution subnuclear localization studies using highly specific antibodies and nucleic acid probes. High quality (confocal) light microscopy is essential for quantitative analysis and provides three-dimensional information. Ultimately, light microscopy on living cells, applying suitable fluorescent probes, will be essential for understanding important dynamics in the cell nucleus.

REFERENCES

Abulafia, R., Ben-Ze'ev, A., Hay, N., and Aloni, Y. (1984). Control of late Simian Virus 40 transcription by the attenuation mechanism and transcriptionally active ternary complexes are associated with the nuclear matrix. *J. Mol. Biol.* **172,** 467–487.

Amero, S. A., Raychaudhuri, G., Cass, C. L., van Venrooij, W. J., Habets, W. J., Krainer, A. R., and Beyer, A. L. (1992). Independent deposition of heterogeneous nuclear ribonucleoproteins and small nuclear ribonucleoproteins at sites of transcription. *Proc. Natl. Sci. U.S.A.* **89,** 8409–8413.

Andreeva, M., Markova, D., Loidl, P., and Djondjurov, L. (1992). Intranuclear compartmentalization of transcribed and nontranscribed c-myc sequences in Namalva-S cells. *Eur. J. Biochem.* **207,** 887–894.

Antoniou, M., Carmo-Fonseca, M., Ferreira, J., and Lamond, A. I. (1993). Nuclear organization of splicing snRNPs during differentiation of murine erythroleukemia cells in vitro. *J. Cell Biol.* **123,** 1055–1068.

Archambault, J., and Friesen, J. D. (1993). Genetics of eukaryotic RNA polymerases I, II, and III. *Microbiol. Rev.* **57,** 703–724.

Bachellerie, J.-P., Puvion, E., and Zalta, J.-P. (1975). Ultrastructural organization and biochemical characterization of chromatin-RNA-protein complexes isolated from mammalian cell nuclei. *Eur. J. Biochem.* **58,** 327–337.

Barrack, E. R. (1987). Steroid hormone receptor localization in the nuclear matrix: Interaction with acceptor sites. *J. Steroid Biochem.* **27,** 115–121.

Baurén, G., and Wieslander, L. (1994). Splicing of Balbiani ring 1 gene pre-mRNA occurs simultaneously with transcription. *Cell (Cambridge, Mass.)* **76,** 183–192.

Bernhard, W. (1969). A new staining procedure for electron microscopical cytology. *J. Ultrastruct. Res.* **27,** 250–265.

Beyer, A. L., and Osheim, Y. N. (1991). Visualization of RNA transcription and processing. *Semin. Cell Biol.* **2,** 131–140.

Bidwell, J. P., van Wijnen, A. J., Fey, E. G., Dworetzky, S., Penman, S., Stein, J. L., Lian, J. B., and Stein, G. S. (1993). Osteocalcin gene promoter-binding factors are tissue-specific nuclear matrix components. *Proc. Natl. Acad. Sci. U.S.A.* **90,** 3162–3166.

Bona, M., Scheer, U., and Bautz, E. K. F. (1981). Antibodies to RNA polymerase II (B) inhibit transcription in lampbrush chromosomes after microinjection into living amphibian oocytes. *J. Mol. Biol.* **151,** 81–99.

Bridge, E., Carmo-Fonseca, M., Lamond, A., and Pettersson, U. (1993). Nuclear organization of splicing small nuclear ribonucleoproteins in adenovirus-infected cells. *J. Virol.* **67,** 5792–5802.

Buttyan, R., and Olsson, C. A. (1986). Prediction of transcriptional activity based on gene association with the nuclear matrix. *Biochem. Biophys. Res. Commun.* **138,** 1334–1340.

Carmo-Fonseca, M., Pepperkok, R., Sproat, B. S., Ansorge, W., Swanson, M. S., and Lamond, A. I. (1991). In vivo detection of snRNP-rich organelles in the nuclei of mammalian cells. *EMBO J.* **10,** 1863–1873.

Carmo-Fonseca, M., Pepperkok, R., Carvalho, M. T., and Lamond, A. I. (1992). Transcription-dependent colocalization of the U1, U2, U4/U6, and U5 snRNPs in coiled bodies. *J. Cell Biol.* **117,** 1–14.

Carter, K. C., and Lawrence, J. B. (1991). DNA and RNA within the nucleus: How much sequence-specific spatial organization? *J. Cell. Biochem.* **47,** 124–129.

Carter, K. C., Taneja, K. L., and Lawrence, J. B. (1991). Discrete nuclear domains of poly(A) RNA and their relationship to the functional organization of the nucleus. *J. Cell Biol.* **115,** 1191–1202.

Carter, K. C., Bowman, D., Carrington, W., Fogarty, K., McNeil, J. A., Fay, F. S., and Lawrence, J. B. (1993). A three-dimensional view of precursor messenger RNA metabolism within the mammalian nucleus. *Science* **259,** 1330–1335.

Ciejek, A. M., Nordstrom, J. L., Tsai, M.-J., and O'Malley, B. W. (1982). Ribonucleic acid precursors are associated with the chick oviduct nuclear matrix. *Biochemistry* **21,** 4945–4953.

Ciejek, E. M., Tsai, M.-J., and O'Malley, B. W. (1983). Actively transcribed genes are associated with the nuclear matrix. *Nature* (*London*) **306,** 607–609.

Clark, R. F., Cho, K. W. Y., Weinmann, R., and Hamkalo, B. A. (1991). Preferential distribution of active RNA polymerase II molecules in the nuclear periphery. *Gene Expression* **1,** 61–70.

Cook, P. R. (1988). The nucleoskeleton: Artifact, passive framework or active site? *J. Cell Sci.* **90,** 1–6.

Cook, P. R. (1989). The nucleoskeleton and the topology of transcription. *Eur. J. Biochem.* **185,** 487–501.

de Graaf, A., van Hemert, F., Linnemans, W. A. M., Brakenhoff, G. J., de Jong, L., van Renswoude, J., and van Driel, R. (1990). Three-dimensional distribution of DNase I-sensitive chromatin regions in interphase nuclei of embryonal carcinoma cells. *Eur. J. Cell Biol.* **52,** 135–141.

de Jong, L., van Driel, R., Stuurman, N., Meijne, A. M. L., and van Renswoude, J. (1990). Principles of nuclear organization. *Cell Biol. Int. Rep.* **14,** 1051–1074.

Dundr, M., and Raška, I. (1993). Nonisotopic ultrastructural mapping of transcription sites within the nucleolus. *Exp. Cell Res.* **208,** 275–281.

Fakan, S. (1986). Structural support for RNA synthesis in the cell nucleus. *Methods Achiev. Exp. Pathol.* **12,** 105–140.

Fakan, S. (1994). Perichromatin fibrils are *in situ* forms of nascent transcripts. *Trends Cell Biol.* **4,** 86–90.

Fakan, S., and Bernhard, W. (1971). Localization of rapidly and slowly labelled nuclear RNA as visualized by high resolution autoradiography. *Exp. Cell Res.* **67,** 129–141.

Fakan, S., and Hancock, R. (1974). Localization of newly-synthesized DNA in a mammalian cell as visualized by high resolution autoradiography. *Exp. Cell Res.* **83,** 95–102.

Fakan, S., and Nobis, P. (1978). Ultrastructural localization of transcription sites and of RNA distribution during the cell cycle of synchronized cells. *Exp. Cell Res.* **113,** 327–337.

Fakan, S., and Puvion, E. (1980). The ultrastructural visualization of nuclear and extranucleolar RNA synthesis and distribution. *Int. Rev. Cytol.* **65,** 255–299.

Fakan, S., Puvion, E., and Spohr, G. (1976). Localization and characterization of newly synthesized nuclear RNA in isolated rat hepatocytes. *Exp. Cell Res.* **99,** 155–164.

Fakan, S., Leser, G., and Martin, T. E. (1984). Ultrastructural distribution of nuclear ribonucleoproteins as visualized by immunocytochemistry on thin sections. *J. Cell Biol.* **98,** 358–363.

Fakan, S., Leser, G., and Martin, T. E. (1986). Immunoelectron microscope visualization of nuclear ribonucleoprotein antigens within spread transcription complexes. *J. Cell Biol.* **103,** 1153–1157.

Fisher, P. A., Lin, L., McConnell, M., Greenleaf, A., Lee, J.-M., and Smith, D. E. (1989). Heat shock-induced appearance of RNA polymerase II in karyoskeletal protein-enriched (nuclear "matrix") fractions correlates with transcriptional shutdown in *Drosophila melanogaster*. *J. Biol. Chem.* **264,** 3463–3469.

Fu, X.-D., and Maniatis, T. (1990). Factor required for mammalian spliceosome assembly is localized to discrete regions in the nucleus. *Nature* (*London*) **343,** 437–441.

Goldman, M. A. (1988). The chromatin domain as a unit of gene regulation. *BioEssays* **9,** 50–55.

Hassan, A. B., Errington, R. J., White, N. S., Jackson, D. A., and Cook, P. R. (1994). Replication and transcription sites are colocalized in human cells. *J. Cell Sci.* **107,** 425–434.

He, D., Nickerson, J. A., and Penman, S. (1990). Core filaments of the nuclear matrix. *J. Cell Biol.* **110,** 569–580.

Hozák, P., Hassan, A. B., Jackson, D. A., and Cook, P. R. (1993). Visualization of replication factories attached to a nucleoskeleton. *Cell (Cambridge, Mass.)* **73,** 361–373.

Hozák, P., Cook, P. R., Schöfer, C., Mosgöller, W., and Wachtler, F. (1994). Site of transcription of ribosomal RNA and intranucleolar structure in HeLa cells. *J. Cell Sci.* **107,** 639–648.

Huang, S., and Spector, D. L. (1991). Nascent pre-mRNA transcripts are associated with nuclear regions enriched in splicing factors. *Genes Dev.* **5,** 2288–2302.

Hutchison, N., and Weintraub, H. (1985). Localization of DNAse I-sensitive sequences to specific regions of interphase nuclei. *Cell (Cambridge, Mass.)* **43,** 471–482.

Jackson, D. A., and Cook, P. R. (1985). Transcription occurs at a nucleoskeleton. *EMBO J.* **4,** 919–925.

Jackson, D. A., Hassan, A. B., Errington, R. J., and Cook, P. R. (1993). Visualization of focal sites of transcription within human nuclei. *EMBO J.* **12,** 1059–1065.

Jiménez-García, L. F., and Spector, D. L. (1993). In vivo evidence that transcription and splicing are coordinated by a recruiting mechanism. *Cell (Cambridge, Mass.)* **73,** 47–59.

Jost, J.-P., and Seldran, M. (1984). Association of transcriptionally active vitellogenin II gene with the nuclear matrix of chicken liver. *EMBO J.* **3,** 2005–2008.

Kirov, N., Djondjurov, L., and Tsanev, R. (1984). Nuclear matrix and transcriptional activity of the mouse α-globin gene. *J. Mol. Biol.* **180,** 601–614.

Klempnauer, K.-H. (1988). Interaction of myb proteins with nuclear matrix in vitro. *Oncogene* **2,** 545–551.

Kopczynski, C. C., and Muskavitch, M. A. T. (1992). Introns excised from the *Delta* primary transcript are localized near sites of *Delta* transcription. *J. Cell Biol.* **119,** 503–512.

Kramer, J., Zachar, Z., and Bingham, P. M. (1994). Nuclear pre-mRNA metabolism: Channels and tracks. *Trends Cell Biol.* **4,** 35–37.

Krystosek, A., and Puck, T. T. (1990). The spatial distribution of exposed nuclear DNA in normal, cancer, and reverse-transformed cells. *Proc. Natl. Acad. Sci. U.S.A.* **87,** 6560–6564.

Lamond, A. I., and Carmo-Fonseca, M. (1993a). Localisation of splicing snRNPs in mammalian cells. *Mol. Biol. Rep.* **18,** 127–133.

Lamond, A. I., and Carmo-Fonseca, M. (1993b). The coiled body. *Trends Cell Biol.* **3,** 198–204.

Lawrence, J. B., and Singer, R. H. (1991). Spatial organization of nucleic acid sequences within cells. *Semin. Cell Biol.* **2,** 83–101.

Lawrence, J. B., Villnave, C. A., and Singer, R. H. (1988). Sensitive, high-resolution chromatin and chromosome mapping in situ: Presence and orientation of two closely integrated copies of EBV in a lymphoma line. *Cell (Cambridge, Mass.)* **52,** 51–61.

LeMaire, M. F., and Thummel, C. S. (1990). Splicing precedes polyadenylation during *Drosophila* E74A transcription. *Mol. Cell. Biol.* **10,** 6059–6063.

Levitt, A., Axel, R., and Cedar, H. (1979). Nick translation of active genes in intact nuclei. *Dev. Biol.* **69,** 496–505.

Manders, E. M. M., Stap, J., Brakenhoff, G. J., van Driel, R., and Aten, J. A. (1992). Dynamics of three-dimensional replication patterns during the S-phase, analysed by double labelling of DNA and confocal microscopy. *J. Cell Sci.* **103,** 857–862.

Manders, E. M. M., Verbeek, F. J., and Aten, J. A. (1993). Measurement of co-localization of objects in dual-colour confocal images. *J. Microsc. (Oxford)* **169,** 375–382.

Mariman, E. C. M., van Eekelen, C. A. G., Reinders, R. J., Berns, A. J. M., and van Venrooij, W. J. (1982). Adenoviral heterogeneous nuclear RNA is associated with the host nuclear matrix during splicing. *J. Mol. Biol.* **154,** 103–119.

Martin, T. E., Barghusen, S. C., Leser, G. P., and Spear, P. G. (1987). Redistribution of nuclear ribonucleoprotein antigens during herpes simplex virus infection. *J. Cell Biol.* **105,** 2069–2082.

Matunis, E. L., Matunis, M. J., and Dreyfuss, G. (1993). Association of individual hnRNP proteins and snRNPs with nascent transcripts. *J. Cell Biol.* **121,** 219–228.

Nakamura, H., Morita, T., and Sato, C. (1986). Structural organization of replicon domains during DNA synthetic phase in the mammalian nucleus. *Exp. Cell Res.* **165,** 291–297.

Nakayasu, H., and Berezney, R. (1989). Mapping replicational sites in the eucaryotic cell nucleus. *J. Cell Biol.* **108,** 1–11.

Nickerson, J. A., Krochmalnic, G., Wan, K. M., and Penman, S. (1989). Chromatin architecture and nuclear RNA. *Proc. Natl. Acad. Sci. U.S.A.* **86,** 177–181.

O'Keefe, R. T., Henderson, S. C., and Spector, D. L. (1992). Dynamic organization of DNA replication in mammalian cell nuclei: Spatially and temporally defined replication of chromosome-specific α-satellite DNA sequences. *J. Cell Biol.* **116,** 1095–1110.

O'Keefe, R. T., Mayeda, A., Sadowski, C. L., Krainer, A. R., and Spector, D. L. (1994). Disruption of pre-mRNA splicing in vivo results in reorganization of splicing factors. *J. Cell Biol.* **124,** 249–260.

Raška, I., Andrade, L. E. C., Ochs, R. L., Chan, E. K. L., Chang, C.-M., Roos, G., and Tan, E. M. (1991). Immunological and ultrastructural studies of the nuclear coiled body with autoimmune antibodies. *Exp. Cell Res.* **195,** 27–37.

Razin, S. V., and Yarovaya, O. V. (1985). Initiated complexes of RNA polymerase II are concentrated in the nuclear skeleton associated DNA. *Exp. Cell Res.* **158,** 273–275.

Razin, S. V., Yarovaya, O. V., and Georgiev, G. P. (1985). Low ionic strength extraction of nuclease-treated nuclei destroys the attachment of transcriptionally active DNA to the nuclear skeleton. *Nucleic Acids Res.* **13,** 7427–7444.

Roberge, M., Dahmus, M. E., and Bradbury, E. M. (1988). Chromosomal loop/nuclear matrix organization of transcriptionally active and inactive RNA polymerases in HeLa nuclei. *J. Mol. Biol.* **201,** 545–555.

Robinson, S. I., Nelkin, B. D., and Vogelstein, B. (1982). The ovalbumin gene is associated with the nuclear matrix of chicken oviduct cells. *Cell (Cambridge, Mass.)* **28,** 99–106.

Rosbash, M., and Singer, R. H. (1993). RNA travel: Tracks from DNA to cytoplasm. *Cell (Cambridge, Mass.)* **75,** 399–401.

Sass, H. (1982). RNA polymerase B in polytene chromosomes: Immunofluorescent and autoradiographic analysis during stimulated and repressed RNA synthesis. *Cell (Cambridge, Mass.)* **28,** 269–278.

Sentenac, A. (1985). Eukaryotic RNA polymerases. *CRC Crit. Rev. Biochem.* **18,** 31–90.

Shafit-Zagardo, B., Brown, F. L., Zavodny, P. J., and Maio, J. J. (1983). Transcription of the *Kpn*I families of long interspersed DNAs in human cells. *Nature (London)* **304,** 277–280.

Sibon, O. C. M., Cremers, F. F. M., Boonstra, J., Humbel, B. M., and Verkleij, A. J. (1993). Localisation of EGF-receptor mRNA in the nucleus of A431 cells by light microscopy. *Cell Biol. Int. Rep.* **17,** 1–11.

Sibon, O. C. M., Humbel, B. M., de Graaf, A., Verkleij, A. J., and Cremers, F. F. M. (1994). Ultrastructural localization of epidermal growth factor (EGF)-receptor transcripts in the cell nucleus using pre-embedding in situ hybridization in combination with ultra-small gold probes and silver enhancement. *Histochemistry* **101,** 223–232.

Sibon, O. C. M., Cremers, F. F. M., Humbel, B. M., Boonstra, J., and Verkley, A. J. (1995). Localization of nuclear RNA by pre- and post-embedding in situ hybridization using different gold probes. *Histochem. J.* **27,** 35–45.

Sierakowska, H., Shukla, R. R., Dominski, Z., and Kole, R. (1989). Inhibition of pre-mRNA splicing by 5′-fluoro, 5-chloro, and 5-bromouridine. *J. Biol. Chem.* **264,** 19185–19191.

Sollner-Webb, B., and Mougey, E. B. (1991). News from the nucleolus: rRNA gene expression. *Trends Biochem. Sci.* **16,** 58–62.

Spector, D. L. (1993). Macromolecular domains within the cell nucleus. *Annu. Rev. Cell Biol.* **9,** 265–315.

Spector, D. L., Fu, X.-D., and Maniatis, T. (1991). Associations between distinct pre-mRNA splicing components and the cell nucleus. *EMBO J.* **10,** 3467–3481.

Spector, D. L., Lark, G., and Huang, S. (1992). Differences in snRNP localization between transformed and nontransformed cells. *Mol. Biol. Cell.* **3,** 555–569.

Stein, G. S., Lian, J. B., Dworetzky, S. I., Owen, T. A., Bortell, R., Bidwell, J. P., and van Wijnen, A. J. (1991). Regulation of transcription-factor activity during growth and differentiation involvement of the nuclear matrix in concentration and localization of promoter binding proteins. *J. Cell. Biochem.* **47,** 300–305.

Stuurman, N., de Jong, L., and van Driel, R. (1992). Nuclear frameworks: Concepts and operational definitions. *Cell Biol. Int. Rep.* **16,** 837–852.

Taneja, K. L., Lifshitz, L. M., Fay, F. S., and Singer, R. H. (1992). Poly(A) RNA codistribution with microfilaments—evaluation by in situ hybridization and quantitative digital imaging microscopy. *J. Cell Biol.* **119,** 1245–1260.

Testillano, P. S., Gorab, E., and Risueño, M. C. (1994). A new approach to map transcription sites at the ultrastructural level. *J. Histochem. Cytochem.* **42,** 1–10.

Thompson, N. E., Steinberg, T. H., Aronson, D. B., and Burgess, R. R. (1989). Inhibition of *in vivo* and *in vitro* transcription by monoclonal antibodies prepared against wheat germ RNA polymerase II that react with the heptapeptide repeat of eukaryotic RNA polymerase II. *J. Biol. Chem.* **264,** 11511–11520.

van Driel, R., Humbel, B., and de Jong, L. (1991). The nucleus: A black box being opened. *J. Cell. Biochem.* **47,** 1–6.

van Eekelen, C. A. G., and van Venrooij, W. J. (1981). HnRNA and its attachment to a nuclear protein matrix. *J. Cell Biol.* **88,** 554–563.

van Steensel, B., Jenster, G., Damm, K., Brinkman, A. O., and van Driel, R. (1995). Domains of the human androgen receptor and glucocorticoid receptor involved in binding to the nuclear matrix. *J. Cell. Biochem.* **57,** 465–478.

van Wijnen, A. J., Bidwell, J. P., Fey, E. G., Penman, S., Lian, J. B., Stein, J. L., and Stein, G. S. (1993). Nuclear matrix association of multiple sequence-specific DNA binding activities related to SP-1, ATF, CCAAT, C/EBP, OCT-1, and AP-1. *Biochemistry* **32,** 8397–8402.

Verheijen, R., van Venrooij, W., and Ramaekers, F. (1988). The nuclear matrix: Structure and composition. *J. Cell Sci.* **90,** 11–36.

Visa, N., Puvion-Dutilleul, F., Harper, F., Bachellerie, J.-P., and Puvion, E. (1993). Intranuclear distribution of poly(A) RNA determined by electron microscope *in situ* hybridization. *Exp. Cell Res.* **208,** 19–34.

Waitz, W., and Loidl, P. (1991). Cell cycle dependent association of c-myc protein with the nuclear matrix. *Oncogene* **6,** 29–35.

Wansink, D. G., Schul, W., van der Kraan, I., van Steensel, B., van Driel, R., and de Jong, L. (1993). Fluorescent labeling of nascent RNA reveals transcription by RNA polymerase II in domains scattered throughout the nucleus. *J. Cell Biol.* **122,** 283–293.

Wansink, D. G., Motley, A. M., van Driel, R., and de Jong, L. (1994a). Fluorescent labeling of nascent RNA in the cell nucleus using 5-bromouridine 5′-triphosphate. *In* "Cell Biology: A Laboratory Handbook" (J. E. Celis, ed.), Vol. 2, pp. 368–374. Academic Press, San Diego, CA.

Wansink, D. G., Manders, E. E. M., van der Kraan, I., Aten, J. A., van Driel, R., and de Jong, L. (1994b). RNA polymerase II transcription is concentrated outside replication domains throughout S-phase. *J. Cell Sci.* **107,** 1449–1456.

Wansink, D. G., Nelissen, R. L. H., and de Jong, L. (1994c). In vitro splicing of pre-mRNA containing bromouridine. *Mol. Biol. Rep.* **19,** 109–113.

Wansink, D. G., Sibon, O. C. M., Cremers, F. M. M., van Driel, R., and de Jong, L. (1996). Ultrastructural localization of active genes in nuclei of A341 cells. *J. Cell. Biochem.* **62,** 10–18.

Weintraub, H., and Groudine, M. (1976). Chromosomal subunits in active genes have an altered conformation. *Science* **193,** 848–856.

Willis, I. M. (1993). RNA polymerase III. Genes, factors and transcriptional specificity. *Eur. J. Biochem.* **212,** 1–11.

Xing, Y., and Lawrence, J. B. (1991). Preservation of specific RNA distribution within the chromatin-depleted nuclear substructure demonstrated by in situ hybridization coupled with biochemical fractionation. *J. Cell Biol.* **112,** 1055–1063.

Xing, Y., and Lawrence, J. B. (1993). Nuclear RNA tracks: Structural basis for transcription and splicing? *Trends Cell Biol.* **3,** 346–353.

Xing, Y., Johnson, C. V., Dobner, P. R., and Lawrence, J. B. (1993). Higher level organization of individual gene transcription and RNA splicing. *Science* **259,** 1326–1330.

Young, R. A. (1991). RNA polymerase II. *Annu. Rev. Biochem.* **60,** 689–715.

Zhang, M., Zamore, P. D., Carmo-Fonseca, M., Lamond, A. I., and Green, M. R. (1992). Cloning and intracellular localization of the U2 small nuclear ribonucleoprotein auxiliary factor small subunit. *Proc. Natl. Acad. Sci. U.S.A.* **89,** 8769–8773.

IV

Nuclear mRNP Complexes

8

Protein Deposition on Nascent Pre-mRNA Transcripts

SALLY A. AMERO AND KENNETH C. SORENSEN

Stritch School of Medicine
Loyola University Chicago
Maywood, Illinois

I. INTRODUCTION

The messenger hypothesis first proposed the existence of short-lived gene products that become associated with ribosomes to direct protein

NUCLEAR STRUCTURE
AND GENE EXPRESSION

synthesis (Jacob and Monod, 1961), and such messengers were shown to be RNA (Brenner *et al.,* 1961; Gros *et al.,* 1961). Because these studies were conducted on bacterial systems, the additional discoveries awaiting studies of eukaryotic premessenger RNA (pre-mRNA) transcripts could not have been imagined. Pre-mRNA molecules in eukaryotes later were found to undergo numerous maturation processes in the nucleus prior to their participation in protein synthesis: 5′-terminal capping (Green *et al.,* 1983; Konarska *et al.,* 1984; Edery and Sonenberg, 1985; Shatkin, 1985; Mizumoto and Kaziro, 1987), removal of intervening sequences (Berget *et al.,* 1977; Berk and Sharp, 1977, 1978, Breathnach *et al.,* 1977; Chow *et al.,* 1977; Gelinas and Roberts, 1977; Jeffreys and Flavell, 1977; Klessig, 1977), 3′-end cleavage and polyadenylation (Edmonds and Caramela, 1969; Edmonds *et al.,* 1971; Molloy *et al.,* 1972; Wahle and Keller, 1992, 1993; Manley and Proudfoot, 1994), nucleotide base modifications (Benne, 1994), and transport through the nuclear pores (Stevens and Swift, 1966; Monneron and Bernhard, 1969; Newmeyer, 1993).

The fact that nascent eukaryotic pre-mRNA exists in a ribonucleoprotein (RNP) context was recognized even before many of these significant discoveries were reported. The colocalization of nascent RNA and protein in lateral loops of newt lampbrush chromosomes (reviewed in Callan, 1987) was demonstrated by cytochemical tests coupled with enzymatic treatments (Gall, 1954, 1956; Callan and Macgregor, 1958; Maundrell, 1975), incorporation of radiolabeled uridine and phenylalanine (Gall and Callan, 1962; Macgregor and Callan, 1962; Miller, 1965), actinomycin D treatment (Snow and Callan, 1969), and ultrastructural visualization of dense 300- to 400-Å RNP granules within the loop matrices (Gall, 1956; Sommerville, 1973). Similar large (0.1–0.3 μm) RNP granules and numerous smaller ones (280–320 Å) were visualized in particular puff sites of insect polytene chromosomes (Swift *et al.,* 1964; Derksen, 1976); as expected of messenger RNP particles, the granules were found in nucleoplasm and at nuclear pores (Beermann and Bahr, 1954; Stevens and Swift, 1966; Derksen *et al.,* 1973). In the mammalian nucleus, perichromatin fibrils (400–450 Å) and granules (30–200 Å) were visualized through special staining techniques (Monneron and Bernhard, 1969) and shown to represent sites of heterogeneous nuclear RNA (hnRNA) synthesis through rapid labeling with [^{3}H]uridine (Fakan and Bernhard, 1971, 1973; Bachellerie *et al.,* 1975; Fakan *et al.,* 1976) and sensitivity to ribonuclease (RNase) (Monneron and Bernhard, 1969; reviewed in Fakan and Puvion, 1980). Interchromatin granules in mammalian nuclei, on the other hand, did not become labeled rapidly with [^{3}H]uridine but were RNase-sensitive (Fakan and Nobis, 1978).

The assembly of nonribosomal nascent transcripts into RNP complexes was also revealed in early ultrastructural observations of active chromatin where nascent RNA displays a granular appearance arising from 30-nm RNP particles (Miller, 1965; Miller and Bakken, 1972; Angelier and Lacroix, 1975; Mott and Callan, 1975; Foe *et al.,* 1976; Laird and Chooi, 1976; Laird *et al.,* 1976; McKnight and Miller, 1976, 1979; Scheer *et al.,* 1976; Malcolm and Sommerville, 1977; Busby and Bakken, 1979; reviewed in Scheer, 1987). In numerous studies, the micrographs were interpreted to represent the aggregation of the 30-nm RNP particles into larger granules, typical in size to the RNP granules seen in lampbrush loop matrices, Balbiani rings, and perichromatin granules. Interestingly, discrete 30S RNP particles isolated from newt oocytes (Sommerville, 1973; Sommerville and Hill, 1973) and rat liver nuclei (Samarina *et al.,* 1968; Georgiev and Samarina, 1971; Lukanidin *et al.,* 1972a,b; Albrecht and Van Zyl, 1973) were shown to contain DNA-like RNA—as determined by base composition—and to resemble the smaller RNP particles seen at the ultrastructural level.

This chapter addresses the molecular mechanisms directing protein deposition on nascent RNA polymerase II (Pol II) transcripts *in vivo,* with a view to interpreting the cytological data and the significance of these deposition processes to gene expression. First, models concerning the primary signals directing deposition of the major RNP complexes onto nascent pre-mRNA transcripts in eukaryotes will be reviewed, and other reviews will be referenced, where possible, for discussions of their functions. The major transcript-binding proteins will be divided for convenience into three categories: the hnRNP (heterogeneous nuclear ribonucleoprotein) proteins, the snRNP (small nuclear ribonucleoprotein) particles and snRNP particle-associated proteins, and the SR (serine- and arginine-rich) splicing factors. The reader is referred to other reviews (Wahle and Keller, 1992, 1993; Manley and Proudfoot, 1994) for discussions concerning the deposition of cleavage and polyadenylation factors on nascent pre-mRNA transcripts. Second, the timing and mode of protein deposition on nascent pre-mRNA will be addressed; evidence for a recruitment and cotranscriptional deposition model will be presented, and the available information will be interpreted to predict the precise timing of the earliest deposition event. Two contrasting models of cotranscriptional protein deposition—the independent and unitary deposition models—will be analyzed. Third, the location of RNA processing events in the nucleus—including contrasting models for cotranscriptional and extrachromosomal splicing—will be discussed. Finally, a question that we find of particular interest will be explored: Do RNP complexes on nascent pre-mRNA transcripts interact with their transcription complexes?

II. PRIMARY DETERMINANTS FOR DEPOSITION OF TRANSCRIPT-BINDING PROTEINS

A. hnRNP Complexes

1. hnRNP Proteins

The hnRNP proteins (reviewed in Dreyfuss *et al.,* 1990, 1993) are abundant, conserved nuclear proteins that were first described in mammalian cell extracts on the basis of their cosedimentation in sucrose density gradients (Samarina *et al.,* 1968; Faiferman *et al.,* 1971; Georgiev and Samarina, 1971; Lukanidin *et al.,* 1972a,b; Albrecht and Van Zyl, 1973; Gallinaro-Matringe and Jacob, 1973; Sommerville, 1973; Sommerville and Hill, 1973; Pederson, 1974; Kinniburgh and Martin, 1976; Beyer *et al.,* 1977; Malcolm and Sommerville, 1977), cesium sulfate gradients (Mayrand and Pederson, 1981; Mayrand *et al.,* 1981), or cesium chloride gradients (Samarina *et al.,* 1968; Niessing and Sekeris, 1971) with rapidly labeled hnRNA in 30–50S RNP complexes. In subsequent studies, the association of hnRNP proteins with nuclear pre-mRNA was established through cross-linking studies (Mayrand and Pederson, 1981; van Eekelen *et al.,* 1981; Economidis and Pederson, 1983; Mayrand *et al.,* 1981; Dreyfuss *et al.,* 1984) or immunoprecipitation with hnRNP protein-specific antibodies (Choi and Dreyfuss, 1984a). Six major proteins in the size range of 32–44 kDa were designated group A, B, and C hnRNP proteins in single-dimension sodium dodecyl sulfate (SDS)–polyacrylamide gels; each group constitutes a doublet, invoking the nomenclature A1, A2, and so on (Beyer *et al.,* 1977; Economides and Pederson, 1983). More recently, two-dimensional gel electrophoresis of immunopurified hnRNP complexes from human and *Drosophila* cells revealed over 20 hnRNP proteins ranging from 32 to 120 kDa, designated hnRNP proteins A through U (Leser and Martin, 1987; Piñol-Roma *et al.,* 1988; Matunis *et al.,* 1992a,b); the RNA fragments in these complexes range up to 10 kb in length (Choi and Dreyfuss, 1984a).

Three models have been proposed to explain the deposition of hnRNP proteins on nascent pre-mRNA *in vivo*. First, the ribonucleosome model proposes the indiscriminate deposition of hnRNP core particles, rather than that of individual hnRNP proteins, along the length of nascent transcripts. Second, the sequence-specific model proposes the binding of individual hnRNP proteins according to the RNA sequence of the transcript. Finally, a composite view proposes both the general deposition of hnRNP proteins on nascent pre-mRNA transcripts to serve roles in packaging and protecting transcripts, and the sequence-specific deposition of hnRNP proteins to render important RNA sequences accessible to other *trans*-acting factors. The deposition of the hnRNP U protein is discussed in Section V,B,3.

2. *Ribonucleosome Model*

The traditional view of hnRNP deposition on nascent pre-mRNA suggests the uniform binding of hnRNP complexes to RNA without sequence specificity as nonspecific RNA-packaging proteins (Georgiev and Samarina, 1971; Chung and Wooley, 1986; Huang *et al.*, 1994). The ribonucleosome model predicts a uniform protein complement in isolated hnRNP complexes, a uniform ultrastructural appearance for nascent pre-mRNA transcripts, and a general packaging function for hnRNP proteins.

Fairly uniform hnRNP complexes can be isolated from rapidly dividing HeLa cells (or many other eukaryotic sources) by sucrose density gradient centrifugation. These "particles" display a spherical, particulate appearance (diameter of ≈20 nm) in the electron microscope and sediment in sucrose gradients in a broad peak as 30–50S particles (Sommerville, 1973; Sommerville and Hill, 1973; Samarina *et al.*, 1968; Georgiev and Samarina, 1971; Lukanidin *et al.*, 1972a,b; Albrecht and Van Zyl, 1973; Beyer *et al.*, 1977). Such purified hnRNP complexes possess a fixed stoichiometry of the "core" hnRNP proteins (Lothstein *et al.*, 1985; Barnett *et al.*, 1989, 1990, 1991; Huang *et al.*, 1994) and fragments of pre-mRNA ranging in length from 400 to 900 nucleotides. In the presence of RNase inhibitors, however, larger (≈200S) RNP particles can be recovered from the gradients; these larger RNP particles resemble beads on a string under the electron microscope and can be converted to the smaller 30–50S form by limited RNase digestion of the pre-mRNA strands connecting the monomer particles (Samarina *et al.*, 1968; Beyer *et al.*, 1977; Spann *et al.*, 1989). It is presumed that immunoprecipitation allows the isolation of hnRNP complexes with even larger RNA fragments (see the earlier discussion) due to limited exposure to RNases.

Visualizations of active genes by the Miller spreading technique fit, in part, the prediction for uniform protein deposition. At the ultrastructural level, protein deposition is visible on very early transcripts from *Drosophila* embryos, giving rise to a repeating beaded appearance on a ≈5-nm fibril; occasionally larger ≈25-nm RNP particles are seen at nonrandom locations on the transcripts (Beyer *et al.,* 1980). That some of these 25-nm RNP complexes represent hnRNP complexes was assumed originally from similarities in their appearance to isolated or reconstituted 40S hnRNP complexes (reviewed in Beyer and Osheim, 1990; Dreyfuss *et al.,* 1990) and by immunoelectron microscopy studies placing hnRNP proteins on nascent transcripts of mouse tissue culture cells (Fakan *et al.,* 1986). A later evaluation of the ultrastructural evidence, in light of the reconstitution studies to be described, leads to the conclusion that hnRNP proteins are localized predominantly in the 5-nm RNP fibril of nascent transcripts but may also be present within the larger 25-nm particles (Beyer and Osheim, 1990).

The results of *in vitro* reconstitution assays also support a packaging function for hnRNP proteins. hnRNP complexes assembled *in vitro* at high protein : RNA stoichiometries, in low ionic strength buffers and in the absence of competitor RNA sequences, are indistinguishable from the 20- to 25-nm particles viewed in the electron microscope (Wilk *et al.,* 1983). These reconstituted complexes are also indistinguishable from 40S particles isolated from sucrose gradients (Pullman and Martin, 1983; Conway *et al.,* 1988) and from those isolated following ultraviolet (UV) cross-linking *in vivo* (Mayrand and Pederson, 1981; Mayrand *et al.,* 1981). At lower protein : RNA ratios, however, reconstituted hnRNP complexes resemble thin fibrils in electron micrographs that may correspond to the ≈5-nm RNP fibril conformation displayed by nascent transcripts in spread chromatin (Malcolm and Sommerville, 1977; Beyer and Osheim, 1990). These observations are consistent with the sequence-independent interactions of hnRNP proteins with many single-stranded nucleic acids (reviewed in Merrill and Williams, 1990).

3. Sequence-Specific Model

The sequence-specific model of hnRNP deposition, in its simplest form, suggests that hnRNP proteins bind nascent pre-mRNA *in vivo* individually—rather than as a complex—according to their affinity for

RNA-binding sites [reviewed by Dreyfuss *et al.*, 1993; an early version of the sequence-specific deposition model proposed the existence of cell-specific hnRNP protein populations (Pederson, 1974)]. The sequence-specific model predicts the nonrandom association of hnRNP complexes with nascent pre-mRNA transcripts, sequence-specific RNA-binding activity for individual hnRNP proteins, deposition of different protein complexes on different transcripts, and specific roles for hnRNP proteins in pre-mRNA maturation.

The nonrandom, sequence-dependent deposition of hnRNP proteins on nascent pre-mRNA transcripts is consistent with the following experiments. When isolated 30–50S hnRNP complexes were subjected to additional RNase digestion *in vitro,* specific RNA sequences from different target transcripts remained nuclease-resistant and therefore presumably protected by hnRNP proteins (Steitz and Kamen, 1981; Ohlsson *et al.,* 1982; Patton *et al.,* 1985). Significantly, UV cross-linking *in vivo* of hnRNA and associated proteins also revealed a nonrandom association of the hnRNP C proteins on RNA sequences in late adenoviral transcripts (van Eekelen *et al.,* 1982).

Perhaps the most compelling demonstration of sequence-specific binding by hnRNP proteins is the selection/amplification of a consensus high-affinity binding site from pools of random-sequence RNA by the mammalian hnRNP protein A1 (Burd and Dreyfuss, 1994b). Consistent with sequence specificity is the preferential and avid *in vitro* binding displayed by numerous hnRNP proteins to certain homoribopolymer affinity columns (Swanson and Dreyfuss, 1988b; see Table I) and the transcript-specific assembly *in vitro* of early splicing complexes on pre-mRNA templates (Bennett *et al.,* 1992b). Also, when RNP complexes were assembled *in vitro* on defined pre-mRNA substrates, hnRNP proteins A1, C, and D were found to protect preferentially the 3′ ends of introns in the pre-mRNA transcripts from nuclease digestion (Swanson and Dreyfuss, 1988b). In similar studies, the hnRNP C proteins became cross-linked preferentially to sequences downstream of polyadenylation cleavage sites on pre-mRNA templates (Moore *et al.,* 1988; Wilusz *et al.,* 1988), although this interaction is sensitive to the presence of an upstream intron (Stolow and Berget, 1990). A list of known binding sites for mammalian hnRNP proteins is provided in Table I.

The expectation that hnRNP proteins perform specific roles in nuclear RNA metabolism is consistent with the inhibition of pre-mRNA splicing observed *in vitro* on addition to splicing extracts of antibodies specific for core hnRNP proteins or on immunodepletion of splicing extracts for

TABLE I

Human hnRNP Proteins

hnRNP protein	Molecular mass (kDa)	Binding specificity	Comments
A1	34	3′ End of intron[a]	Increases RNA–RNA base pairing *in vitro* (RNA annealing activity)[g]
		3′ Splice site[b–d]	ASF/SF2 antagonist for 5′ splice site selection *in vitro*[h] and *in vivo*[i]
		5′ Splice site[b]	Shuttles between nucleus and cytoplasm[j]
		d(TTAGGG)$_n$ telomeric sequence[c]	
		3′-UTR (uridylate sequences)[e]	
		5′ Ends of U2, U4 snRNA[f]	
A2/B1	36/38	3′ Splice site[k]	A2 and B1 are alternatively spliced products[l]
		d(TTAGGG)$_n$ telomeric sequence[k]	
C1/C2	41/43	3′ End of intron[m]	Phosphorylated[p]
		Poly(U)[n]	Immunodepletion blocks splicing *in vitro*[q]
		3′-UTR (uridylate sequences)[a]	Increases RNA–RNA base pairing *in vitro* (RNA annealing activity)[r]
			C1 and C2 are alternatively spliced products[s]
D group	44–48	3′ End of intron[t]	
		3′ Splice site[u]	
		d(TTAGGG)$_n$ telomeric sequence[u]	
E group	36–43	Poly(G)[v]	
		d(TTAGGG)$_n$ telomeric sequence[w]	
		3′ Splice site[w]	
F/H	53/56	Poly(G)[x]	F and H are immunologically related[y]
I	59	Polypyrimidine tract[z]	Identical to the PTB

K/J	62/68	Poly(C)[aa]	K and J are immunologically related[aa]
			K protein regulates transcription of c-*myc in vitro*[bb]
L	68		Associated with giant loops of newt lampbrush chromosomes[cc]
M	68	Poly(G)[dd]	
		Poly(U)[ee]	
P	72	Poly(A)[ff]	Identical to the poly(A) binding protein[gg]
U	120	Poly(G)[hh]	Identical to the scaffold attachment factor-A[ii]
		Poly(U)[hh]	Increases RNA–RNA base pairing *in vitro* (RNA annealing activity)[jj]
		DNA (scaffold attachment region)[ii]	

[a] Swanson and Dreyfuss (1988b).
[b] Burd and Dreyfuss (1994b).
[c] Ishikawa *et al.* (1993).
[d] Buvoli *et al.* (1990).
[e] Hamilton *et al.*, (1993).
[f] Buvoli *et al.*, (1992).
[g] Portman and Dreyfuss (1994).
[h] Mayeda and Krainer (1992).
[i] Cáceres *et al.* (1994).
[j] Piñol-Roma and Dreyfuss (1992).
[k] Ishikawa *et al.* (1993); McKay and Cooke (1992).
[l] Burd *et al.* (1989).
[m] Swanson and Dreyfuss (1988b).
[n] Swanson and Dreyfuss (1988a); Görlach *et al.* (1992).
[o] Wilusz and Shenk (1990); Hamilton *et al.* (1993); Stolow and Berget (1990).
[p] Piñol-Roma and Dreyfuss (1993); Mayraud *et al.* (1993).
[q] Choi *et al.* (1986).
[r] Portman and Dreyfuss (1994).
[s] Burd *et al.* (1989).
[t] Swanson and Dreyfuss (1988b).
[u] Ishikawa *et al.* (1993).
[v] Swanson and Dreyfuss (1988a).
[w] Ishikawa *et al.* (1993).
[x] Matunis *et al.* (1994b).
[y] Swanson and Dreyfuss (1988a).
[z] Ghetti *et al.* (1992).
[aa] Matunis *et al.* (1992c); Swanson and Dreyfuss (1988a).
[bb] Takimoto *et al.* (1993).
[cc] Piñol-Roma *et al.* (1989).
[dd] Datar *et al.* (1993).
[ee] Swanson and Dreyfuss (1988a).
[ff] Swanson and Dreyfuss (1988a).
[gg] Adam *et al.* (1986).
[hh] Swanson and Dreyfuss (1988a); Kiledjian and Dreyfuss (1992).
[ii] Romig *et al.* (1992); von Kries *et al.* (1994); Fackelmayer and Richter (1994a–c).
[jj] Portman and Dreyfuss (1994).

the hnRNP protein C (Choi *et al.,* 1986; Sierakowska *et al.,* 1986). In particular, the mammalian hnRNP proteins A1 and U may perform essential roles in pre-mRNA splicing because they possess RNA annealing activity *in vitro* (Portman and Dreyfuss, 1994). The mammalian hnRNP protein I, on the other hand, may play a different role in pre-mRNA splicing because it is identical to the polypyrimidine-tract binding (PTB) protein (Ghetti *et al.,* 1992). The PTB/hnRNP I, in association with another protein splicing factor (PSF), binds specifically to polypyrimidine tracts in pre-mRNA templates *in vitro* (García-Blanco *et al.,* 1989; Wang and Pederson, 1990; Gil *et al.,* 1991; Patton *et al.,* 1993), presumably as a constitutive splicing factor. In addition, PTB/hnRNP I interacts preferentially with sequences involved in alternative splicing of β-tropomyosin pre-mRNA through an exon blockage mechanism (Mulligan *et al.,* 1992). Numerous other studies of alternative pre-mRNA splicing, discussed in Section II,B,4, further suggest the involvement of particular hnRNP proteins in splice site decisions.

The prediction for transcript-specific hnRNP complex deposition is evidenced by locus-specific ratios of different hnRNP proteins at chromosomal sites of transcription. On polytene chromosomes from severely heat shocked larvae, certain hnRNP proteins are concentrated preferentially on the heat shock-induced puff at locus 93D but not on other conspicuous heat shock puffs (Dangli *et al.,* 1983; Amero *et al.,* 1991; Hovemann *et al.,* 1991). Moreover, direct comparisons of the *Drosophila* protein on ecdysone puffs (PEP) protein (Amero *et al.,* 1991) and the *Drosophila* hnRNA-binding (HRB) proteins on the same chromosome (Raychaudhuri *et al.,* 1992) revealed varying ratios of immunofluorescent signals from the two antibodies at neighboring chromosomal sites, thereby revealing qualitatively similar, but nonidentical, puff-preferential staining patterns for these hnRNP proteins (Amero *et al.,* 1993). Similar double staining experiments likewise revealed locus-preferential deposition for other individual hnRNP proteins on polytene chromosomes. That is, pairwise comparisons of deposition patterns for the *Drosophila* hrp36, hrp40, and hrp48 proteins using directly conjugated monoclonal antibodies demonstrated nonidentical, transcript-specific staining patterns for these core hnRNP proteins on developmental and heat shock puffs (Matunis *et al.,* 1993).

4. Composite Model

A composite model of hnRNP deposition accommodates the two models just discussed. This model suggests that the specific RNA sequences recognized by individual hnRNP proteins *in vitro* represent high-affinity binding sites, whereas those involved in hnRNP complex reconstitution experiments represent less stringent binding sites (Beyer and Osheim, 1990; Dreyfuss *et al.,* 1990). This low-affinity binding may enable hnRNP proteins to "scan" the nascent pre-mRNA molecule for a high-affinity binding site such as an intron sequence; under conditions of hnRNP protein abundance, 30S hnRNP complexes may also form. Thus, it is possible that hnRNP proteins perform both a general RNA-packaging role and sequence-dependent, specific roles in the maturation of pre-mRNA.

B. Spliceosome Assembly

1. snRNP Particles and Pre-mRNA Splicing

The snRNP particles are the major factors involved in pre-mRNA splicing (reviewed in Maniatis and Reed, 1987; Lührmann *et al.,* 1990; Green, 1991; Rosbash and Séraphin, 1991; Ruby and Abelson, 1988; Guthrie, 1991; Steitz, 1992; Baserga and Steitz, 1993; Lamond, 1993; Moore *et al.,* 1993). These particles were first recognized from their antigenicity for certain autoimmune sera called Sm and RNP from systemic lupus erythematosus patients (Lerner and Steitz, 1979; reviewed in Craft, 1992). Each snRNP particle is an intricate complex of a single small nuclear RNA molecule—designated U1, U2, U4, U5, U6 snRNA—and protein (reviewed in Steitz *et al.,* 1988); the exception is the double snRNP particle formed by U4 and U6 snRNP particles in solution (Hashimoto and Steitz, 1984). All snRNP particles share a core of common proteins (Lerner and Steitz, 1979), and each possesses several snRNP-specific proteins.

The term spliceosome denotes the macromolecular complex that contains pre-mRNA, snRNP particles, and other *trans*-acting splicing factors (such as the SR protein family) and that performs the spicing reaction. The studies presented in Sections II,B,2–6 concerning deposition of these components on nascent pre-mRNA templates *in vitro* suggest a sequential pathway for spliceosome assembly. An early model suggested that spliceosome assembly and splice site selection depend primarily on RNA–RNA interactions between snRNA and pre-mRNA molecules as the primary determinants in splice site selection. Additional studies to be discussed suggest a more recent model that highlights the importance of RNA–protein and protein–protein interactions in the assembly process; this model is also applicable to examples of alternative splice site selection. In Section III, studies addressing deposition of these components on nascent pre-mRNA *in vivo* will be presented.

2. *Ordered Spliceosome Assembly* in Vitro

An ordered pathway of spliceosome assembly has been established from analyses of the complexes that assemble on pre-mRNA templates incubated in mammalian splicing extracts (Brody and Abelson, 1985; Frendewey and Keller, 1985; Grabowski *et al.,* 1985; Ruskin and Green, 1985; Konarska and Sharp, 1986, 1987; Bindereif and Green, 1986, 1987; Jamison *et al.,* 1992, reviewed in Lamond, 1993; Moore *et al.,* 1993). A comparable pathway has been elucidated in yeast (reviewed in Rosbash and Séraphin, 1991; Ruby and Abelson, 1991).

In each study, a small (≈20–30S) RNP complex appears first, regardless of the presence of intron sequences in the pre-mRNA template; on intron-containing templates, one or two larger RNP particles then appear at the expense of the first. The initial complex in mammalian systems can be fractionated into two early, adenosine triphosphate (ATP)-independent complexes: the H complex, composed of hnRNP proteins and pre-mRNA (Bennett *et al.,* 1992a), is not committed to splicing; the E complex contains U1 and unstably bound U2 snRNP particles (Michaud and Reed, 1991, 1993) and loosely associated hnRNP proteins (Bennett *et al.,* 1992b). Formation of the H and E complexes is followed by formation of three ATP-dependent complexes: the A complex contains U1 and stably bound U2 snRNP particles, probably due to an interaction with SR proteins (see Section II,B,4); the B complex results from addition of the U4/U5/U6 tri-snRNP particle and ensuing rearrangements in the spliceosome; and the C complex arises concomitantly with the first and second steps in the splicing reaction. Because the E complex is the first to be committed to splicing, it can be chased sequentially into A, B, and C complexes on addition of ATP (Michaud and Reed, 1991).

3. *RNA/RNA-Mediated Interactions*

Defined RNA sequences are important determinants directing the deposition of snRNP particles on simple pre-mRNA templates (reviewed in Steitz, 1992; McKeown, 1993; Moore *et al.,* 1993). The term simple pre-mRNA template here refers to one containing a single intron and limited exon sequences. An early model of spliceosome assembly suggested that base-pairing interactions between U snRNA molecules and conserved sequences in pre-mRNA templates determine splice site selection. This model predicts preferential binding between U snRNP particles and conserved pre-mRNA sequences; moreover, given the necessity of U1 and U2 snRNP particles for splicing *in vitro* (Yang *et al.,* 1981; Padgett *et al.,* 1983; Black *et al.,* 1985; Krainer and Maniatis, 1985), the model predicts that genetic disruption of the snRNA/pre-mRNA base-pairing interactions should result in inhibition of splicing *in vivo.*

The primary determinants for snRNP particle deposition on pre-mRNA transcripts were determined initially through nuclease protection and immunoprecipitation experiments: U1 snRNP particles recognize the 5′ splice site (Mount *et al.,* 1983), U2 snRNP particles recognize the branchpoint (Black *et al.,* 1985), and U5 snRNP particles recognize the 3′ splice site [Chabot *et al.,* 1985; in mammalian systems the deposition of the U2 snRNP particle requires the U2AF factor (Ruskin *et al.,* 1988) which binds to the polypyrimidine tract/3′ splice site region of pre-mRNA (Zamore and Green, 1989)]. The requirement for base-pairing interactions between pre-mRNA and the 5′ ends of U snRNA molecules was elucidated in the following experiments. First, oligo-mediated RNase H degradation of U1 and U2 snRNA 5′-terminal sequences in nuclear extracts abolishes splicing activity *in vitro* (Krämer *et al.,* 1984; Black *et al.,* 1985; Krainer and Maniatis, 1985). In these experiments, short single-stranded DNA primers complementary in sequence to the 5′ end of target snRNA molecules were added to splicing extracts to form DNA/RNA hybrids; these hybrids form specific targets for RNase H digestion.

The best evidence for pre-mRNA–snRNA base-pairing interactions is provided by genetic experiments wherein splicing defects *in vivo* that arise from point mutations in 5′ or 3′ splice sites are rescued by compensatory base changes in snRNA that restore complementarity (Zhuang and Weiner, 1986, 1989; Parker *et al.,* 1987; Zhuang *et al.,* 1987; Séraphin *et al.,* 1988; Siliciano and Guthrie, 1988; Wu and Manley, 1989). Additional genetic experiments have revealed other RNA–RNA interactions that occur during spliceosome assembly. In yeast, sequences in the 3′ splice site interact specifically with U1 snRNA (Ruby and Abelson, 1988; Reich *et al.,* 1992)—an interaction that may be necessary, but not sufficient, for stable binding of the U2 snRNP particle to the pre-mRNA branch site (Nelson and Green, 1989; Barabino *et al.,* 1990). At a later step in the splicing reaction, U2 snRNA interacts with U6 snRNA (Madhani and Guthrie, 1992), probably as the U4 snRNP particle leaves the spliceosome.

Determinants of deposition for the remaining splicing snRNP particles are not clearly defined, but they probably enter the spliceosome as a U4/5/6 tri-snRNP complex (Bindereif and Green, 1987; Cheng and Abelson, 1987; Konarska and Sharp, 1987; Utans *et al.,* 1992). Both the 5′ and 3′ splice sites are likely to interact with the U4/5/6 tri-snRNP particle in the spliceosome because the 3′ splice site is required for interactions with a mammalian U5 snRNP-associated protein (Gerke and Steitz, 1986; Tazi *et al.,* 1986), both 5′ and 3′ exon sequences interact with U5 snRNA (Newman and Norman, 1992; Wassarman and Steitz, 1992), and mammalian U6 snRNA interacts with the 5′ splice site region (Sawa and Shimura, 1992).

4. Protein–RNA Interactions

A recently defined family of proteins, the SR proteins, may also participate in snRNP deposition by recognizing conserved sequences in mammalian and *Drosophila* pre-mRNA transcripts (reviewed in Horowitz and Krainer, 1994). Each SR protein is distinguished by two sequence motifs (Krainer *et al.,* 1991; Zahler *et al.,* 1992)—a serine- and arginine-rich (SR) domain and a RNA recognition motif (RRM; Burd and Dreyfuss, 1994a); each SR protein is recognized by a monoclonal antibody directed against a shared, phosphorylated epitope (Roth *et al.,* 1990, 1991), and each can restore splicing activity on simple pre-mRNA templates to S100 extracts (Krainer and Maniatis, 1985; Mayeda *et al.,* 1992; Fu *et al.,* 1992). In addition, individual SR proteins affect alternative splice site choices (see Section II,B,5). Table II lists SR proteins and their observed activities. SR-rich domains are found in other proteins with roles in pre-mRNA splicing, including the U1 snRNP 70K protein (Query *et al.,* 1989), the mammalian U2AF splicing factor (Zamore *et al.,* 1992; Kanaar *et al.,* 1993), and the product of the *suppressor-of-white-apricot [su(w^a)]* gene in *Drosophila* (Chou *et al.,* 1987; Zachar *et al.,* 1987).

Specific recognition of conserved pre-mRNA sequences by SR proteins is an important determinant in constitutive splice site selection. U2AF and the U1 snRNP 70K protein are components of E complexes (see Section II,B,2) and interact specifically with the SR proteins SC35 and

SF2/ASF to effect A complex formation *in vitro* (Fu and Maniatis, 1990). SC35 probably interacts with the pre-mRNA 3′ splice site and with U1 and U2 snRNP particles (Fu and Maniatis, 1992), enhancing their binding to the pre-mRNA template, whereas SF2/ASF (Krainer and Maniatis, 1985; Ge and Manley, 1990) recognizes the 5′ splice site in pre-mRNA transcripts (Zuo and Manley, 1994) to stimulate binding of the U1 snRNP particle (Kohtz *et al.*, 1994). Although the exact timing of SR protein deposition on pre-mRNA templates is not known, it is clear that certain SR proteins are required for the commitment of simple pre-mRNA transcripts to splicing (Fu, 1993) (see Section II,B,2).

SR domain-mediated interactions may function further to bring splice sites together in the spliceosome because both SC35 and SF2/ASF have been shown to interact specifically with the U1 snRNP 70K protein and the small subunit of U2AF (Wu and Maniatis, 1993; Kohtz *et al.*, 1994). This model is consistent with the juxtaposition of the 5′ and 3′ splice sites observed in E complexes (Michaud and Reed, 1993).

SnRNP deposition on nascent pre-mRNA transcripts may also depend on interactions between hnRNP proteins and snRNP particles. The cross-linking *in vitro* of hnRNP proteins in nuclear extracts to pre-mRNA templates apparently requires U1 and U2 snRNP particles because the cross-linkable interactions of hnRNP protein A1 with β-globin pre-mRNA templates and of hnRNP C proteins with adenoviral pre-mRNA templates are sensitive to RNase H-mediated degradation of U1 and U2 snRNA in splicing extracts (Mayrand and Pederson, 1990). In addition, hnRNP A1 protein binds specifically to the 5′ ends of U2 and U4 snRNA in snRNP particles *in vitro,* leading to the suggestion that hnRNP A1 protein promotes spliceosome assembly through its RNA annealing activity (Buvoli *et al.*, 1992). However, the story of RNP assembly on nascent transcripts remains incomplete because purified mammalian spliceosomes contain numerous additional proteins (Patton *et al.*, 1993), including those called spliceosome-associated proteins (SAPs) that are not yet characterized (Reed, 1990; Bennett *et al.*, 1992a). The reader is referred to other descriptions of these factors (Baserga and Steitz, 1993; Moore *et al.*, 1993).

5. Protein–Protein Interactions

The SR domain-mediated model of snRNP deposition may also apply to alternative splice site selection in complex pre-mRNA transcripts (Wu and Maniatis, 1993). Complex transcripts are those containing either

TABLE II

SR Proteins

Protein	Molecular mass (kDa)	Effect	Antagonist	Binding site
SF2/ASF	33[a]	Constitutive human β-globin pre-mRNA splicing[b–f]		
SRp30a (calf thymus)		Early spliceosome assembly on human β-globin pre-mRNA and HIV-1 pre-mRNA[d]		Nonspecific pre-mRNA sequences[d]
		Annealing of HIV-1 pre-mRNA[d]		
		Activation of proximal 5′ splice sites in β-thalassemic allele of human β globin pre-mRNA,[c,g] in SV40 early pre-mRNA,[g,f,j,k] duplicated human β-globin pre-mRNA,[c] and adenovirus E1a pre-mRNA[f,k]	hnRNP A1[e,g–l]	
		Promotion of exon inclusion or skipping in model pre-mRNAs[h]		
		Stimulation of upstream intron (D) splicing in bovine growth hormone pre-mRNA[i]		Splicing enhancer in last exon in bovine growth hormone pre-mRNA[i]
		Commitment of human HIV-1 *tat* pre-mRNA to splicing[l]		
		Increase in strength of U1 snRNP binding to all 5′ splice sites in model pre-mRNAs from rabbit β-globin IV2[m]	SF7[m]	
		Promotion of use of proximal 3′ splice site in D115 pre-mRNA[e]		
		Binding to SV40 t intron and adenovirus E1a pre-mRNA containing 5′ slice sites[n]		5′ Splice sites[n]
		Stimulation of binding of U1 snRNP to 5′ splice sites in adenovirus major late transcription unit derivative[o]	DSF (distal splicing factor)[f]	U1 70K snRNP protein[o]
SC35	35[p]	Commitment of human β-globin pre-mRNA to splicing[q,t]	hnRNP A1[e]	Competed by 5′ splice site[q]
SRp30b (calf thymus)[f]		Promotion of use of proximal 5′ splice site in model pre-mRNA and of proximal 3′ splice site in 3′ D115 pre-mRNA[e]		
		Assembly of 60S spliceosomal complex on human β-globin pre-mRNA[p,s]		

		Activation of splicing β-globin pre-mRNA[f]	
		Strong promotion of proximal 5′ splice site[f]	
		Strong promotion of proximal 5′ splice site[f]	
SRp40 (calf thymus)	40	Activation of β-globin pre-mRNA splicing[f]	
		Weak promotion of proximal 5′ splice site in adenovirus E1a pre-mRNA[f]	
B52 (*Drosophila*)	52 (app.)		Boundaries of transcriptionally active chromatin[t,u]
	43 (calc.)		
SRp55 (*Drosophila*)	55 (app.)	Activation of human β-globin pre-mRNA splicing[v]	Lateral loops on amphibian lampbrush chromosomes[w]
	39 (calc.)	Switch from distal to proximal 5′ splice site in β-thalassemic human pre-mRNA[z]	
SRp55 (calf thymus)		Activation of β-globin pre-mRNA splicing[c]	
		Promotion of proximal 5′ splice site in adenovirus E1a pre-mRNA and in SV40 pre-mRNA[f]	
SRp70 (calf thymus)	70 (app.)	Activation of β-globin pre-mRNA splicing[f]	
		Weak promotion of proximal 5′ splice site in adenovirus E1a pre-mRNA[f]	

[a] Krainer *et al.* (1991).
[b] Krainer and Meniatis (1985).
[c] Krainer *et al.* (1990a).
[d] Krainer *et al.* (1990b).
[e] Fu *et al.* (1992).
[f] Zahler *et al.* (1993).
[g] Mayeda and Krainer (1992).
[h] Mayeda *et al.* (1993).
[i] Sun *et al.* (1993).
[j] Ge and Manley (1990).
[k] Harper and Manley (1991).
[l] Fu (1993).
[m] Eperon *et al.* (1993).
[n] Zuo and Manley (1994).
[o] Kohtz *et al.* (1994).
[p] Fu and Maniatis (1992).
[q] Fu (1993).
[r] Spector *et al.* (1991).
[s] Fu and Maniatis (1990).
[t] Champlin *et al.* (1991).
[u] Champlin and Lis (1994).
[v] Mayeda *et al.* (1992).
[w] Roth *et al.* (1990).

multiple competing splice sites or multiple introns. One mechanism of alternative splicing apparently involves "exon bridging" between SR domains to regulate exon inclusion/exclusion choices, a proposal similar to the exon definition model of splice site selection (Robberson *et al.*, 1990). In one example, inclusion of the alternative exon E4 in transcripts of the rat preprotachykinin gene requires facilitated binding of U1 snRNP particles to the 5′ splice site adjacent to E4 (Kuo *et al.*, 1991). Because the adjacent upstream intron possesses a weak polypyrimidine tract, enhanced deposition of U1 snRNP particles to a strong 5′ splice site sequence of E4 facilitates binding of U2AF to the 3′ splice site region upstream across the exon (Hoffman and Grabowski, 1992), presumably enhancing deposition of U2 snRNP particles on this intron. Thus, SR domain-mediated interactions may span the alternative exon to compensate for a weak upstream polypyrimidine tract.

A similar mechanism may be utilized by the purine-rich "splicing enhancers" (Tian and Maniatis, 1994). For example, the SR protein SF2/ASF binds preferentially to a purine-rich splicing enhancer in the last exon of bovine growth hormone transcripts, stimulating splicing of the upstream intron (Sun *et al.*, 1993). Similarly, interactions between a purine-rich splicing enhancer in the ED1 exon of the human fibronectin gene and SR proteins enhances deposition of U2 snRNP particles to the ED1 3′ splice site and thereby stimulates inclusion of the ED1 exon (Lavigueur *et al.*, 1993).

Another mechanism of SR domain-mediated alternative splice site selection may work to enhance the proximity effect (Eperon *et al.*, 1993). In the case of adjacent, competing 5′ splice sites, the proximal 5′ splice site normally is preferred over a distal 5′ site if the two are separated by a distance greater than 40 nucleotides (Reed and Maniatis, 1986). SF2/ASF promotes use of the proximal 5′ splice site (Ge and Manley, 1990; Krainer *et al.*, 1990a) but is antagonized by increased abundance of hnRNP A1 (Mayeda and Krainer, 1992; Cáceres *et al.*, 1994), which leads to preferred splicing of the distal 5′ site. Because SF2/ASF seems to increase the binding of U1 snRNP particles to adjacent but separated 5′ splice sites indiscriminately (Eperon *et al.*, 1993), one wonders if the SF2/ASF-mediated enhancement of proximal splice site selection involves interactions with SR domains in other proteins, such as U2AF, situated downstream in the intron.

Both positive and negative SR domain-mediated mechanisms of alternative 3′ splice site selection are evident in studies of sex-specific splicing in *Drosophila* (reviewed in Baker, 1989; Mattox *et al.*, 1992). In males, spliced transcripts of the *transformer* (*tra*) gene contain an in-frame stop codon and encode a truncated protein; in females, nearly half of the *tra* transcripts undergo a splicing event that removes the stop codon (McKeown *et al.,* 1987). The difference arises from alternative splicing of competing 3′ splice sites and is regulated by the product of the *Sex lethal* (*Sxl*) gene (Boggs *et al.,* 1987), an RRM-containing protein that lacks an SR domain (Bell *et al.,* 1988). Binding of Sxl protein in females to the non-sex-specific 3′ splice site in *tra* transcripts (Inoue *et al.,* 1990) prevents the binding of U2AF to this site (Valcárcel *et al.,* 1993). This blockage mechanism leads to inactivation of the Sxl-engaged splice site, due to the absence of an SR splicing domain, and concomitantly to activation of the female-specific site by U2AF. Interestingly, this event leads to additional consequences concerning alternative 3′ splice site selection in the sex determination pathway because the *tra* gene encodes another protein with an SR domain (Amrein *et al.,* 1988). The *tra* protein, in conjunction with the non-sex-specific protein TRA-2, regulates sex-specific splicing of the *doublesex* (*dsx*) gene (Burtis and Baker, 1989; Hoshijima *et al.,* 1991). In females, TRA and TRA-2 proteins recognize an exon sequence downstream of the regulated 3′ splice site in *dsx* transcripts (Nagoshi and Baker, 1990; Hedley and Maniatis, 1991; Ryner and Baker, 1991; Inoue *et al.,* 1992) and recruit members of the SR family to the site (Tian and Maniatis, 1993).

6. *Composite Model*

In sum, constitutive spliceosome assembly on simple pre-mRNA templates *in vitro* involves numerous RNA–RNA base-pairing interactions, protein–RNA sequence recognition, and protein–protein interactions. The earliest RNP complex formed on pre-mRNA templates *in vitro* contains hnRNP proteins, some of which probably bind with sequence specificity. Initial deposition of U1 and U2 snRNP particles onto pre-mRNA templates *in vitro* requires precise snRNA–pre-mRNA base-pairing interactions, whereas tight binding of U1 and U2 snRNP particles to pre-mRNA transcripts requires SR domain-mediated protein interactions. These interactions may also bring splice sites into juxtaposition. Moreover, the examples presented here point to a general mechanism for alternative splice site selection on complex pre-mRNA transcripts whereby protein–protein interactions redirect the deposition of general splicing factors on pre-mRNA templates by bridging from a strong binding site to a weak binding site or by blocking normal binding sites for constitutive factors. It will be interesting to see if SR domain-mediated interactions are responsible for other known instances of regulated alternative splice site selection (Black, 1992; Gallego *et al.*, 1992; Guo and Helfman, 1993; Siebel *et al.*, 1994; Tanaka *et al.*, 1994).

III. TIMING OF PROTEIN DEPOSITION ON NASCENT PRE-mRNA TRANSCRIPTS

The precise timing of protein deposition on nascent pre-mRNA transcripts *in vivo* is unknown. One model suggests that each class of transcript-binding factor—hnRNP proteins, snRNP particles, and SR proteins—is recruited from nuclear storage sites to chromosomal sites of transcription, presumably to bind nascent pre-mRNA transcripts cotranscriptionally. The recruitment model predicts the presence of transcript-binding factors at chromosomal sites of transcription, plus the detectable relocation of such factors in response to induction of transcription or the inhibition of splicing; the model further predicts the occurrence of some cotranscriptional RNA processing reactions. An alternative view suggests that some protein deposition occurs in discrete processing centers in the nucleus and predicts the passage of nascent transcripts through defined nuclear paths. A third model proposes the channeled diffusion of nascent transcripts through extrachromosomal regions, with most intron removal occurring cotranscriptionally.

A. Recruitment Model

1. Sites of Transcription

Abundant immunocytological evidence demonstrates the presence of transcript-binding proteins at chromosomal sites of transcription. On lampbrush chromosomes, with a few notable exceptions (see the following discussion), most hnRNP proteins tested—the A and B group hnRNP proteins (Scott and Sommerville, 1974; Lacroix *et al.*, 1985; Wu *et al.*, 1991) and presumed hnRNP proteins of 90 kDa and 120 kDa (Roth and Gall, 1987), plus the major spliceosomal snRNP particles (Gall and Callan, 1989; Wu *et al.*, 1991; Jantsch and Gall, 1992) and the SR protein

SC35 (Roth *et al.*, 1990; Wu *et al.*, 1991)—are present on lampbrush chromosome loops. As expected of transcript-binding proteins, the staining patterns reflect the characteristic "thin-to-thick" morphology of loop RNP matrices. The hnRNP antigens just listed are not detectable, however, on the giant loops of chromosome 2 or on the sequentially labeling loops on chromosome 11, although the giant loops on chromosome 2 are preferentially stained with antibodies specific for the hnRNP L protein (Piñol-Roma *et al.*, 1989). The distribution of hnRNP proteins and snRNP particles in "snurposomes" is discussed in Section IV,A.

Immunostaining experiments on polytene chromosomes are consistent with cotranscriptional protein deposition on nascent pre-mRNA transcripts. Numerous hnRNP proteins are localized in polytene chromosome puff sites (Risau *et al.*, 1983; Hovemann *et al.*, 1991; Amero *et al.*, 1991, 1993; Matunis *et al.*, 1992b, 1993; Weeks *et al.*, 1993); as expected for transcript-bound proteins, the retention of certain hnRNP proteins in *Drosophila* chromosomal puffs is RNase-sensitive (Amero *et al.*, 1993; Matunis *et al.*, 1992b, 1993). Other transcript-binding factors found in puff sites include the U1 and U2 snRNP particles (Sass and Pederson, 1984; Vazquez-Nin *et al.*, 1990; Amero *et al.*, 1992; Matunis *et al.*, 1993; Weeks *et al.*, 1993), the SR proteins B52 (Champlin *et al.*, 1991) and RBP1 (Kim *et al.*, 1992), and the Sxl protein (Samuels *et al.*, 1994). (The B52 protein also displays a "puff bracketing" pattern that is related to a DNA-binding function; see Section V,B.)

In mammalian cells, definitive localization of the hnRNP C proteins on nascent pre-mRNA was provided by immunocytochemistry (Fakan *et al.*, 1984, 1986; Puvion *et al.*, 1984; Leser *et al.*, 1989; Spector *et al.*, 1991). The hnRNP proteins were identified in perichromatin fibrils and interchromatin granules, structures implicated previously as sites of Pol II transcript synthesis and maturation, respectively (Fakan *et al.*, 1984, 1986). However, these are not the only nuclear structures associated with hnRNP proteins, and not all hnRNP proteins display identical nuclear distributions. Numerous visualizations of mammalian interphase nuclei stained with antibodies specific for the hnRNP A1, C, G, and U proteins reveal homogeneous nucleoplasmic distributions that exclude chromatin and nucleoli (Choi and Dreyfuss, 1984b; Dreyfuss *et al.*, 1984; Leser *et al.*, 1989; Carmo-Fonseca *et al.*, 1991; Soulard *et al.*, 1993). Moreover, several hnRNP proteins display peculiarities in addition to their general, diffuse presence throughout the nucleoplasm; the hnRNP I protein is concentrated in a discrete perinucleolar structure (Ghetti *et al.*, 1992), the hnRNP K and L proteins are found in two or three unidentified, brightly stained loci in each nucleus (Piñol-Roma *et al.*, 1989; Siomi *et al.*, 1993), and the hnRNP F and H proteins are concentrated in numerous discrete regions, or "speckles," in the nucleoplasm (Matunis *et al.*, 1994b).

Spliceosomal snRNP particles and the SR protein SC35 also were identified on perichromatin fibrils and interchromatin granules by immunoelectron microscopy (Fakan *et al.*, 1984, 1986; Puvion *et al.*, 1984; Leser *et al.*, 1989; Spector *et al.*, 1991). However, early immunostaining experiments utilizing sera from autoimmune patients revealed three types of nuclear staining patterns: nucleolar, homogeneous, and a speckled pattern superimposed on a low-level, diffuse pattern throughout the nucleus (Beck, 1961; Lachmann and Kunkel, 1961). The speckled pattern is characteristic of the Sm antisera (Tan and Kunkel, 1966; Tan, 1967; Lerner *et al.*, 1981), which recognize proteins in spliceosomal snRNP particles (Lerner and Steitz, 1979). It is now clear that numerous snRNP particle components and splicing factors are colocalized in nuclear speckles [including U1 and U2 snRNP particles (Spector, 1984; Reuter *et al.*, 1984; Nyman *et al.*, 1986; Verheijen *et al.*, 1986; Habets *et al.*, 1989; Huang and Spector, 1992) and the SR protein SC35 (Fu and Maniatis, 1990)]. Thus, one interpretation of the data suggests that speckles represent interchromatin granules, or storage areas for transcript-binding factors. In some cells, the spliceosomal snRNP particles and the SR domain-containing protein U2AF—but not SC35—are also concentrated in coiled bodies within the nucleus (Fakan *et al.*, 1984; Zamore and Green, 1991; Carmo-Fonseca *et al.*, 1992; Matera and Ward, 1993), but the role of coiled bodies in nuclear RNA metabolism is not yet known.

2. *Induction of Transcription*

The best evidence that transcript-binding factors are recruited to active sites of transcription exists in the following observations. First, the absence of snRNP particle deposition prior to induction of puffing on *Chironomus* chromosomes (Sass and Pederson, 1984) has been interpreted to indicate that transcript-binding factors are not poised at active sites prior to transcriptional activation. Second, the SR protein SC35, snRNP particles, RNA polymerase II, and hnRNP proteins C1 and C2 are seen to move to, and colocalize on, sites of adenoviral-specific transcription in HeLa cells (Jiménez-García and Spector, 1993). Circumstantial evidence for the recruitment model exists in the correlation observed between deposition patterns of the phosphorylated form of the large subunit of RNA polymerase II—and therefore probably the elongating form—on polytene chromosomes and the deposition patterns of hnRNP proteins and snRNP proteins on most major puffs (Weeks *et al.*, 1993).

3. *Inhibition of Splicing*

Another prediction of the recruitment model is the redistribution of nuclear transcript-binding factors in response to inhibition of pre-mRNA splicing *in vivo*. This prediction was confirmed directly by the microinjection into living cells of oligonucleotides or antibodies that inhibit pre-mRNA splicing *in vitro;* following such treatment, the interchromatin granules decreased in number, but those remaining were larger and more concentrated in splicing factors than granules in control cells (O'Keefe *et al.*, 1994). A similar phenomenon occurs in response to herpes simplex virus infection and may reflect partial inhibition of host pre-mRNA splicing (Martin *et al.*, 1987). Consistent with these results are the changes in interchromatin granules observed following the incubation of cells in the transcriptional inhibitors DRB (5,6-dichloro-1-β-D-ribofuranosylbenzimidazole; Spector *et al.*, 1983) or α-amanitin (Carmo-Fonseca *et al.*, 1992), treatments that would also result in decreased pre-mRNA splicing activity.

4. *Cotranscriptional RNA Processing*

The recruitment model predicts the cotranscriptional processing of nascent pre-mRNA molecules, and the cytological evidence (discussed in Sections III,A,4,a–c) demonstrates that both folding and splicing of nascent pre-mRNA transcripts can be cotranscriptional events.

a. RNA Folding. The distinctive RNP granules that appear in Balbiani rings of *Chironomus tentans* polytene chromosomes (Beermann and Bahr, 1954; Stevens and Swift, 1966) present an unequaled opportunity to follow the synthesis and maturation of a single transcript through the nucleus (Skoglund *et al.*, 1983). These giant puffs produce long transcripts (35–40 kb; Case and Daneholt, 1978) encoding large secretory proteins (Höög *et al.*, 1986, 1988). Mature nascent transcripts in sectioned material assume a characteristic morphology consisting of a granule ($\approx$50 nm in diameter) connected to a looped chromatin axis by a short stalk ($\approx$20 nm in diameter and $\approx$90 nm in length; Lamb and Daneholt, 1979;

Andersson *et al.*, 1980). The mature 50-nm granules share a fairly uniform structure, with an RNP ribbon (30–60 nm broad and 10–15 nm thick) bent into a ring; four discernible domains in the granule differ in ribbon width (Skoglund *et al.*, 1986). The 50-nm granules are abundant in the nucleoplasm, allowing their biochemical purification (Wurtz *et al.*, 1990b), and they can be followed during transit through nuclear pores (Stevens and Swift, 1966; Mehlin *et al.*, 1991, 1992).

A defined pathway for early, orderly cotranscriptional folding of the Balbiani ring transcript has been proposed. Circumstantial evidence for the pathway exists in serial sections, where most of the loop containing a Balbiani ring transcription unit can be visualized. The "stalks" but not the 50-nm granules are present on the proximal side of the loop containing immature transcripts; the granules do appear on distal portions of the loop that contain more mature transcripts (Andersson *et al.*, 1980; Olins *et al.*, 1983; Skoglund *et al.*, 1983). The basic structure of the RNP particle is thought to be a 7-nm fibril that thickens and bends to form the granule because 50-nm granules in low ionic strength buffers dissociate into 7-nm RNP fibrils (Lönnroth *et al.*, 1992). Such a fibril is also visible in chromatin dispersed by the Miller spreading technique, where nascent transcripts in Balbiani rings appear as growing RNP fibrils ($\approx$10 nm in diameter) in a single length gradient per loop (Lamb and Daneholt, 1979); the Miller spreading technique probably induces unfolding of the 50-nm granule into the 10-nm fibril (Andersson *et al.*, 1980). Also, because intact 50-nm granules are sensitive to RNase digestion (Wurtz *et al.*, 1990a), the pre-mRNA molecule probably is folded around a protein core initially to form the 7-nm fibril.

An analogous assembly pathway has been proposed for folding of perichromatin fibrils into perichromatin granules during transcript maturation in mammalian cells (reviewed in Puvion-Dutilleul and Puvion, 1981). In rat liver chromatin spread in the presence of Photo Flo detergent, perichromatin granules appear as dense particles (33–55 nm in diameter) connected to chromatin threads through short RNP fibrils (9–12 nm in diameter) (Monneron and Bernhard, 1969; Puvion-Dutilleul and Puvion, 1981). The smaller fibrils resemble perichromatin fibrils seen in more nearly denatured chromatin spreads (Bachellerie *et al.*, 1975), suggesting that the fibrils may become folded into perichromatin granules. Both perichromatin fibrils and perichromatin granules are rapidly labeled *in vivo* with [^{3}H]uridine (Fakan and Bernhard, 1971, 1973; Bachellerie *et al.*, 1975; Fakan *et al.*, 1976; Puvion-Dutilleul and

Puvion, 1981), indicating that both structures contain nascent pre-mRNA transcripts.

Partially denatured pre-mRNA transcripts in *Drosophila* chromatin also assume a granular morphology that may arise from the folding of a precursor RNP fibril. Rather than using milder detergents to achieve partial denaturation of chromatin, the pH of the water used for dispersal can be lowered from 8.5–9.0 to 7.5–8.0 (Beyer and Osheim, 1988). Transcripts prepared from *Drosophila* embryos in "more physiological" water appear as 50- to 60-nm granules attached to the DNA template (Beyer and Osheim, 1988). At intermediate stages of dispersal, the granules dissociate partially into the 25-nm particles and 5-nm fibrils seen in fully dispersed chromatin, producing a "rosette" on each transcript (Beyer and Osheim, 1990).

b. Timing of Cotranscriptional Protein Deposition. The exact timing of early protein deposition on nascent pre-mRNA transcripts is unknown, but it probably occurs more quickly than can be measured biochemically by conventional cross-linking techniques—given the average elongation rate *in vivo* for eukaryotic RNA polymerase II of 1.1–1.4 kb per minute (Kafatos, 1972; Thummel *et al.,* 1990; Shermoen and O'Farrell, 1991; O'Brien and Lis, 1993). However, one study estimated that a "window" of approximately 100 nucleotides exists *in vivo* between RNA synthesis and protein deposition (Eperon *et al.,* 1988). These experiments introduced sequences capable of forming stable hairpin loops at fixed distances along a transcription unit to disrupt alternative splicing *in vivo.*

Studies of paused transcription complexes may be used to refine further the window for protein deposition. Numerous genes in *Drosophila* and other organisms are regulated at the level of transcription elongation (Lis and Wu, 1993), and regulation of the *Drosophila hsp 70* gene is particularly well elucidated. An RNA polymerase II molecule resides at the promoter region of the *hsp 70* gene in uninduced (Gilmour and Lis, 1986) and induced (heat-shocked) cells, as detected by UV cross-linking (Giardina *et al.,* 1992). The polymerase molecule pauses over a promoter-proximal region approximately 20 bp wide (Rasmussen and Lis, 1993), having synthesized approximately 25 nucleotides of nascent RNA, but is transcriptionally engaged, poised for rapid elongation on induction by heat shock (Rougvie and Lis, 1988; O'Brien and Lis, 1991). The short RNA transcript associated with the paused polymerase undergoes 5′ cap formation *in vivo* during the progression of polymerase between its pause sites (Rasmussen and Lis, 1993). However, it is unlikely that short nascent transcripts on the paused polymerase are associated with RNA-binding proteins because RNA of this length observed in *in vitro* elongation assays is intimately associated with RNA polymerase II (Rice *et al.,* 1991) and probably occluded from transacting factors. Moreover, immunopurified complexes assembled on the *hsp 70* gene promoter contain predominantly transcription factor IID (TFIID), with no indication of additional proteins such as hnRNP proteins (Emanuel and Gilmour, 1993). We conclude that protein deposition *in vivo* occurs on nascent pre-mRNA between 25 and 100 nucleotides in length.

c. Pre-mRNA Splicing. Visualization of active genes by the Miller spreading technique clearly demonstrates cotranscriptional RNA processing *in vivo* (Beyer and Osheim, 1991), as predicted by the recruitment model. At the ultrastructural level, nascent pre-mRNA transcripts from *Drosophila* embryos have a beaded appearance reflecting nonrandom RNP particles (Beyer *et al.,* 1980). Correlation of particle deposition with the *Drosophila* chorion gene sequence shows 25-nm RNP particles on splice sites, and on more mature transcripts these particles occasionally coalesce into 40-nm particles (Osheim *et al.,* 1985). On unidentified *Drosophila* transcription units, particle assembly occurs at the bases of loops that are to be excised with kinetics that are faster than, but consistent with, predictions from *in vitro* splicing reactions (Beyer and Osheim, 1988). These processes of particle coalescence and loop excision no doubt represent cotranscriptional spliceosome assembly and intron removal from certain transcripts *in vivo.*

Biochemical experiments prove that pre-mRNA splicing can occur cotranscriptionally. Molecular analyses of products from the ecdysone-responsive *E74* transcription unit in *Drosophila* showed a defined splicing pathway that initiates on incomplete transcripts prior to transcription termination (LeMaire and Thummel, 1990). These analyses utilized the rapid inducibility and unusual length (60 kb) of the *E74* transcription unit to analyze nascent *E74* transcripts at defined time intervals following transcriptional initiation. Similar analyses of nascent transcripts purified from microdissected Balbiani ring chromatin showed a graded disappearance of introns from the immature transcripts (Baurén and Wieslander, 1994).

B. Extrachromosomal RNA Processing

1. *Tracking Model*

Not all pre-mRNA splicing occurs cotranscriptionally. Studies in human diploid fibroblasts suggest that nuclear speckles are coincident with transcript domains, or areas of the nucleus defined by *in situ* hybridization as being enriched in poly(A) RNA (Carter *et al.*, 1991, 1993). The transcript domains also are enriched in SC35 and overlap regions in rat fibroblast nuclei where fibronectin gene transcripts are concentrated and undergo intron removal; these areas of fibronectin transcript concentration have been termed tracks (Xing *et al.*, 1993). The fact that the intron signals extend further along the track than would be predicted from Miller spreads suggests that some intron removal occurs after transcript release from the chromosome.

One interpretation of the track data suggests that nuclear speckles are active RNA processing centers, as opposed to storage sites for splicing factors, and that pre-mRNA transcripts move through these centers during maturation. Consistent with this view are experiments involving the microinjection of rhodamine-labeled β-globin pre-mRNA transcripts into the nuclei of cultured rat kidney epithelial cells. Splicing-competent pre-mRNA transcripts became concentrated in speckles and were colocalized in these sites with snRNP particles and SC35, while splicing-defective pre-mRNA transcripts did not (Wang *et al.*, 1991). The literature concerning nuclear sites of pre-mRNA splicing, as well as compartmentalization of snRNP particles and splicing factors in the eukaryotic nucleus, has been reviewed (Spector, 1993a,b).

2. *Channeled Diffusion Model*

Studies in polytene nuclei of *Drosophila* suggest an alternative interpretation of the tracking phenomenon. Integration of a *LacZ* reporter construct containing the *Drosophila suppressor-of-white-apricot* gene into polytene chromosomes permitted comparative localization of chromosomes and $su(w^a)$ transcripts through combined cytochemistry and *in situ* hybridization; the evidence suggests that transcript diffusion through channels between chromosomes is responsible for the appearance of concentrated transcript tracks (Zacher *et al.*, 1993). In addition, constitutive removal of the third intron from *su*(w^a) pre-mRNA transcripts occurred at or close to the site of synthesis, whereas regulated removal of at least one other (upstream) intron occurred more slowly—apparently at a location in the extrachromosomal space removed from the site of synthesis (Zachar *et al.*, 1993).

In sum, substantial experimental evidence exists to suggest that each class of major transcript-binding factor—hnRNP proteins, snRNP particles, and SR proteins—is associated with chromosomal sites of transcription, as viewed on loops of lampbrush chromosomes, on puffs of polytene chromosomes, and on perichromatin fibrils in mammalian nuclei; these factors are also concentrated in extrachromosomal nuclear sites. The earliest protein deposition may occur on nascent pre-mRNA transcripts between 25 and 100 nucleotides in length. The presence of snRNP particles and SR proteins on sites of transcription is consistent with ultrastructural observations and molecular analyses demonstrating the occurrence of RNA folding and pre-mRNA splicing during transcript elongation, but may signify prespliceosome assembly and commitment to splicing of pre-mRNA transcripts that will be spliced subsequently at extrachromosomal locations, possibly in nuclear speckles.

IV. INTERDEPENDENCE AMONG RIBONUCLEOPROTEIN COMPLEXES ON PRE-mRNA TRANSCRIPTS

A separate question concerning protein deposition on nascent pre-mRNA transcripts concerns the interdependence of the major transcript-binding factors. One model for cotranscriptional protein deposition on nascent pre-mRNA transcripts *in vivo* suggests the preassembly of transcript-binding proteins—including hnRNP proteins, snRNP particles, and SR proteins—into a unitary complex prior to their deposition. The unitary deposition model predicts that the major transcript-binding factors should be colocalized in extrachromosomal nuclear complexes, undergo extrachromosomal assembly, and exist in a fixed ratio at chromosomal sites of transcription. An alternative model of protein deposition *in vivo* suggests the independent deposition of individual transcript-binding factors, although several variations of the independent model can be envisioned. A broad definition of the independent deposition model, used here, refers to hnRNP complex formation and prespliceosome assembly as separate processes. The independent deposition model predicts locus-specific ratios of hnRNP proteins and snRNP particles at chromosomal sites of transcription (the question of unitary versus independent deposition was addressed in the context of hnRNP complexes in Section II,A).

A. Unitary Deposition Model

The unitary deposition model arose from studying the thousands of nucleoplasmic granules in amphibian oocytes (Gall and Callan, 1989; Gall, 1991). The granules, or snurposomes, are classified on the basis of immunofluorescence and *in situ* hybridization reactivities into three classes (Wu *et al.*, 1991): the A snurposomes contain U1 snRNP particles; the B snurposomes (1–4 μm in diameter) contain hnRNP proteins, the splicing snRNAs and the SR protein SC35; and the C snurposomes (or sphere organelles, ≤ 1 μm to 10 μm in diameter) contain U7 snRNA but not the "splicing" snRNAs (Wu and Gall, 1993). B snurposomes exist on the surface of, and within, C snurposomes (Wu *et al.*, 1991); C snurposomes are also attached to lampbrush chromosomes at specific loci, the sphere organizers (Gall and Callan, 1989). A pathway for snurposome assembly proposed that B snurposomes are precursor unitary processing particles and that A snurposomes are storage sites, possibly analogous to interchromatin granules in mammalian nuclei (Gall, 1991; Wu *et al.*, 1991).

The predictions from the unitary deposition model for the extrachromosomal assembly of a processing particle and the simultaneous deposition of transcript-binding factors on nascent pre-mRNA transcripts are fulfilled in the following observations. The time course of [^{3}H]uridine incorporation into B and C snurposomes in cultured *Xenopus* oocytes involves the appearance of labeled B snurposomes in preparations from oocytes incubated for 12 or 24 hr, but labeled C snurposomes appear in preparations from oocytes incubated for 48 hr; thus, RNA—and probably snRNA—may pass from B to C snurposomes during their maturation (Callan and Gall, 1991). Interestingly, however, the injection of pre-mRNA templates encoding *myc*-tagged, U1 snRNP-specific C protein into amphibian oocytes demonstrated the simultaneous association of the nascent C protein with all U1 snRNP-containing structures (Jantsch and Gall, 1992). Once recruited to lampbrush loops, possibly through C snurposomes, both hnRNP proteins and snRNP proteins show coincident immunofluorescent labeling along loop lengths, even at the "front end" of the transcription unit where short transcripts are unlikely to contain intron sequences (Wu *et al.,* 1991).

B. Independent Deposition Model

The prediction from the independent model for locus-specific ratios of hnRNP proteins and snRNP particle components is fulfilled by the results of two different studies performed on *Drosophila* polytene chromosomes. In one study, comparative localization of a U1 and U2 snRNP-specific protein and the *Drosophila* "core" HRB proteins revealed generally coincident distribution patterns on most developmental puffs for the two antigen types, but qualitative differences in the patterns were evidenced by reproducible, locus-specific variations in the two signal intensities (Amero *et al.,* 1992). The correlation observed between puff size and HRB-specific patterns was interpreted to reflect a correlation between hnRNP protein deposition and general RNA mass, whereas a similar correlation was not observed in the snRNP-specific pattern. In the other study, antibodies specific for individual hnRNP proteins and for the Sm epitope of splicing snRNP particle proteins also produced locus-specific signals suggestive of transcript-specific RNP complexes (Matunis *et al.,* 1993). The results of both studies are consistent with the pathway of spliceosome assembly *in vitro,* where hnRNP proteins become associated first with pre-mRNA transcripts; snRNP particle and SR protein deposition follow in subsequent steps (see Section II,B,2).

Direct tests of the unitary and independent deposition models should now be feasible with the advent of *Drosophila* mutagenesis to produce null mutations in genes encoding transcript-binding proteins (Kelley, 1993; Matunis *et al.,* 1994a). It will be interesting to see if the deposition of all transcript-binding proteins is affected by the depletion of individual factors.

V. INTERACTIONS WITH TRANSCRIPTION COMPLEXES

Protein deposition on nascent pre-mRNA transcripts is thought to be independent of transcription machineries because transcription and

RNA processing are not interdependent processes per se. For instance, plasmid DNA is transcribed efficiently when injected into splicing-deficient *Xenopus* oocytes that have been depleted of snRNA by RNAse H-mediated cleavage (Pan and Prives, 1988, 1989; Hamm *et al.,* 1989). Although injection of the oligonucleotides to create substrates for RNAse H produces a general collapse of lampbrush chromosome loops and shutdown of Pol II transcription, the oocytes do resume transcription after ≈12 hr in the absence of detectable U2 snRNA (Tsvetkov *et al.,* 1992). Further evidence that the processes of transcription and pre-mRNA splicing can be uncoupled is provided by the accurate splicing observed on injection into *Xenopus* oocytes of pre-mRNA templates synthesized *in vitro* (Green *et al.,* 1983; Kedes and Steitz, 1988). Nonetheless, interactions between transcription complexes and transcript-binding factors are apparent in the examples described in Sections V,A–C, suggesting that additional determinants for deposition of transcript-binding proteins may reside in DNA sequences or in DNA-binding proteins.

A. *Tat*

The mechanism of transcriptional enhancement utilized by the human immunodeficiency virus-1 (HIV-1) exemplifies the influence of transcript-binding factors on transcription complexes (reviewed in Jones, 1989; Sharp and Marciniak, 1989). The virally encoded Tat protein upregulates transcription from the viral LTR sequence by binding specifically to RNA hairpin sequences—the *trans*-activating region (TAR) element—in the 5′ end of the viral transcript (Berkhout *et al.,* 1989; Selby and Peterlin, 1990; Southgate *et al.,* 1990; Calnan *et al.,* 1991). In addition, Tat affects Pol II elongation when bound to TAR, promoting read-through and production of full-length transcripts (Kao *et al.,* 1987; Selby *et al.,* 1989; Feinberg *et al.,* 1991; Marciniak and Sharp, 1991; Kato *et al.,* 1992).

The purpose of Tat–TAR interactions is to recruit the Tat activation domain into the vicinity of LTR-engaged transcription complexes. The Tat activation domain functions normally when fused to either the DNA-binding domain from the yeast GAL4 protein (Southgate and Green, 1991) or the RNA-binding domain from the bacteriophage MS2 coat protein (Selby and Peterlin, 1990), provided that the target LTR sequences contain either a GAL4 binding site in the DNA sequence or the operator of the bacteriophage replicase gene in the RNA sequence, respectively. Tat-mediated activation may be mediated through direct interactions with the TATA-binding protein in the TFIID complex (Kashanchi *et al.,* 1994).

B. B52/SRp55

A curious example of interplay between transcription complexes and transcript-binding factors exists in the properties of the 52-kDa SR protein from *Drosophila,* designated either B52 (Champlin *et al.,* 1991) or SRp55 (Roth *et al.,* 1991). The protein is both an essential splicing factor and an alternative splicing factor (Mayeda *et al.,* 1992). As expected of an RNA-binding protein, at intermediate levels of heat shock transcription the protein is distributed uniformly over the transcription units containing the *hsp70* genes, as determined by immunofluorescence mapping and *in vivo* UV cross-linking (Champlin and Lis, 1994). However, deposition of the protein on polytene chromosomes from severely heat-shocked larvae reveals an unusual pattern on heat shock puffs that brackets the pattern produced by antibodies to RNA polymerase II (Champlin *et al.,* 1991), and the protein can be cross-linked *in vivo* to specific DNA sequences flanking the heat shock genes (Champlin *et al.,* 1991; Champlin and Lis, 1994). These observations indicate an inverse correlation between protein deposition on nascent pre-mRNA transcripts and decondensation of the chromatin template. One proposal further suggests the interaction of the SR domain with the C-terminal repeat of the RNA polymerase II large subunit to link RNA processing to transcription (Greenleaf, 1993).

C. hnRNP U/SAF-A

The identification of the mammalian 120-kDa hnRNP U protein as the nuclear matrix/scaffold attachment protein (Fackelmayer and Richter, 1994a; von Kries *et al.,* 1994) creates the intriguing possibility that the primary deposition signal for this hnRNP protein may reside in specific DNA sequences. The protein binds specifically to DNA MAR/SAR sequences (matrix attachment regions/scaffold attachment regions) and possesses sequence-preferential RNA-binding characteristics *in vitro*

(Kiledjian and Dreyfuss, 1992; Romig *et al.,* 1992; Tsutsui *et al.,* 1993; Fackelmayer and Richter, 1994b; Fackelmayer *et al.,* 1994; von Kries *et al.,* 1994). It will be interesting to learn if this protein is required for establishment of proper chromatin structure and for transport of nascent pre-mRNA molecules through the nucleus.

VI. CONCLUSIONS

A simple model is proposed for protein deposition on nascent pre-mRNA transcripts *in vivo.* Most hnRNP proteins are envisioned to bind nascent pre-mRNA immediately following passage of the transcription bubble, possibly within 100 nucleotides of the polymerase complex, and for certain proteins through interactions with chromatin. The generalized nucleoplasmic distribution exhibited by most hnRNP proteins suggests that these proteins are abundant and free to diffuse through the nucleus. The strength of the hnRNP protein binding site may dictate whether certain RNA sequences remain accessible for binding of other *trans*-acting factors or become folded into compact hnRNP complexes. Those sequences rendered accessible by high-affinity hnRNP protein binding sites then become associated with snRNP particles and SR proteins to commit the nascent pre-mRNA transcript to splicing. The snRNP particles and SR proteins may be recruited from concentrated nuclear storage sites. The rates of protein recruitment, transcription, and intron removal may dictate whether removal of a particular intron occurs cotranscriptionally or following release of the transcript from the DNA template. Thus, we envision a fairly simple extension of the messenger hypothesis to accommodate the steps involved in maturation of eukaryotic pre-mRNA transcripts.

ACKNOWLEDGMENTS

The authors are grateful to Drs. Ann Beyer, Allen Frankfater, Hans-Martin Jäck, and Adrian Krainer, and to members of the laboratory for critical comments on the manuscript, to Dr. J. G. Gall for helpful discussions, and to Ms. Bridget Borja and Mrs. Janet Flores for secretarial assistance. S.A.A. is the recipient of a Career Advancement Award from the National Science Foundation.

REFERENCES

Adam, S. A., Nakagawa, T., Swanson, M. S., Woodruff, T. K., and Dreyfuss, G. (1986). mRNA polyadenylate-binding protein: Gene isolation and sequencing and identification of a ribonucleoprotein consensus sequence. *Mol. Cell. Biol.* **6,** 2932–2943.

Albrecht, C., and Van Zyl, I. M. (1973). A comparative study of the protein components of ribonucleoprotein particles isolated from rat liver and hepatoma nuclei. *Exp. Cell Res.* **76,** 8–14.

Amero, S. A., Elgin, S. C. R., and Beyer, A. L. (1991). A unique zinc finger protein is associated preferentially with active ecdysone-responsive loci in *Drosophila. Genes Dev.* **5,** 188–200.

Amero, S. A., Raychaudhuri, G., Cass, C. L., van Venrooij, W. J., Habets, W. J., Krainer, A. R., and Beyer, A. L. (1992). Independent deposition of heterogeneous nuclear ribonucleoproteins and small nuclear ribonucleoprotein particles at sites of transcription. *Proc. Natl. Acad. Sci. U.S.A.* **89,** 8409–8413.

Amero, S. A., Matunis, M. J., Matunis, E. L., Hockensmith, J. W., Raychaudhuri, G., and Beyer, A. L. (1993). A unique ribonucleoprotein complex assembles preferentially on ecdysone-responsive sites in *Drosophila melanogaster. Mol. Cell. Biol.* **13,** 5323–5330.

Amrein, H., Gorman, M., and Nöthiger, R. (1988). The sex-determining gene *tra*-2 of *Drosophila* encodes a putative RNA binding protein. *Cell* (*Cambridge, Mass.*) **55,** 1025–1035.

Andersson, K., Björkroth, B., and Daneholt, B. (1980). The *in situ* structure of the active 75S RNA genes in Balbiani rings of *Chironomus tentans. Exp. Cell Res.* **130,** 313–326.

Angelier, N., and Lacroix, J. C. (1975). Complexes de transcription d'origines nucléolaire et chromosomique d'ovocytes de *Pleurodeles waltlii* et *P. poireti* (Amphibiens, Urodèles). *Chromosoma* **51,** 323–335.

Bachellerie, J., Puvion, E., and Zalta, J. (1975). Ultrastructural organization and biochemical characterization of chromatin-RNA-protein complexes isolated from mammalian cell nuclei. *Eur. J. Biochem.* **58,** 327–337.

Baker, B. S. (1989). Sex in flies: The splice of life. *Nature* (*London*) **340,** 521–524.

Barabino, S. M. L., Blencowe, B. J., Ryder, U., Sproat, B. S., and Lamond, A. I. (1990). Targeted snRNP depletion reveals an additional role for mammalian U1 snRNP in spliceosome assembly. *Cell* (*Cambridge, Mass.*) **63,** 293–302.

Barnett, S. F., Friedman, D. L., and LeStourgeon, W. M. (1989). The C proteins of HeLa 40S nuclear ribonucleoprotein particles exist as anisotropic tetramers of $(C1)_3$ C2. *Mol. Cell. Biol.* **9,** 492–498.

Barnett, S. F., Northington, S. J., and LeStourgeon, W. M. (1990). Isolation and *in vitro* assembly of nuclear ribonucleoprotein particles and purification of core particle proteins. *In* "Methods in Enzymology" (J. E. Dahlberg and J. N. Abelson, eds.), Vol. 181, pp. 293–307. Academic Press, San Diego, CA.

Barnett, S. F., Theiry, T. A., and LeStourgeon, W. M. (1991). The core proteins A2 and B1 exist as $(A2)_3$ B1 tetramers in 40S nuclear ribonucleoprotein particles. *Mol. Cell. Biol.* **11,** 864–871.

Baserga, S. J., and Steitz, J. A. (1993). The diverse world of small ribonucleoproteins. *In* "The RNA World: The Nature of Modern RNA Suggests a Prebiotic RNA World" (R. F. Gesteland and J. F. Atkins, eds.), pp. 359–381. Cold Spring Harbor Lab. Press, Cold Spring Harbor, NY.

Baurén, G., and Wieslander, L. (1994). Splicing of Balbiani Ring 1 gene pre-mRNA occurs simultaneously with transcription. *Cell (Cambridge, Mass.)* **76,** 183–192.

Beck, J. S. (1961). Variations in the morphological patterns of "autoimmune" nuclear fluorescence. *Lancet* **1,** 1203–1205.

Beermann, W., and Bahr, G. F. (1954). The submicroscopic structure of the Balbiani Ring. *Exp. Cell Res.* **6,** 195–201.

Bell, L. R., Maine, E. M., Schedl, P., and Cline, T. W. (1988). *Sex-lethal,* a *Drosophila* sex determination switch gene, exhibits sex-specific RNA splicing and sequence similarity to RNA binding proteins. *Cell (Cambridge, Mass.)* **55,** 1037–1046.

Benne, R. (1994). RNA editing in trypanosomes. *Eur. J. Biochem.* **221,** 9–23.

Bennett, M., Michaud, S., Kingston, J., and Reed, R. (1992a). Protein components specifically associated with prespliceosome and spliceosome complexes. *Genes Dev.* **6,** 1986–2000.

Bennett, M., Piñol-Roma, S., Staknis, D., Dreyfuss, G., and Reed, R. (1992b). Differential binding of heterogeneous nuclear ribonucleoproteins to mRNA precursors prior to spliceosome assembly *in vitro. Mol. Cell. Biol.* **12,** 3165–3175.

Berget, S. M., Moore, C., and Sharp, P. A. (1977). Spliced segments at the 5′ terminus of adenovirus 2 late mRNA. *Proc. Natl. Acad. Sci. U.S.A.* **74,** 3171–3175.

Berk, A. J., and Sharp, P. A. (1977). Sizing and mapping of early adenovirus mRNAs by gel electrophoresis of S1 endonuclease-digested hybrids. *Cell (Cambridge, Mass.)* **12,** 721–732.

Berk, A. J., and Sharp, P. A. (1978). Structure of the adenovirus 2 early mRNAs. *Cell (Cambridge, Mass.)* **14,** 695–711.

Berkhout, B., Silverman, R. H., and Jeang, K. T. (1989). Tat *trans*-activates the human immunodeficiency virus through a nascent RNA target. *Cell (Cambridge, Mass.)* **59,** 273–282.

Beyer, A. L., and Osheim, Y. N. (1988). Splice site selection, rate of splicing, and alternative splicing on nascent transcripts. *Genes Dev.* **2,** 754–765.

Beyer, A. L., and Osheim, Y. N. (1990). Ultrastructural analysis of the ribonucleoprotein substrate for pre-mRNA processing. *In* "The Eukaryotic Nucleus: Molecular Biochemical and Macromolecular Assemblies" (P. R. Strauss and S. H. Wilson, eds.), Vol. 2, pp. 1–21. Telford Press, Caldwell, NJ.

Beyer, A. L., and Osheim, Y. N. (1991). Visualization of RNA transcription and processing. *Semin. Cell Biol.* **2,** 131–140.

Beyer, A. L., Christensen, M. E., Walker, B. W., and LeStourgeon, W. M. (1977). Identification and characterization of the packaging proteins of core 40S hnRNP particles. *Cell (Cambridge, Mass.)* **11,** 127–138.

Beyer, A. L., Miller, O. L., Jr., and McKnight, S. L. (1980). Ribonucleoprotein structure in nascent hnRNA is nonrandom and sequence-dependent. *Cell (Cambridge, Mass.)* **20,** 75–84.

Bindereif, A., and Green, M. R. (1986). Ribonucleoprotein complex formation during pre-mRNA splicing *in vitro. Mol. Cell. Biol.* **6,** 2582–2592.

Bindereif, A., and Green, M. R. (1987). An ordered pathway of snRNP binding during mammalian pre-mRNA splicing complex assembly. *EMBO J.* **6,** 2415–2424.

Black, D. L. (1992). Activation of c-*src* neuron-specific splicing by an unusual RNA element *in vivo* and *in vitro. Cell (Cambridge, Mass.)* **69,** 795–807.

Black, D. L., Chabot, B., and Steitz, J. A. (1985). U2 as well as U1 small nuclear ribonucleoproteins are involved in premessenger RNA splicing. *Cell (Cambridge, Mass.)* **42,** 737–750.

Boggs, R. T., Gregor, P., Idriss, S., Belote, J. M., and McKeown, M. (1987). Regulation of sexual differentiation in *D. melanogaster* via alternative splicing of RNA from the *transformer* gene. *Cell (Cambridge, Mass.)* **50,** 739–747.

Breathnach, R., Mandel, J. L., and Chambon, P. (1977). Ovalbumin gene is split in chicken DNA. *Nature (London)* **270,** 314–319.

Brenner, S., Jacob, F., and Meselson, M. (1961). An unstable intermediate carrying information from genes to ribosomes for protein synthesis. *Nature (London)* **190,** 576–581.

Brody, E., and Abelson, J. (1985). The "spliceosome": Yeast pre-messenger RNA associates with a 40S complex in a splicing-dependent reaction. *Science* **228,** 963–967.

Burd, C. G., and Dreyfuss, G. (1994a). Conserved structures and diversity of functions of RNA-binding proteins. *Science* **265,** 615–621.

Burd, C. G., and Dreyfuss, G. (1994b). RNA binding specificity of hnRNP A1: Significance of hnRNP A1 high-affinity binding sites in pre-mRNA splicing. *EMBO J.* **13,** 1197–1204.

Burd, C. G., Swanson, M. S., Görlach, M., and Dreyfuss, G. (1989). Primary structures of the heterogeneous nuclear ribonucleoprotein A2, B1, and C2 proteins: A diversity of RNA binding proteins is generated by small peptide inserts. *Proc. Natl. Acad. Sci. U.S.A.* **86,** 9788–9792.

Burtis, K. C., and Baker, B. S. (1989). *Drosophila doublesex* gene controls somatic sexual differentiation by producing alternatively spliced mRNAs encoding related sex-specific polypeptides. *Cell (Cambridge, Mass.)* **56,** 997–1010.

Busby, S., and Bakken, A. (1979). A quantitative electron microscopic analysis of transcription in sea urchin embryos. *Chromosoma* **71,** 249–262.

Buvoli, M., Cobianchi, F., Biamonti, G., and Riva, S. (1990). Recombinant hnRNP protein A1 and its N-terminal domain show preferential affinity for oligodeoxynucleotides homologous to intron/exon acceptor sites. *Nucleic Acids Res.* **18,** 6595–6600.

Buvoli, M., Cobianchi, F., and Riva, S. (1992). Interaction of hnRNP A1 with snRNPs and pre-mRNAs: Evidence for a possible role of A1 RNA annealing activity in the first steps of spliceosome assembly. *Nucleic Acids Res.* **20,** 5017–5025.

Cáceres, J. F., Stamm, S., Helfman, D. M., and Krainer, A. R. (1994). Regulation of alternative splicing *in vivo* by overexpression of antagonistic splicing factors. *Science* **265,** 1706–1709.

Callan, H. G. (1987). Lampbrush chromosomes as seen in historical perspective. *In* "Structure and Function of Eukaryotic Chromosomes" (W. Hennig, ed.), pp. 5–26. Springer-Verlag, Berlin.

Callan, H. G., and Gall, J. G. (1991). Association of RNA with the B and C snurposomes of *Xenopus* oocyte nuclei. *Chromosoma* **101,** 69–82.

Callan, H. G., and Macgregor, H. C. (1958). Action of deoxyribonuclease on lampbrush chromosomes. *Nature (London)* **181,** 1479–1480.

Calnan, B. J., Biancalana, S., Hudson, D., and Frankel, A. D. (1991). Analysis of arginine-rich peptides from the HIV Tat protein reveals unusual features of RNA-protein recognition. *Genes Dev.* **5,** 201–210.

Carmo-Fonseca, M., Pepperkok, R., Sproat, B. S., Ansorge, W., Swanson, M. S., and Lamond, A. I. (1991). *In vivo* detection of snRNP-rich organelles in the nuclei of mammalian cells. *EMBO J.* **10,** 1863–1873.

Carmo-Fonseca, M., Pepperkok, R., Carvalho, M. T., and Lamond, A. I. (1992). Transcription-dependent colocalization of the U1, U2, U4/U6, and U5 snRNPs in coiled bodies. *J. Cell Biol.* **117,** 1–14.

Carter, K. C., Taneja, K. L., and Lawrence, J. B. (1991). Discrete nuclear domains of poly(A) RNA and their relationship to the functional organization of the nucleus. *J. Cell Biol.* **115,** 1191–1202.
Carter, K. C., Bowman, D., Carrington, W., Fogarty, K., McNeil, J. A., Fay, F. S., and Lawrence, J. B. (1993). A three-dimensional view of precursor messenger RNA metabolism within the mammalian nucleus. *Science* **259,** 1330–1335.
Case, S. T., and Daneholt, B. (1978). The size of the transcription unit in Balbiani Ring 2 of *Chironomus tentans* as derived from analysis of the primary transcript and 75 S RNA. *J. Mol. Biol.* **124,** 223–241.
Chabot, B., Black, D. L., LeMaster, D. M., and Steitz, J. A. (1985). The 3′ splice site of pre-messenger RNA is recognized by a small nuclear ribonucleoprotein. *Science* **230,** 1344–1349.
Champlin, D. T., and Lis, J. T. (1994). Distribution of B52 within a chromosomal locus depends on the level of transcription. *Mol. Biol. Cell* **5,** 71–79.
Champlin, D. T., Frasch, M., Saumweber, H., and Lis, J. T. (1991). Characterization of a *Drosophila* protein associated with boundaries of transcriptionally active chromatin. *Genes Dev.* **5,** 1611–1621.
Cheng, S., and Abelson, J. (1987). Spliceosome assembly in yeast. *Genes Dev.* **1,** 1014–1027.
Choi, Y. D., and Dreyfuss, G. (1984a). Isolation of the heterogeneous nuclear RNA-ribonucleoprotein complex (hnRNP): A unique supramolecular assembly. *Proc. Natl. Acad. Sci. U.S.A.* **81,** 7471–7475.
Choi, Y. D., and Dreyfuss, G. (1984b). Monoclonal antibody characterization of the C proteins of heterogeneous nuclear ribonucleoprotein complexes in vertebrate cells. *J. Cell Biol.* **99,** 1997–2004.
Choi, Y. D., Grabowski, P. J., Sharp, P. A., and Dreyfuss, G. (1986). Heterogeneous nuclear ribonucleoproteins: Role in RNA splicing. *Science* **231,** 1534–1539.
Chou, T., Zachar, Z., and Bingham, P. M. (1987). Developmental expression of a regulatory gene is programmed at the level of splicing. *EMBO J.* **16,** 4095–4104.
Chow, L. T., Gelinas, R. E., Broker, T. R., and Roberts, R. J. (1977). An amazing sequence arrangement at the 5′ ends of adenovirus 2 messenger RNA. *Cell* (*Cambridge, Mass.*) **12,** 1–8.
Chung, S. Y., and Wooley, J. (1986). Set of novel, conserved proteins fold pre-messenger RNA into ribonucleosomes. *Proteins* **1,** 195–210.
Conway, G., Wooley, J., Bibring, T., and LeStourgeon, W. M. (1988). Ribonucleoproteins package 700 nucleotides of pre-mRNA into a repeating array of regular particles. *Mol. Cell. Biol.* **8,** 2884–2895.
Craft, J. (1992). Antibodies to snRNPs in systemic lupus erythematosus. *In* "Antinuclear Antibodies" (D. S. Pisetsky, ed.), pp. 311–335. Saunders, Philadelphia.
Dangli, A., Grond, C., Kloetzel, P., and Bautz, E. K. F. (1983). Heat-shock puff 93D from *Drosophila melanogaster:* Accumulation of a RNP-specific antigen associated with giant particles of possible storage function. *EMBO J.* **2,** 1747–1751.
Datar, K. V., Dreyfuss, G., and Swanson, M. S. (1993). The human hnRNP M proteins: Identification of a methionine/arginine-rich repeat motif in ribonucleoproteins. *Nucleic Acids Res.* **21,** 439–446.
Derksen, J., Berendes, H. D., and Willart, E. (1973). Production and release of a locus-specific ribonucleoprotein product in polytene nuclei of *Drosophila hydei. J. Cell Biol.* **59,** 661–668.

Derksen, J. W. M. (1976). Ribonuclear protein formation at locus *2-48BC* in *Drosophila hydei. Nature* (*London*) **263,** 438–439.
Dreyfuss, G., Choi, Y. D., and Adam, S. A. (1984). Characterization of heterogeneous nuclear RNA-protein complexes *in vivo* with monoclonal antibodies. *Mol. Cell. Biol.* **4,** 1104–1114.
Dreyfuss, G., Swanson, M. S., and Piñol-Roma, S. (1990). The composition, structure, and organization of proteins in heterogeneous nuclear ribonucleoprotein complexes. *In* "The Eukaryotic Nucleus: Molecular Biochemistry and Macromolecular Assemblies" (P. R. Strauss and S. H. Wilson, eds.), Vol. 2, pp. 503–517. Telford Press, Caldwell, NJ.
Dreyfuss, G., Matunis, M. J., Piñol-Roma, S., and Burd, C. G. (1993). HnRNP proteins and the biogenesis of mRNA. *Annu. Rev. Biochem.* **62,** 289–321.
Economidis, J. V., and Pederson, T. (1983). Structure of nuclear ribonucleoprotein: Heterogeneous nuclear RNA is complexed with a major sextet of proteins *in vivo. Proc. Natl. Acad. Sci. U.S.A.* **80,** 1599–1602.
Edery, I., and Sonenberg, N. (1985). Cap-dependent RNA splicing in a HeLa nuclear extract. *Proc. Natl. Acad. Sci. U.S.A.* **82,** 7590–7594.
Edmonds, M., and Caramela, M. G. (1969). The isolation and characterization of adenosine monophosphate-rich polynucleotides synthesized by Ehrlich ascites cells. *J. Biol. Chem.* **224,** 1314–1324.
Edmonds, M., Vaughan, M. H., Jr., and Nakazato, H. (1971). Polyadenylic acid sequences in the heterogeneous nuclear RNA and rapidly-labeled polyribosomal RNA of HeLa cells: Possible evidence for a precursor relationship. *Proc. Natl. Acad. Sci. U.S.A.* **68,** 1336–1340.
Emanuel, P. A., and Gilmour, D. S. (1993). Transcription factor TFIID recognizes DNA sequences downstream of the TATA element in the *Hsp*70 heat shock gene. *Proc. Natl. Acad. Sci. U.S.A.* **90,** 8449–8453.
Eperon, I. C., Ireland, D. C., Smith, R. A., Mayeda, A., and Krainer, A. R. (1993). Pathways for selection of 5′ splice sites by U1 snRNPs and SF2/ASF. *EMBO J.* **12,** 3607–3617.
Eperon, L. P., Graham, I. R., Griffiths, A. D., and Eperon, I. C. (1988). Effects of RNA secondary structure on alternative splicing of pre-mRNA: Is folding limited to a region behind the transcribing RNA polymerase? *Cell* (*Cambridge, Mass.*) **54,** 393–401.
Fackelmayer, F. O., and Richter, A. (1994a). HnRNP-U/SAF-A is encoded by two differentially polyadenylated mRNAs in human cells. *Biochim. Biophys. Acta* **1217,** 232–234.
Fackelmayer, F. O., and Richter, A. (1994b). Purification of two isoforms of hnRNP-U and characterization of their nucleic acid binding activity. *Biochemistry* **33,** 10416–10422.
Fackelmayer, F. O., Dahm, K., Renz, A., Ramsperger, U., and Richter, A. (1994). Nucleic-acid-binding properties of hnRNP-U/SAF-A, a nuclear-matrix protein which binds DNA and RNA *in vivo* and *in vitro. Eur. J. Biochem.* **221,** 749–757.
Faiferman, I., Hamilton, M. G., and Pogo, A. O. (1971). Nucleoplasmic ribonucleoprotein particles of rat liver. II. Physical properties and action of dissociating agents. *Biochim. Biophys. Acta* **232,** 685–695.
Fakan, S., and Bernhard, W. (1971). Localisation of rapidly and slowly labelled nuclear RNA as visualized by high resolution autoradiography. *Exp. Cell Res.* **67,** 129–141.
Fakan, S., and Bernhard, W. (1973). Nuclear labelling after prolonged ^{3}H-uridine incorporation as visualized by high resolution autoradiography. *Exp. Cell Res.* **79,** 431–444.
Fakan, S., and Nobis, P. (1978). Ultrastructural localization of transcription sites and of RNA distribution during the cell cycle of synchronized CHO cells. *Exp. Cell Res.* **113,** 327–337.

Fakan, S., and Puvion, E. (1980). The ultrastructural visualization of nucleolar and extra-nucleolar RNA synthesis and distribution. *Int. Rev. Cytol.* **65,** 255–299.

Fakan, S., Puvion, E., and Spohr, G. (1976). Localization and characterization of newly synthesized nuclear RNA in isolated rat hepatocytes. *Exp. Cell Res.* **99,** 155–164.

Fakan, S., Leser, G., and Martin, T. E. (1984). Ultrastructural distribution of nuclear ribonucleoproteins as visualized by immunocytochemistry on thin sections. *J. Cell Biol.* **98,** 358–363.

Fakan, S., Leser, G., and Martin, T. E. (1986). Immunoelectron microscope visualization of nuclear ribonucleoprotein antigens within spread transcription complexes. *J. Cell Biol.* **103,** 1153–1157.

Feinberg, M. B., Baltimore, D., and Frankel, A. D. (1991). The role of Tat in the human immunodeficiency virus life cycle indicates a primary effect on transcriptional elongation. *Proc. Natl. Acad. Sci. U.S.A.* **88,** 4045–4049.

Foe, V. E., Wilkinson, L. E., and Laird, C. D. (1976). Comparative organization of active transcription units in *Oncopeltus fasciatus. Cell (Cambridge, Mass.)* **9,** 131–146.

Frendewey, D., and Keller, W. (1985). Stepwise assembly of a pre-mRNA splicing complex requires U-snRNPs and specific intron sequences. *Cell (Cambridge, Mass.)* **42,** 355–367.

Fu, X. (1993). Specific commitment of different pre-mRNAs to splicing by single SR proteins. *Nature (London)* **365,** 82–85.

Fu, X., and Maniatis, T. (1990). Factor required for mammalian spliceosome assembly is localized to discrete regions in the nucleus. *Nature (London)* **343,** 437–441.

Fu, X., and Maniatis, T. (1992). The 35-kDa mammalian splicing factor SC35 mediates specifc interactions between U1 and U2 small nuclear ribonucleoprotein particles at the 3′ splice site. *Proc. Natl. Acad. Sci. U.S.A.* **89,** 1725–1729.

Fu, X., Mayeda, A., Maniatis, T., and Krainer, A. R. (1992). General splicing factors SF2 and SC35 have equivalent activities *in vitro,* and both affect alternative 5′ and 3′ splice site selection. *Proc. Natl. Acad. Sci. U.S.A.* **89,** 11224–11228.

Gall, J. G. (1954). Lampbrush chromosomes from oocyte nuclei of the newt. *J. Morphol.* **94,** 283–339.

Gall, J. G. (1956). On the submicroscopic structure of chromosomes. *Brookhaven Symp. Biol.* **8,** 17–32.

Gall, J. G. (1991). Spliceosomes and snurposomes. *Science* **252,** 1499–1500.

Gall, J. G., and Callan, H. G. (1962). H^3 uridine incorporation in lampbrush chromosomes. *Proc. Natl. Acad. Sci. U.S.A.* **48,** 562–570.

Gall, J. G., and Callan, H. G. (1989). The sphere organelle contains small nuclear ribonucleoproteins. *Proc. Natl. Acad. Sci. U.S.A.* **86,** 6635–6639.

Gallego, M. E., Balvay, L., and Brody, E. (1992). *Cis*-acting sequences involved in exon selection in the chicken β-tropomyosin gene. *Mol. Cell. Biol.* **12,** 5415–5425.

Gallinaro-Matringe, H., and Jacob, M. (1973). Nuclear particles from rat brain: Complexity of the major proteins and their phosphorylation *in vivo. FEBS Lett.* **36,** 105–108.

García-Blanco, M. A., Jamison, S. F., and Sharp, P. A. (1989). Identification and purification of a 62,000-dalton protein that binds specifically to the polypyrimidine tract of introns. *Genes Dev.* **3,** 1874–1886.

Ge, H., and Manley, J. L. (1990). A protein factor, ASF, controls cell-specific alternative splicing of SV40 early pre-mRNA *in vitro. Cell (Cambridge, Mass.)* **62,** 25–34.

Gelinas, R. E., and Roberts, R. J. (1977). One predominant 5′-undecanucleotide in adenovirus 2 late messenger RNAs. *Cell (Cambridge, Mass.)* **11,** 533–544.

Georgiev, G. P., and Samarina, O. P. (1971). D-RNA containing ribonucleoprotein particles. *In* "Advances in Cell Biology" (L. Goldstein, D. M. Prescott, and E. McConkey, eds.), pp. 47–110. Appleton-Century-Crofts, New York.

Gerke, V., and Steitz, J. A. (1986). A protein associated with small nuclear ribonucleoprotein particles recognizes the 3′ splice site of premessenger RNA. *Cell (Cambridge, Mass.)* **47,** 973–984.

Ghetti, A., Piñol-Roma, S., Michael, W. M., Morandi, C., and Dreyfuss, G. (1992). HnRNP I, the polypyrimidine tract-binding protein: Distinct nuclear localization and association with hnRNAs. *Nucleic Acids Res.* **20,** 3671–3678.

Giardina, C., Pérez-Riba, M., and Lis, J. T. (1992). Promoter melting and TFIID complexes on *Drosophila* genes *in vivo. Genes Dev.* **6,** 2190–2200.

Gil, A., Sharp, P. A., Jamison, S. F., and García-Blanco, M. A. (1991). Characterization of cDNAs encoding the polypyrimidine tract-binding protein. *Genes Dev.* **5,** 1224–1236.

Gilmour, D. S., and Lis, J. T. (1986). RNA polymerase II interacts with the promoter region of the noninduced *hsp*70 gene in *Drosophila melanogaster* cells. *Mol. Cell. Biol.* **6,** 3984–3989.

Görlach, M., Wittekind, M., Beckman, R. A., Mueller, L., and Dreyfuss, G. (1992). Interaction of the RNA-binding domain of the hnRNP C proteins with RNA. *EMBO J.* **11,** 3289–3295.

Grabowski, P. J., Seiler, S. R., and Sharp, P. A. (1985). A multicomponent complex is involved in the splicing of messenger RNA precursors. *Cell (Cambridge, Mass.)* **42,** 345–353.

Green, M. R. (1991). Biochemical mechanisms of constitutive and regulated pre-mRNA splicing. *Annu. Rev. Cell Biol.* **7,** 559–599.

Green, M. R., Maniatis, T., and Melton, D. A. (1983). Human β-globin pre-mRNA synthesized *in vitro* is accurately spliced in *Xenopus* oocyte nuclei. *Cell (Cambridge, Mass.)* **32,** 681–694.

Greenleaf, A. L. (1993). Positive patches and negative noodles: Linking RNA processing to transcription. *Trends Prochem. Sci.* **18,** 117–119.

Gros, F., Hiatt, H., Gilbert, W., Kurland, C. G., Risebrough, R. W., and Watson, J. D. (1961). Unstable ribonucleic acid revealed by pulse labelling of *Escherichia coli. Nature (London)* **190,** 581–585.

Guo, W., and Helfman, D. M. (1993). *Cis*-elements involved in alternative splicing in the rat beta-tropomyosin gene: The 3′-splice site of the skeletal muscle exon 7 is the major site of blockage in nonmuscle cells. *Nucleic Acids Res.* **21,** 4762–4768.

Guthrie, C. (1991). Messenger RNA splicing in yeast: Clues to why the spliceosome is a ribonucleoprotein. *Science* **253,** 157–163.

Habets, W. J., Hoet, M. H., de Jong, B. A. W., van der Kemp, A., and van Venrooij, W. J. (1989). Mapping of B cell epitopes on small nuclear ribonucleoproteins that react with human autoantibodies as well as with experimentally-induced mouse monoclonal antibodies. *J. Immunol.* **143,** 2560–2566.

Hamilton, B. J., Nagy, E., Malter, J. S., Arrick, B. A., and Rigby, W. F. C. (1993). Association of heterogeneous nuclear ribonucleoprotein A1 and C proteins with reiterated AUUUA sequences. *J. Biol. Chem.* **268,** 8881–8887.

Hamm, J., Dathan, N. A., and Mattaj, I. W. (1989). Functional analysis of mutant *Xenopus* U2 snRNAs. *Cell (Cambridge, Mass.)* **59,** 159–169.

Harper, J. E., and Manley, J. L. (1991). A novel protein factor is required for use of distal alternative 5′ splice sites *in vitro. Mol. Cell. Biol.* **11,** 5945–5953.

Hashimoto, C., and Steitz, J. A. (1984). U4 and U6 RNAs coexist in a single small nuclear ribonucleoprotein particle. *Nucleic Acids Res.* **12,** 3283–3293.

Hedley, M. L., and Maniatis, T. (1991). Sex-specific splicing and polyadenylation of *dsx* pre-mRNA requires a sequence that binds specifically to *tra*-2 protein *in vitro. Cell (Cambridge, Mass.)* **65,** 579–586.

Hoffman, B. E., and Grabowski, P. J. (1992). U1 snRNP targets an essential splicing factor, U2AF65, to the 3′ splice site by a network of interactions spanning the exon. *Genes Dev.* **6,** 2554–2568.

Höög, C., Engberg, C., and Wieslander, L. (1986). A BR 1 gene in *Chironomus tentans* has a composite structure: A large repetitive core block is separated from a short unrelated 3′-terminal domain by a small intron. *Nucleic Acids Res.* **14,** 703–719.

Höög, C., Daneholt, B., and Wieslander, L. (1988). Terminal repeats in long repeat arrays are likely to reflect the early evolution of Balbiani Ring genes. *J. Mol. Biol.* **200,** 655–664.

Horowitz, D. S., and Krainer, A. R. (1994). Mechanisms for selecting 5′ splice sites in mammalian pre-mRNA splicing. *Trends Genet.* **10,** 100–106.

Hoshijima, K., Inoue, K., Higuchi, I., Sakamoto, H., and Shimura, Y. (1991). Control of *doublesex* alternative splicing by *transformer* and *transformer-2* in *Drosophila. Science* **252,** 833–836.

Hovemann, B. T., Dessen, E., Mechler, H., and Mack, E. (1991). *Drosophila* snRNP associated protein P11 which specifically binds to heat shock puff 93D reveals strong homology with hnRNP core protein A1. *Nucleic Acids Res.* **19,** 4909–4914.

Huang, M., Rech, J. E., Northington, S. J., Flicker, P. F., Mayeda, A., Krainer, A. R., and LeStourgeon, W. M. (1994). The C-protein tetramer binds 230 to 240 nucleotides of pre-mRNA and nucleates the assembly of 40S heterogeneous nuclear ribonucleoprotein particles. *Mol. Cell. Biol.* **14,** 518–533.

Huang, S., and Spector, D. L. (1992). U1 and U2 small nuclear RNAs are present in nuclear speckles. *Proc. Natl. Acad. Sci. U.S.A.* **89,** 305–308.

Inoue, K., Hoshijima, K., Sakamoto, H., and Shimura, Y. (1990). Binding of the *Drosophila Sex-lethal* gene product to the alternative splice site of *transformer* primary transcript. *Nature (London)* **344,** 461–463.

Inoue, K., Hoshijima, K., Higuchi, I., Sakamoto, H., and Shimura, Y. (1992). Binding of the *Drosophila transformer* and *transformer-2* proteins to the regulatory elements of *doublesex* primary transcript for sex-specific RNA processing. *Proc. Natl. Acad. Sci. U.S.A.* **89,** 8092–8096.

Ishikawa, F., Matunis, M. J., Dreyfuss, G., and Cech, T. R. (1993). Nuclear proteins that bind the pre-mRNA 3′ splice site sequence r(UUAG/G) and the human telomeric DNA sequence $d(TTAGGG)_n$. *Mol. Cell. Biol.* **13,** 4301–4310.

Jacob, F., and Monod, J. (1961). Genetic regulatory mechanisms in the synthesis of proteins. *J. Mol. Biol.* **3,** 318–356.

Jamison, S. F., Crow, A., and Carcía-Blanco, M. A. (1992). The spliceosome assembly pathway in mammalian extracts. *Mol. Cell. Biol.* **12,** 4279–4287.

Jantsch, M. F., and Gall, J. G. (1992). Assembly and localization of the U1-specific snRNP C protein in the amphibian oocyte. *J. Cell Biol.* **119,** 1037–1046.

Jeffreys, A. J., and Flavell, R. A. (1977). The rabbit β-globin gene contains a large insert in the coding sequence. *Cell (Cambridge, Mass.)* **12,** 1097–1108.

Jiménez-García, L. F., and Spector, D. L. (1993). *In vivo* evidence that transcription and splicing are coordinated by a recruiting mechanism. *Cell (Cambridge, Mass.)* **73,** 47–59.

Jones, K. A. (1989). HIV trans-activation and transcription control mechanisms. *New Biol.* **1,** 127–135.

Kafatos, F. C. (1972). The cocoonase zymogen cells of silk moths: A model of terminal cell differentiation for specific protein synthesis. *Curr. Top. Dev. Biol.* **7,** 125–191.

Kanaar, R., Roche, S. E., Beall, E. L., Green, M. R., and Rio, D. C. (1993). The conserved pre-mRNA splicing factor U2AF from *Drosophila:* Requirement for viability. *Science* **262,** 569–573.

Kao, S. Y., Calman, A. F., Luciw, P. A., and Peterlin, B. M. (1987). Anti-termination of transcription within the long terminal repeat of HIV-1 by *tat* gene product. *Nature* (*London*) **330,** 489–493.

Kashanchi, F., Piras, G., Radonovich, M. F., Duvall, J. F., Fattaey, A., Chiang, C. M., Roeder, R. G., and Brady, J. N. (1994). Direct interaction of human TFIID with the HIV-1 transactivator *Tat. Nature* (*London*) **367,** 295–299.

Kato, H., Sumimoto, H., Pognonec, P., Chen, C. H., Rosen, C. A., and Roeder, R. G. (1992). HIV-1 *Tat* acts as a processivity factor *in vitro* in conjunction with cellular elongation factors. *Genes Dev.* **6,** 655–666.

Kedes, D. H., and Steitz, J. A. (1988). Correct *in vivo* splicing of the mouse immunoglobulin kappa light-chain pre-mRNA is dependent on 5′ splice-site position even in the absence of transcription. *Genes Dev.* **2,** 1448–1459.

Kelley, R. L. (1993). Initial organization of the *Drosophila* dorsoventral axis depends on an RNA-binding protein encoded by the *squid* gene. *Genes Dev.* **7,** 948–960.

Kiledjian, M., and Dreyfuss, G. (1992). Primary structure and binding activity of the human hnRNP U protein: Binding RNA through RGG box. *EMBO J.* **11,** 2655–2664.

Kim, Y. J., Zuo, P., Manley, J. L., and Baker, B. S. (1992). The *Drosophila* RNA-binding protein RBP1 is localized to transcriptionally active sites of chromosomes and shows a functional similarity to human splicing factor ASF/SF2. *Genes Dev.* **6,** 2569–2579.

Kinniburgh, A. J., and Martin, T. E. (1976). Detection of mRNA sequences in nuclear 30S ribonucleoprotein subcomplexes. *Proc. Natl. Acad. Sci. U.S.A.* **73,** 2725–2729.

Klessig, D. F. (1977). Two adenovirus mRNAs have a common 5′ terminal leader sequence encoded at least 10 kb upstream from their main coding regions. *Cell* (*Cambridge, Mass.*) **12,** 9–21.

Kohtz, J. D., Jamison, S. F., Will, C. L., Zuo, P., Lührmann, R., García-Blanco, M. A., and Manley, J. L. (1994). Protein-protein interactions and 5′-splice-site recognition in mammalian mRNA precursors. *Nature* (*London*) **368,** 119–124.

Konarska, M. M., and Sharp, P. A. (1986). Electrophoretic separation of complexes involved in the splicing of precursors to mRNAs. *Cell* (*Cambridge, Mass.*) **46,** 845–855.

Konarska, M. M., and Sharp, P. A. (1987). Interactions between small nuclear ribonucleoprotein particles in formation of spliceosomes. *Cell* (*Cambridge, Mass.*) **49,** 763–774.

Konarska, M. M., Padgett, R. A., and Sharp, P. A. (1984). Recognition of cap structure in splicing *in vitro* of mRNA precursors. *Cell* (*Cambridge, Mass.*) **38,** 731–736.

Krainer, A. R., and Maniatis, T. (1985). Multiple factor including the small nuclear ribonucleoproteins U1 and U2 are necessary for pre-mRNA splicing *in vitro. Cell* (*Cambridge, Mass.*) **42,** 725–736.

Krainer, A. R., Conway, G. C., and Kozak, D. (1990a). The essential pre-mRNA splicing factor SF2 influences 5′ splice site selection by activating proximal sites. *Cell* (*Cambridge, Mass.*) **62,** 35–42.

Krainer, A. R., Conway, G. C., and Kozak, D. (1990b). Purification and characterization of pre-mRNA splicing factor SF2 from HeLa cells. *Genes Dev.* **4,** 1158–1171.

Krainer, A. R., Mayeda, A., Kozak, D., and Binns, G. (1991). Functional expression of cloned human splicing factor SF2: Homology to RNA-binding proteins, U1 70K, and *Drosophila* splicing regulators. *Cell (Cambridge, Mass.)* **66,** 383–394.

Krämer, A., Keller, W., Appel, B., and Lührmann, R. (1984). The 5′ terminus of the RNA moiety of U1 small nuclear ribonucleoprotein particles is required for the splicing of messenger RNA precursors. *Cell (Cambridge, Mass.)* **38,** 299–307.

Kuo, H. C., Nasim, F. H., and Grabowski, P. J. (1991). Control of alternative splicing by the differential binding of U1 small nuclear ribonucleoprotein particle. *Science* **251,** 1045–1050.

Lachmann, P. J., and Kunkel, H. G. (1961). Correlation of antinuclear antibodies and nuclear staining patterns. *Lancet* **2,** 436–437.

Lacroix, J. C., Azzouz, R., Boucher, D., Abbadie, C., Pyne, C. K., and Charlemagne, J. (1985). Monoclonal antibodies to lampbrush chromosome antigens of *Pleurodeles waltlii. Chromosoma* **92,** 69–80.

Laird, C. D., and Chooi, W. Y. (1976). Morphology of transcription units in *Drosophila melanogaster. Chromosoma* **58,** 193–218.

Laird, C. D., Wilkinson, L. E., Foe, V. E., and Chooi, W. Y. (1976). Analysis of chromatin-associated fiber arrays. *Chromosoma* **58,** 169–192.

Lamb, M. M., and Daneholt, B. (1979). Characterization of active transcription units in Balbiani Rings of *Chironomus tentans. Cell (Cambridge, Mass.)* **17,** 835–848.

Lamond, A. I. (1993). The spliceosome. *BioEssays* **15,** 595–603.

Lavigueur, A., La Branche, H. L., Kornblihtt, A. R., and Chabot, B. (1993). A splicing enhancer in the human fibronectin alternate ED1 exon interacts with SR proteins and stimulates U2 snRNP binding. *Genes Dev.* **7,** 2405–2417.

LeMaire, M. F., and Thummel, C. S. (1990). Splicing precedes polyadenylation during *Drosophila E74A* transcription. *Mol. Cell. Biol.* **10,** 6059–6063.

Lerner, E. A., Lerner, M. R., Janeway, C. A., Jr., and Steitz, J. A. (1981). Monoclonal antibodies to nucleic acid-containing cellular constituents: Probes for molecular biology and autoimmune disease. *Prob. Natl. Acad. Sci. U.S.A.* **78,** 2737–2741.

Lerner, M. R., and Steitz, J. A. (1979). Antibodies to small nuclear RNAs complexed with proteins are produced by patients with systemic lupus erythematosus. *Proc. Natl. Acad. Sci. U.S.A.* **76,** 5495–5499.

Leser, G. P., and Martin, T. E. (1987). Changes in heterogeneous nuclear RNP core polypeptide complements during the cell cycle. *J. Cell Biol.* **105,** 2083–2094.

Leser, G. P., Fakan, S., and Martin, T. E. (1989). Ultrastructural distribution of ribonucleoprotein complexes during mitosis: snRNP antigens are contained in mitotic granule clusters. *Eur. J. Cell Biol.* **50,** 376–389.

Lis, J., and Wu, C. (1993). Protein traffic on the heat shock promoter: Parking, stalling, and trucking along. *Cell (Cambridge, Mass.)* **74,** 1–4.

Lönnroth, A., Alexciev, K., Mehlin, H., Wurtz, T., Skoglund, U., and Daneholt, B. (1992). Demonstration of a 7-nm RNP fiber as the basic structural element in a premessenger RNP particle. *Exp. Cell Res.* **199,** 292–296.

Lothstein, L., Arenstorf, H. P., Chung, S., Walker, B. W., Wooley, J. C., and LeStourgeon, W. M. (1985). General organization of protein in HeLa 40S nuclear ribonucleoprotein particles. *J. Cell Biol.* **100,** 1570–1581.

Lührmann, R., Kastner, B., and Bach, M. (1990). Structure of spliceosomal snRNPs and their role in pre-mRNA splicing. *Biochim. Biophys. Acta* **1087,** 265–292.

Lukanidin, E. M., Zalmanzon, E. S., Komaromi, L., Samarina, O. P., and Georgiev, G. P. (1972a). Structure and function of informofers. *Nature (London), New Biol.* **238,** 193–197.

Lukanidin, E. M., Olsnes, S., and Phil, A. (1972b). Antigenic difference between informofers and protein bound to polyribosomal mRNA from rat liver. *Nature (London), New Biol.* **240,** 90–92.

Macgregor, H. C., and Callan, H. G. (1962). The actions of enzymes on lampbrush chromosomes. *J. Microsc. Sci.* **103,** 173–203.

Madhani, H. D., and Guthrie, C. (1992). A novel base-pairing interaction between U2 and U6 snRNAs suggests a mechanism for the catalytic activation of the spliceosome. *Cell (Cambridge, Mass.)* **71,** 803–817.

Malcolm, D. B., and Sommerville, J. (1977). The structure of nuclear ribonucleoprotein of amphibian oocytes. *J. Cell Sci.* **24,** 143–165.

Maniatis, T., and Reed, R. (1987). The role of small nuclear ribonucleoprotein particles in pre-mRNA splicing. *Nature (London)* **325,** 673–678.

Manley, J. L., and Proudfoot, N. J. (1994). RNA 3′ ends: Formation and function-meeting review. *Genes Dev.* **8,** 259–264.

Marciniak, R. A., and Sharp, P. A. (1991). HIV-1 *Tat* protein promotes formation of more-processive elongation complexes. *EMBO J.* **10,** 4189–4196.

Martin, T. E., Barghusen, S. C., Leser, G. P., and Spear, P. G. (1987). Redistribution of nuclear ribonucleoprotein antigens during herpes simplex virus infection. *J. Cell Biol.* **105,** 2069–2082.

Matera, A. G., and Ward, D. C. (1993). Nucleoplasmic organization of small nuclear ribonucleoproteins in cultured human cells. *J. Cell Biol.* **121,** 715–727.

Mattox, W., Ryner, L., and Baker, B. S. (1992). Autoregulation and multifunctionality among *trans*-acting factors that regulate alternative pre-mRNA processing. *J. Biol. Chem.* **267,** 19023–19026.

Matunis, M. J., Matunis, E. L., and Dreyfuss, G. (1992a). Isolation of hnRNP complexes from *Drosophila melanogaster. J. Cell Biol.* **116,** 245–255.

Matunis, E. L., Matunis, M. J., and Dreyfuss, G. (1992b). Characterization of the major hnRNP proteins from *Drosophila melanogaster. J. Cell Biol.* **116,** 257–269.

Matunis, M. J., Michael, W. M., and Dreyfuss, G. (1992c). Characterization and primary structure of the poly(C)-binding heterogeneous nuclear ribonucleoprotein complex K protein. *Mol. Cell. Biol.* **12,** 164–171.

Matunis, E. L., Matunis, M. J., and Dreyfuss, G. (1993). Association of individual hnRNP proteins and snRNPs with nascent transcripts. *J. Cell Biol.* **121,** 219–228.

Matunis, E. L., Kelley, R., and Dreyfuss, G. (1994a). Essential role for a heterogeneous nuclear ribonucleoprotein (hnRNP) in oogenesis: hrp40 is absent from the germ line in the dorsoventral mutant squid[1]. *Proc. Natl. Acad. Sci. U.S.A.* **91,** 2781–2784.

Matunis, M. J., Xing, J., and Dreyfuss, G. (1994b). The hnRNP F protein: Unique primary structure, nucleic acid-binding properties, and subcellular localization. *Nucleic Acids Res.* **22,** 1059–1067.

Maundrell, K. (1975). Proteins of the newt oocyte nucleus: Analysis of the nonhistone proteins from lampbrush chromosomes, nucleoli and nuclear sap. *J. Cell Sci.* **17,** 579–588.

Mayeda, A., and Krainer, A. R. (1992). Regulation of alternative pre-mRNA splicing by hnRNP A1 and splicing factor SF2. *Cell (Cambridge, Mass.)* **68,** 365–375.

Mayeda, A., Zahler, A. M., Krainer, A. R., and Roth, M. B. (1992). Two members of a conserved family of nuclear phosphoproteins are involved in pre-mRNA splicing. *Proc. Natl. Acad. Sci. U.S.A.* **89,** 1301–1304.

Mayeda, A., Helfman, D. M., and Krainer, A. R. (1993). Modulation of exon skipping and inclusion by heterogeneous nuclear ribonucleoprotein A1 and pre-mRNA splicing factor SF2/ASF. *Mol. Cell. Biol.* **13,** 2993–3001.

Mayrand, S., and Pederson, T. (1981). Nuclear ribonucleoprotein particles probed in living cells. *Proc. Natl. Acad. Sci. U.S.A.* **78,** 2208–2212.

Mayrand, S. H., and Pederson, T. (1990). Crosslinking of hnRNP proteins to pre-mRNA requires U1 and U2 snRNPs. *Nucleic Acids Res.* **18,** 3307–3318.

Mayrand, S., Setyono, B., Greenberg, J. R., and Pederson, T. (1981). Structure of nuclear ribonucleoprotein: Identification of proteins in contact with poly $(A)^+$ heterogeneous nuclear RNA in living HeLa cells. *J. Cell Biol.* **90,** 380–384.

Mayrand, S. H., Dwen, P., and Pederson, T. (1993). Serine/threonine phosphorylation regulates binding of C hnRNP proteins to pre-mRNA. *Proc. Natl. Acad. Sci. U.S.A.* **90,** 7764–7768.

McKay, S. J., and Cooke, H. (1992). HnRNP A2/B1 binds specifically to single stranded vertebrate telomeric repeat $TTAAGGG_n$. *Nucleic Acids Res.* **20,** 6461–6464.

McKeown, M. (1993). The role of small nuclear RNAs in RNA splicing. *Curr. Opin. Cell Biol.* **5,** 448–454.

McKeown, M., Belote, J. M., and Baker, B. S. (1987). A molecular analysis of *transformer,* a gene in *Drosophila melanogaster* that controls female sexual differentiation. *Cell* (*Cambridge, Mass.*) **48,** 489–499.

McKnight, S. L., and Miller, O. L., Jr. (1976). Ultrastructural patterns of RNA synthesis during early embryogenesis of *Drosophila melanogaster. Cell* (*Cambridge, Mass.*) **8,** 305–319.

McKnight, S. L., and Miller, O. L., Jr. (1979). Post-replicative nonribosomal transcription units in *D. melanogaster* embryos. *Cell* (*Cambridge, Mass.*) **17,** 551–563.

Mehlin, H., Skoglund, U., and Daneholt, B. (1991). Transport of Balbiani Ring granules through nuclear pores in *Chironomus tentans. Exp. Cell Res.* **193,** 72–77.

Mehlin, H., Daneholt, B., and Skoglund, U. (1992). Translocation of a specific premessenger ribonucleoprotein particle through the nuclear pore: Studies with electron microscope tomography. *Cell* (*Cambridge, Mass.*) **69,** 605–613.

Merrill, B. M., and Williams, K. R. (1990). Structure/function relationships in hnRNP proteins. *In* "The Eukaryotic Nucleus: Molecular Biochemistry and Macromolecular Assemblies" (P. R. Strauss, and S. H. Wilson, eds.), Vol. 2, pp. 579–604. Telford Press, Caldwell, NJ.

Michaud, S., and Reed, R. (1991). An ATP-independent complex commits pre-mRNA to the mammalian spliceosome assembly pathway. *Genes Dev.* **5,** 2534–2546.

Michaud, S., and Reed, R. (1993). A functional association between the 5′ and 3′ splice sites is established in the earliest prespliceosome complex (E) in mammals. *Genes Dev.* **7,** 1008–1020.

Miller, O. L., Jr. (1965). Fine structure of lampbrush chromosomes. *Natl. Cancer Inst. Monogr.* **18,** 79–99.

Miller, O. L., Jr., and Bakken, A. H. (1972). Morphological studies of transcription. *Acta Endocrinol.* (*Copenhagen*) **168,** 155–177.

Mizumoto, K., and Kaziro, Y. (1987). Messenger RNA capping enzymes from eukaryotic cells. *Prog. Nucleic Acid Res. Mol. Biol.* **34,** 1–28.

Molloy, G. R., Sporn, M. B., Kelley, D. E., and Perry, R. P. (1972). Localization of polyadenylic acid sequences in messenger ribonucleic acid of mammalian cells. *Biochemistry* **11,** 3256–3260.

Monneron, A., and Bernhard, W. (1969). Fine structural organization of the interphase nucleus in some mammalian cells. *J. Ultrastruct. Res.* **27,** 266–288.

Moore, C. L., Chen, J., and Whoriskey, J. (1988). Two proteins crosslinked to RNA containing the adenovirus L3 poly(A) site require the AAUAAA sequence for binding. *EMBO J.* **7,** 3159–3169.

Moore, M. J., Query, C. C., and Sharp, P. A. (1993). Splicing of precursors to mRNAs by the spliceosome. *In* "The RNA World: The Nature of Modern RNA Suggests a Prebiotic RNA World" (R. F. Gesteland, and J. R. Atkins, eds.), pp. 303–357. Cold Spring Harbor Lab. Press, Cold Spring Harbor, NY.

Mott, M. R., and Callan, H. G. (1975). An electron-microscope study of the lampbrush chromosomes of the newt *Triturus cristatus. J. Cell Sci.* **17,** 241–261.

Mount, S. M., Pettersson, I., Hinterberger, M., Karmas, A., and Steitz, J. A. (1983). The U1 small nuclear RNA-protein complex selectively binds a 5′ splice site *in vitro. Cell (Cambridge, Mass.)* **33,** 509–518.

Mulligan, G. J., Guo, W., Wormsley, S., and Helfman, D. M. (1992). Polypyrimidine tract binding protein interacts with sequences involved in alternative splicing of β-tropomyosin pre-mRNA. *J. Biol. Chem.* **267,** 25480–25487.

Nagoshi, R. N., and Baker, B. S. (1990). Regulation of sex-specific RNA splicing at the *Drosophila doublesex* gene: *cis*-acting mutations in exon sequences alter sex-specific RNA splicing patterns. *Genes Dev.* **4,** 89–97.

Nelson, K. K., and Green, M. R. (1989). Mammalian U2 snRNP has a sequence-specific RNA-binding activity. *Genes Dev.* **3,** 1562–1571.

Newman, A. J., and Norman, C. (1992). U5 snRNA interacts with exon sequences at 5′ and 3′ splice sites. *Cell (Cambridge, Mass.)* **68,** 743–754.

Newmeyer, D. D. (1993). The nuclear pore complex and nucleocytoplasmic transport. *Curr. Opin. Cell Biol.* **5,** 395–407.

Niessing, J., and Sekeris, C. E. (1971). Further studies on nuclear ribonucleoprotein particles containing DNA-like RNA from rat liver. *Biochim. Biophys. Acta* **247,** 391–403.

Nyman, U., Hallman, H., Hadlaczky, G., Pettersson, I., Sharp, G., and Ringertz N. R. (1986). Intranuclear localization of snRNP antigens. *J. Cell Biol.* **102,** 137–144.

O'Brien, T., and Lis, J. T. (1991). RNA polymerase II pauses at the 5′ end of the transcriptionally induced *Drosophila hsp70* gene. *Mol. Cell. Biol.* **11,** 5285–5290.

O'Brien, T., and Lis. J. T. (1993). Rapid changes in *Drosophila* transcription after an instantaneous heat shock. *Mol. Cell. Biol.* **13,** 3456–3463.

Ohlsson, R. I., van Eekelen, C., and Philipson, L. (1982). Non-random localization of ribonucleoprotein (RNP) structures within an adenovirus mRNA precursor. *Nucleic Acids Res.* **10,** 3053–3068.

O'Keefe, R. T., Mayeda, A., Sadowski, C. L., Krainer, A. R., and Spector, D. L. (1994). Disruption of pre-mRNA splicing *in vivo* results in reorganization of splicing factors. *J. Cell Biol.* **124,** 249–260.

Olins, D. E., Olins, A. L., Levy, H. A., Durfee, R. C., Margle, S. M., Tinnel, E. P., and Dover, S. D. (1983). Electron microscope tomography: Transcription in three dimensions. *Science* **220,** 498–500.

Osheim, Y. N., Miller, O. L., Jr., and Beyer, A. L. (1985). RNP particles at splice junction sequences on *Drosophila* chorion transcripts. *Cell (Cambridge, Mass.)* **43,** 143–151.

Padgett, R. A., Mount, S. M., Steitz, J. A., and Sharp, P. A. (1983). Splicing of messenger RNA precursors is inhibited by antisera to small nuclear ribonucleoprotein. *Cell (Cambridge, Mass.)* **35,** 101–107.

Pan, Z. Q., and Prives, C. (1988). Assembly of functional U1 and U2 human-amphibian hybrid snRNPs in *Xenopus laevis* oocytes. *Science* **241,** 1328–1331.

Pan, Z. Q., and Prives, C. (1989). U2 snRNA sequences that bind U2-specific proteins are dispensible for the function of U2 snRNP in splicing. *Genes Dev.* **3,** 1887–1898.

Paker, R., Siliciano, P. G., and Guthrie, C. (1987). Recognition of the TACTAAC box during mRNA splicing in yeast involves base pairing to the U2-like snRNA. *Cell (Cambridge, Mass.)* **49,** 229–239.

Patton, J. G., Porro, E. B., Galcéran, J., Tempst, P., and Nadal-Ginard, B. (1993). Cloning and characterization of PSF, a novel pre-mRNA splicing factor. *Genes Dev.* **7,** 393–406.

Patton, J. R., Ross, D. A., and Chae, C. (1985). Specific regions of β-globin RNA are resistant to nuclease digestion in RNA-protein complexes in chicken reticulocyte nuclei. *Mol. Cell. Biol.* **5,** 1220–1228.

Pederson, T. (1974). Proteins associated with heterogeneous nuclear RNA in eukaryotic cells. *J. Mol. Biol.* **83,** 163–183.

Piñol-Roma, S., and Dreyfuss, G. (1992). Shuttling of pre-mRNA binding proteins between nucleus and cytoplasm. *Nature (London)* **355,** 730–732.

Piñol-Roma, S., and Dreyfuss, G. (1992). Cell cycle-regulated phosphorylation of the pre-mRNA-binding (heterogeneous nuclear ribonucleoprotein) C proteins. *Mol. Cell. Biol.* **13,** 5762–5770.

Piñol-Roma, S., Choi, Y. D., Matunis M. J., and Dreyfuss, G. (1988). Immunopurification of heterogeneous nuclear ribonucleoprotein particles reveals an assortment of RNA-binding proteins. *Genes Dev.* **2,** 215–227.

Piñol-Roma, S., Swanson, M. S., Gall, J. G., and Dreyfuss, G. (1989). A novel heterogeneous nuclear RNP protein with a unique distribution on nascent transcripts. *J. Cell Biol.* **109,** 2575–2587.

Portman, D. S., and Dreyfuss, G. (1994). RNA annealing activities in HeLa nuclei. *EMBO J.* **13,** 213–221.

Pullman, J. M., and Martin, T. E. (1983). Reconstitution of nucleoprotein complexes with mammalian heterogeneous nuclear ribonucleoprotein (hnRNP) core proteins. *J. Cell Biol.* **97,** 99–111.

Puvion, E., Viron, A., Assens, C., Leduc, E. H., and Jeanteur, P. (1984). Immunocytochemical identification of nuclear structures containing snRNPs in isolated rat liver cells. *J. Ultrastruct. Res.* **87,** 180–189.

Puvion-Dutilleul, F., and Puvion, E. (1981). Relationship between chromatin and perichromatin granules in cadmium-treated isolated hepatocytes. *J. Ultrastruct. Res.* **74,** 341–350.

Query, C. C., Bentley, R. C., and Keene, J. D. (1989). A common RNA recognition motif identified within a defined U1 RNA binding domain of the 70K U1 snRNP protein. *Cell (Cambridge, Mass.)* **57,** 89–101.

Rasmussen, E. B., and Lis, J. T. (1993). *In vivo* transcriptional pausing and cap formation on three *Drosophila* heat shock genes. *Proc. Natl. Acad. Sci. U.S.A.* **90,** 7923–7927.

Raychaudhuri, G., Haynes, S. R., and Beyer, A. L. (1992). Heterogeneous nuclear ribonucleoprotein complexes and proteins in *Drosophila melanogaster. Mol. Cell. Biol.* **12,** 847–855.

Reed, R. (1990). Protein composition of mammalian spliceosomes assembled *in vitro*. *Proc. Natl. Acad. Sci. U.S.A.* **87,** 8031–8035.

Reed, R., and Maniatis, T. (1986). A role for exon sequences and splice-site proximity in splice-site selection. *Cell (Cambridge, Mass.)* **46,** 681–690.

Reich, C. I., VanHoy, R. W., Porter, G. L., and Wise, J. A. (1992). Mutations at the 3′ splice site can be suppressed by compensatory base changes in U1 snRNA in fission yeast. *Cell (Cambridge, Mass.)* **69,** 1159–1169.

Reuter, R., Appel, B., Bringmann, P., Rinke, J., and Lührmann, R. (1984). 5′-Terminal caps of snRNAs are reactive with antibodies specific for 2,2,7-trimethylguanosine in whole cells and nuclear matrices. *Exp. Cell Res.* **154,** 548–560.

Rice, G. A., Kane, C. M., and Chamberlin, M. J. (1991). Footprinting analysis of mammalian RNA polymerase II along its transcript: An alternative view of transcription elongation. *Proc. Natl. Acad. Sci. U.S.A.* **88,** 4245–4249.

Risau, W., Symmons, P., Saumweber, H., and Frasch, M. (1983). Nonpackaging and packaging proteins of hnRNA in *Drosophilia melanogaster. Cell (Cambridge, Mass.)* **33,** 529–541.

Robberson, B. L., Cote, G. J., and Berget, S. M. (1990). Exon definition may facilitate splice site selection in RNAs with multiple exons. *Mol. Cell. Biol.* **10,** 84–94.

Romig, H., Fackelmayer, F. O., Renz, A., Ramsperger, U., and Richter, A. (1992). Characterization of SAF-A, a novel nuclear SNA binding protein from HeLa cells with high affinity for nuclear matrix/scaffold attachment DNA elements. *EMBO J.* **11,** 3431–3440.

Rosbash, M., and Séraphin, B. (1991). Who's on first? The U1 sRNP-5′ splice site interaction and splicing. *Trends Biochem. Sci.* **16,** 187–190.

Roth, M. B., and Gall, J. G. (1987). Monoclonal antibodies that recognize transcription unit proteins on newt lampbrush chromosomes. *J. Cell Biol.* **105,** 1047–1054.

Roth, M. B., Murphy, C., and Gall, J. G. (1990). A monoclonal antibody that recognizes a phosphorylated epitope stains lampbrush chromosome loops and small granules in the amphibian germinal vesicle. *J. Cell Biol.* **111,** 2217–2223.

Roth, M. B., Zahler, A. M., and Stolk, J. A. (1991). A conserved family of nuclear phosphoproteins localized to sites of polymerase II transcription. *J. Cell Biol.* **115,** 587–596.

Rougvie, A. E., and Lis, J. T. (1988). The RNA polymerase II molecule at the 5′ end of the uninduced *hsp70* gene of *D. melanogaster* is transcriptionally engaged. *Cell (Cambridge, Mass.)* **54,** 795–804.

Ruby, S. W., and Abelson, J. (1988). An early hierarchic role of U1 small nuclear ribonucleoprotein in spliceosome assembly. *Science* **242,** 1028–1035.

Ruby, S. W., and Abelson, J. (1991). Pre-mRNA splicing in yeast. *Trends Genet.* **7,** 73–85.

Ruskin, B., and Green, M. R. (1985). Specific and stable intron-factor interactions are established early during *in vitro* pre-mRNA splicing. *Cell (Cambridge, Mass.)* **43,** 131–142.

Ruskin, B., Zamore, P. D., and Green, M. R. (1988). A factor, U2AF, is required for U2 snRNP binding and splicing complex assembly. *Cell (Cambridge, Mass.)* **52,** 207–219.

Ryner, L. C., and Baker, B. S. (1991). Regulation of *doublesex* pre-mRNA processing occurs by 3′-splice site activation. *Genes Dev.* **5,** 2071–2085.

Samarina, O. P., Lukanidin, E. M., Molnar, J., and Georgiev, G. P. (1968). Structural organization of nuclear complexes containing DNA-like RNA. *J. Mol. Biol.* **33,** 251–263.

Samuels, M. E., Bopp, D., Colvin, R. A., Roscigno, R. F., García-Blanco, M. A., and Schedl, P. (1994). RNA binding by Sxl proteins in vitro and in vivo. *Mol. Cell. Biol.* **14,** 4975–4990.

Sass, H., and Pederson, T. (1984). Transcription-dependent localization of U1 and U2 small nuclear ribonucleoproteins at major sites of gene activity in polytene chromosomes. *J. Mol. Biol.* **180,** 911–926.

Sawa, H., and Shimura, Y. (1992). Association of U6 snRNA with the 5′-splice site region of pre-mRNA in the spliceosome. *Genes Dev.* **6,** 244–254.

Scheer, U. (1987). Contributions of electron microscopic spreading preparations ("Miller" spreads) to the analysis of chromosome structure. *In* "Structure and Function of Eukaryotic Chromosomes" (W. Hennig, ed.), pp. 147–171. Springer-Verlag, Berlin.

Scheer, U. Franke, W. W., Trendelenburg, M. F., and Spring, H. (1976). Classification of loops of lampbrush chromosomes according to the arrangement of transcriptional complexes. *J. Cell Sci.* **22,** 503–519.

Scott, S. E. M., and Sommerville, J. (1994). Location of nuclear proteins on the chromosomes of newt oocytes. *Nature (London)* **250,** 680–682.

Selby, M. J., and Peterlin, B. M. (1990). *Trans*-activation by HIV-1 *Tat* via a heterologous RNA binding protein. *Cell (Cambridge, Mass.)* **62,** 769–776.

Selby, M. J., Bain, E. S., Luciw, P. A., and Peterlin, B. M. (1989). Structure, sequence, and position of the stem-loop in *tar* determine transcriptional elongation by *tat* through the HIV-1 long terminal repeat. *Genes Dev.* **3,** 547–558.

Séraphin, B., Kretzner, L., and Rosbash, M. (1988). A U1 snRNA: Pre-mRNA base pairing interaction is required early in yeast spliceosome assembly but does not uniquely define the 5′ cleavage site. *EMBO J.* 7, 2533–2538.

Sharp, P. A., and Marciniak, R. A. (1989). HIV TAR: An RNA enhancer? *Cell (Cambridge, Mass.)* **59,** 229–230.

Shatkin, A. J. (1985). mRNA cap binding proteins: Essential factors for initiating translation. *Cell (Cambridge, Mass.)* **40,** 223–224.

Shermoen, A. W., and O'Farrell, P. H. (1991). Progression of the cell cycle through mitosis leads to abortion of nascent transcripts. *Cell (Cambridge, Mass.)* **67,** 303–310.

Siebel, C. W., Kanaar, R., and Rio, D. C. (1994). Regulation of tissue-specific P-element pre-mRNA splicing requires the RNA-binding protein PSI. *Genes Dev.* **8,** 1713–1725.

Sierakowska, H., Szer, W., Furdon, P. J., and Kole, R. (1986). Antibodies to hnRNP core proteins inhibit *in vitro* splicing of human β-globin pre-mRNA. *Nucleic Acids Res.* **14,** 5241–5254.

Siliciano, P. G., and Guthrie, C. (1988). 5′ Splice site selection in yeast: Genetic alterations in base-pairing with U1 reveal additional requirements. *Genes Dev.* **2,** 1258–1267.

Siomi, H., Matunis, M. J., Michael, W. M., and Dreyfuss, G. (1993). The pre-mRNA binding K protein contains a novel evolutionarily conserved motif. *Nucleic Acids Res.* **21,** 1193–1198.

Skoglund, U., Anderson, K., Björkroth, B., Lamb, M. M., and Daneholt, B. (1983). Visualization of the formation and transport of a specific hnRNP particle. *Cell (Cambridge, Mass.)* **34,** 847–855.

Skoglund, U., Andersson, K., Strandberg, B., and Daneholt, B. (1986). Three-dimensional structure of a specific pre-messenger RNP particle established by electron microscope tomography. *Nature (London)* **319,** 560–564.

Snow, M. H. L., and Callan, H. G. (1969). Evidence for a polarized movement of the lateral loops of newt lampbrush chromosomes during oogenesis. *J. Cell Sci.* **5,** 1–25.

Sommerville, J. (1973). Ribonucleoprotein particles derived from the lampbrush chromosomes of newt oocytes. *J. Mol. Biol.* **78,** 487–503.

Sommerville, J., and Hill, R. J. (1973). Proteins associated with heterogeneous nuclear RNA of newt oocytes. *Nature (London)* **245,** 104–106.

Soulard, M., Valle, V. D., Siomi, M. C., Piñol-Roma, S., Codogno, P., Bauvy, C., Bellini, M., Lacroix, J., Monod, G., Dreyfuss, G., and Larsen, C. (1993). HnRNP G: Sequence and characterization of a glycosylated RNA-binding protein. *Nucleic Acids Res.* **21,** 4210–4217.

Southgate, C. D., and Green, M. R. (1991). The HIV-1 *Tat* protein activates transcription from an upstream DNA-binding site: Implications for *Tat* function. *Genes Dev.* **5,** 2496–2507.

Southgate, C., Zapp, M. L., and Green, M. R. (1990). Activation of transcription by HIV-1 *Tat* protein tethered to nascent RNA through another protein. *Nature (London)* **345,** 640–642.

Spann, P., Feinerman, M., Sperling, J., and Sperling, R. (1989). Isolation and visualization of large compact ribonucleoprotein particles of specific nuclear RNAs. *Proc. Natl. Acad. Sci. U.S.A.* **86,** 466–470.

Spector, D. L. (1984). Colocalization of U1 and U2 small nuclear RNPs by immunocytochemistry. *Biol. Cell* **51,** 109–112.

Spector, D. L. (1993a). Macromolecular domains within the cell nucleus. *Annu. Rev. Cell Biol.* **9,** 265–315.

Spector, D. L. (1993b). Nuclear organization of pre-mRNA splicing. *Curr. Opin. Cell Biol.* **5,** 442–448.

Spector, D. L., Schrier, W. H., and Busch, H. (1983). Immunoelectron microscopic localization of snRNPs. *Biol. Cell* **49,** 1–14.

Spector, D. L., Fu, X., and Maniatis, T. (1991). Associations between distinct pre-mRNA splicing components and the cell nucleus. *EMBO J.* **10,** 3467–3481.

Steitz, J. A. (1992). Splicing takes a holliday. *Science* **257,** 888–889.

Steitz, J. A., and Kamen, R. (1981). Arrangement of 30S heterogeneous nuclear ribonucleoprotein on polyoma virus late nuclear transcripts. *Mol. Cell. Biol.* **1,** 21–34.

Steitz, J. A., Black, D. L., Gerke, V., Parker, K. A., Kramer, A., Frendeway, D., and Keller, W. (1988). Functions of the abundant U-snRNPs. *In* "Structure and Function of Major and Minor Small Nuclear Ribonucleoprotein Particles" (M. L. Birnstiel, ed.), pp. 115–154. Springer-Verlag, Berlin.

Stevens, B. J., and Swift, H. (1966). RNA transport from nucleus to cytoplasm in *Chironomus* salivary glands. *J. Cell Biol.* **31,** 55–77.

Stolow, D. T., and Berget, S. M. (1990). UV cross-linking of polypeptides associated with 3′-terminal exons. *Mol. Cell. Biol.* **10,** 5937–5944.

Sun, Q., Mayeda, A., Hampson, R. K., Krainer, A. R., and Rottman, F. M. (1993). General splicing factor SF2/ASF promotes alternative splicing by binding to an exonic splicing enhancer. *Genes Dev.* **7,** 2598–2608.

Swanson, M. S., and Dreyfuss, G. (1988a). Classification and purification of proteins of heterogeneous nuclear ribonucleoprotein particles by RNA-binding specificities. *Mol. Cell. Biol.* **8,** 2237–2241.

Swanson, M. S., and Dreyfuss, G. (1988b). RNA binding specificity of hnRNP proteins: A subset bind to the 3′ end of introns. *EMBO J.* **7,** 3519–3529.

Swift, H., Adams, B. J., and Larsen, K. (1964). Electron microscope cytochemistry of nucleic acids in *Drosophila* salivary glands and *Tetrahymena*. *J. R. Microsc. Soc.* **83,** 161–167.

Takimoto, M., Tomonaga, T., Matunis, M., Avigan, M., Krutzsch, H., Dreyfuss, G., and levens, D. (1993). Specific binding of heterogeneous ribonucleoprotein particle protein κ to the human *c-myc* promoter, *in vitro*. *J. Biol. Chem.* **268,** 18249–18258.

Tan, E. M. (1967). Relationship of nuclear staining patterns with precipitating antibodies in systemic lupus erythematosus. *J. Lab. Clin. Med.* **70,** 800–812.

Tan, E. M., and Kunkel, H. G. (1966). Characteristics of a soluble nuclear antigen precipitating with sera of patients with systemic lupus erythematosus. *J. Immunol.* **96,** 464–471.

Tanaka, K., Watakabe, A., and Shimura, Y. (1994). Polypurine sequences within a downstream exon function as a splicing enhancer. *Mol. Cell. Biol.* **14,** 1347–1354.

Tazi, J., Albert, C., Temsamani, J., Reveillaud, I., Cathala, G., Brunel, C., and Jeanteur, P. (1986). A protein that specifically recognizes the 3′ splice site of mammalian pre-mRNA introns is associated with a small nuclear ribonucleoprotein. *Cell* (*Cambridge, Mass*). **47,** 755–766.

Thummel, C. S., Burtis, K. C., and Hogness, D. S. (1990). Spatial and temporal patterns of *E74* transcription during *Drosophila* development. *Cell* (*Cambridge, Mass.*) **61,** 101–111.

Tian, M., and Maniatis, T. (1993). A splicing enhancer complex controls alternative splicing of *doublesex* pre-mRNA. *Cell* (*Cambridge, Mass.*) **74,** 105–114.

Tian, M., and Maniatis, T. (1994). A splicing enhancer exhibits both constitutive and regulated activities. *Genes Dev.* **8,** 1703–1712.

Tsutsui, K., Okada, S., Watari, S., Seki, S., Yasuda, T., and Shohmori, T. (1993). Identification and characterization of a nuclear scaffold protein that binds the matrix attachment region DNA. *J. Biol. Chem.* **268,** 12886–12894.

Tsvetkov, A., Jantsch, M., Wu, Z., Murphy, C., and Gall, J. G. (1992). Transcription on lampbrush chromosome loops in the absence of U2 snRNA. *Mol. Biol. Cell.* **3,** 249–261.

Utans, U., Behrens, S. E., Lührmann, R., Kole, R., and Krämer, A. (1992). A splicing factor that is inactivated during *in vivo* heat shock is functionally equivalent to the [U4/U6.U5] triple snRNP-specific proteins. *Genes Dev.* 6, 631–641.

Valcárcel, J., Singh, R., Zamore, P. D., and Green, M. R. (1993). The protein *Sex-lethal* antagonizes the splicing factor U2AF to regulate alternate splicing of *transformer* pre-mRNA. *Nature* (*London*) **362,** 171–175.

van Eekelen, C. A., Rieman, T., and van Venrooij, W. J. (1981). Specificity in the interaction of hnRNA and mRNA with proteins as revealed by *in vivo* crosslinking. *FEBS Lett.* **130,** 223–226.

van Eekelen, C., Ohlsson, R., Philipson, L., Mariman, E., van Beek, R., and van Venrooij, W. (1982). Sequence dependent interaction of hnRNP proteins with late adenoviral transcripts. *Nucleic Acids Res.* **10,** 7115–7131.

Vazquez-Nin, G. H., Echeverria, O. M., Fakan, S., Leser, G., and Martin, T. E. (1990). Immunoelectron microscope localization of snRNPs in the polytene nucleus of salivary glands of *Chironomus thummi. Chromosoma* **99,** 44–51.

Verheijen, R., Kuijpers, H., Vooijs, P., van Venrooij, W., and Ramaekers, F. (1986). Distribution of the 70K U1 RNA-associated protein during interphase and mitosis. *J. Cell Sci.* **86,** 173–190.

von Kries, J. P., Buck, F., and Strätling, W. H. (1994). Chicken MAR binding protein p120 is identical to human heterogeneous nuclear ribonucleoprotein (hnRNP) U. *Nucleic Acids Res.* **22,** 1215–1220.

Wahle, E., and Keller, W. (1992). The biochemistry of 3′-end cleavage and polyadenylation of messenger RNA precursors. *Annu. Rev. Biochem.* **61,** 419–440.

Wahle, E., and Keller, W. (1993). RNA-protein interactions in mRNA 3′-end formation. *Mol. Biol. Rep.* **18,** 157–161.

Wang, J., and Pederson, T. (1990). A 62,000 molecular weight spliceosome protein cross-links to the intron polypyrimidine tract. *Nucleic Acids Res.* **18,** 5995–6001.

Wang, J., Cao, L., Wang, Y., and Pederson, T. (1991). Localization of pre-messenger RNA at discrete nuclear sites. *Proc. Natl. Acad. Sci. U.S.A.* **88,** 7391–7395.

Wassarman, D. A., and Steitz, J. A. (1992). Interactions of small nuclear RNA's with precursor messenger RNA during *in vitro* splicing. *Science* **257,** 1918–1925.

Weeks, J. R., Hardin, S. E., Shen, J., Lee, J. M., and Greenleaf, A. L. (1993). Locus-specific variation in phosphorylation state of RNA polymerase II *in vivo:* Correlations with gene activity and transcript processing. *Genes Dev.* **7,** 2329–2344.

Wilk, H., Angeli, G., and Schäfer, K. P. (1983). *In vitro* reconstitution of 35S ribonucleoprotein complexes. *Biochemistry* **22,** 4592–4600.

Wilusz, J., and Shenk, T. (1990). A uridylate tract mediates efficient heterogeneous nuclear ribonucleoprotein C protein-RNA cross-linking and functionally substitutes for the downstream element of the polyadenylation signal. *Mol. Cell. Biol.* **10,** 6397–6407.

Wilusz, J., Feig, D. I., and Shenk, T. (1988). The C proteins of heterogeneous nuclear ribonucleoprotein complexes interact with RNA sequences downstream of polyadenylation cleavage sites. *Mol. Cell. Biol.* **8,** 4477–4483.

Wu, C. H. H., and Gall, J. G. (1993). U7 small nuclear RNA in C snurposomes of *Xenopus* germinal vesicle. *Proc. Natl. Acad. Sci. U.S.A.* **90,** 6257–6259.

Wu, J., and Manley, J. L. (1989). Mammalian pre-mRNA branch site selection by U2 snRNP involves base pairing. *Genes Dev.* **3,** 1553–1561.

Wu, J. Y., and Maniatis, T. (1993). Specific interactions between proteins implicated in splice site selection and regulated alternative splicing. *Cell (Cambridge, Mass.)* **75,** 1061–1070.

Wu, Z., Murphy, C., Callan, H. G., and Gall, J. G. (1991). Small nuclear ribonucleoproteins and heterogeneous nuclear ribonucleoproteins in the amphibian germinal vesicle: Loops, spheres, and snurposomes. *J. Cell Biol.* **113,** 465–483.

Wurtz, T., Lönnroth, A., and Daneholt, B. (1990a). Higher order structure of Balbiani Ring premessenger RNP particles depends on certain RNase A sensitive sites. *J. Mol. Biol.* **215,** 93–101.

Wurtz, T., Lönnroth, A., Ovchinnikov, L., Skoglund, U., and Daneholt, B. (1990b). Isolation and initial characterization of a specific premessenger ribonucleoprotein particle. *Proc. Natl. Acad. Sci. U.S.A.* **87,** 831–835.

Xing, Y., Johnson, C. V., Dobner, P. R., and Lawrence, J. B. (1993). Higher level organization of individual gene transcription and RNA splicing. *Science* **259,** 1326–1330.

Yang, V. W., Lerner, M. R., Steitz, J. A., and Flint, S. J. (1981). A small nuclear ribonucleoprotein is required for splicing of adenoviral early RNA sequences. *Proc. Natl. Acad. Sci. U.S.A.* **78,** 1371–1375.

Zachar, Z., Chou, T., and Bingham, P. M. (1987). Evidence that a regulatory gene autoregulates splicing of its transcript. *EMBO J.* **6,** 4105–4111.

Zachar, Z., Kramer, J., Mims, I. P., and Bingham, P. M. (1993). Evidence for channeled diffusion of pre-mRNAs during nuclear RNA transport in metazoans *J. Cell Biol.* **121,** 729–742.

Zahler, A. M., Lane, W. S., Stolk, J. A., and Roth, M. B. (1992). SR proteins: A conserved family of pre-mRNA splicing factors. *Genes Dev.* **6,** 837–847.

Zahler, A. M., Neugebauer, K. M., Lane, W. S., and Roth, M. B. (1993). Distinct functions of SR proteins in alternative pre-mRNA splicing. *Science* **260,** 219–222.
Zamore, P. D., and Green, M. R. (1989). Identification, purification, and biochemical characterization of U2 small nuclear ribonucleoprotein auxiliary factor. *Proc. Natl. Acad. Sci. U.S.A.* **86,** 9243–9247.
Zamore, P. D., and Green, M. R. (1991). Biochemical characterization of U2 snRNP auxiliary factor: An essential pre-mRNA splicing factor with a novel intranuclear distribution. *EMBO J.* **10,** 207–214.
Zamore, P. D., Patton, J. G., and Green, M. R. (1992). Cloning and domain structure of the mammalian splicing factor U2AF. *Nature (London)* **355,** 609–614.
Zhuang, Y., and Weiner, A. M. (1986). A compensatory base change in U1 snRNA suppresses a 5′ splice site mutation. *Cell (Cambridge, Mass.)* **46,** 827–835.
Zhuang, Y., and Weiner, A. M. (1989). A compensatory base change in human U2 snRNA can suppress a branch site mutation. *Genes Dev.* **3,** 1545–1552.
Zhuang, Y., Leung, H., and Weiner, A. M. (1987). The natural 5′ splice site of Simian Virus 40 large T antigen can be improved by increasing the base complementarity of U1 RNA. *Mol. Cell. Biol.* **7,** 3018–3020.
Zuo, P., and Manley, J. L. (1994). The human splicing factor ASF/SF2 can specifically recognize pre-mRNA 5′ splice sites. *Proc. Natl. Acad. Sci. U.S.A.* **91,** 3363–3367.

NOTE ADDED IN PROOF

The field of pre-mRNA metabolism has advanced rapidly since completion of this review article. The reader is referred to the following review articles for additional information.

HnRNP Proteins

Herschlag, D. (1995). RNA chaperones and the RNA folding problem. *J. Biol. Chem.* **270,** 20871–20874.
Kiledjian, M., Burd, C. G., Görlach, M., Portman, D. S., and Dreyfuss, G. (1994). Structure and function of hnRNP proteins. *In* "RNA-Protein Interactions" (K. Nagai and I. W. Mattaj, eds.), Frontiers in Molecular Biology (B. D. Hames and D. M. Glover, eds.), pp. 127–149. Oxford University Press, New York.
Madhani, H. D., and Guthrie, C. (1994). Dynamic RNA-RNA interactions in the spliceosome. *Annu. Rev. Genet.* **28,** 1–26.
Swanson, M. S. (1995). Functions of nuclear pre-mRNA/mRNA binding proteins. *In* "Pre-mRNA Processing" (A. I. Lamond, ed.), pp. 17–33. R. G. Landes Company, Austin, Texas.

Spliceosome Assembly and Splicing

Ares, M., Jr., and Weiser, B. (1995). Rearrangement of snRNA structure during assembly and function of the spliceosome. *Prog. Nucleic Acid Res. Mol. Biol.* **50,** 131–159.

Berget, S. M. (1995). Exon recognition in vertebrate splicing. *J. Biol. Chem.* **270,** 2411–2414.
Fu, X. D. (1995). The superfamily of arginine/serine-rich splicing factors. *RNA* **1,** 663–680.
Inoue, K., Ohno, M., and Shimura, Y. (1995). Aspects of splice site selection in constitutive and alternative pre-mRNA splicing. *Gene Exp.* **4,** 177–182.
Krämer, A. (1996). The structure and function of proteins involved in mammalian pre-mRNA splicing. *Annu. Rev. Biochem.* **65,** 367–409.
Manley, J. L., and Tacke, R. (1996). SR proteins and splicing control. *Genes Dev.* **10,** 1569–1579.
Umen, J. G., and Guthrie, C. (1995). The second catalytic step of pre-mRNA splicing. *RNA* **1,** 869–885.

Nuclear Architecture

Berezny, R., Mortillaro, M. J., Ma, H., Wei, X., and Samarabandu, J. (1995). The nuclear matrix: A structural milieu for genomic function. *Int. Rev. Cytol.* **162A,** 1–65.
Gall, J. G., Tsvetkov, A., Wu, Z., and Murphy, C. (1995). Is the sphere organelle/coiled body a universal nuclear component? *Dev. Genetics* **16,** 25–35.
Moen, P. T., Jr., Smith, K. P., and Lawrence, J. B. (1995). Compartmentalization of specific pre-mRNA metabolism: An emerging view. *Human Mol. Genetics* **4,** 1779–1789.
Nickerson, J. A., Blencowe, B. J., and Penman, S. (1995). The architectural organization of nuclear metabolism. *Int. Rev. Cytol.* **162A,** 67–123.
van Driel, R., Wansink, D. G., van Steensel, B., Grande, M. A., Schul, W., and de Jong, L. (1995). Nuclear domains and the nuclear matrix. *Int. Rev. Cytol.* **162A,** 151–189.

Tat

Karn, J., Gait, M. J., Churcher, M. J., Mann, D. A., Mikaélian, I., and Pritchard, C. (1994). Control of human immunodeficiency virus gene expression by the RNA-binding proteins tat and rev. *In* "RNA-Protein Interactions" (K. Nagai and I. W. Mattaj, eds.), Frontiers in Molecular Biology (B. D. Hames and D. M. Glover, eds.), pp. 192–220. Oxford University Press, New York.

Index

D

H

I

K

L

M

N

O

P

Q

R